Communications
in Computer and Information Science

2015

Rationale

The CCIS series is devoted to the publication of proceedings of computer science conferences. Its aim is to efficiently disseminate original research results in informatics in printed and electronic form. While the focus is on publication of peer-reviewed full papers presenting mature work, inclusion of reviewed short papers reporting on work in progress is welcome, too. Besides globally relevant meetings with internationally representative program committees guaranteeing a strict peer-reviewing and paper selection process, conferences run by societies or of high regional or national relevance are also considered for publication.

Topics

The topical scope of CCIS spans the entire spectrum of informatics ranging from foundational topics in the theory of computing to information and communications science and technology and a broad variety of interdisciplinary application fields.

Information for Volume Editors and Authors

Publication in CCIS is free of charge. No royalties are paid, however, we offer registered conference participants temporary free access to the online version of the conference proceedings on SpringerLink (http://link.springer.com) by means of an http referrer from the conference website and/or a number of complimentary printed copies, as specified in the official acceptance email of the event.

CCIS proceedings can be published in time for distribution at conferences or as post-proceedings, and delivered in the form of printed books and/or electronically as USBs and/or e-content licenses for accessing proceedings at SpringerLink. Furthermore, CCIS proceedings are included in the CCIS electronic book series hosted in the SpringerLink digital library at http://link.springer.com/bookseries/7899. Conferences publishing in CCIS are allowed to use Online Conference Service (OCS) for managing the whole proceedings lifecycle (from submission and reviewing to preparing for publication) free of charge.

Publication process

The language of publication is exclusively English. Authors publishing in CCIS have to sign the Springer CCIS copyright transfer form, however, they are free to use their material published in CCIS for substantially changed, more elaborate subsequent publications elsewhere. For the preparation of the camera-ready papers/files, authors have to strictly adhere to the Springer CCIS Authors' Instructions and are strongly encouraged to use the CCIS LaTeX style files or templates.

Abstracting/Indexing

CCIS is abstracted/indexed in DBLP, Google Scholar, EI-Compendex, Mathematical Reviews, SCImago, Scopus. CCIS volumes are also submitted for the inclusion in ISI Proceedings.

How to start

To start the evaluation of your proposal for inclusion in the CCIS series, please send an e-mail to ccis@springer.com.

De-Shuang Huang · Prashan Premaratne ·
Changan Yuan
Editors

Applied Intelligence

First International Conference, ICAI 2023
Nanning, China, December 8–12, 2023
Proceedings, Part II

 Springer

Editors
De-Shuang Huang
Eastern Institute of Technology
Zhejiang, China

Prashan Premaratne
University of Wollongong
North Wollongong, NSW, Australia

Changan Yuan
Guangxi Academy of Sciences
Guangxi, China

ISSN 1865-0929   ISSN 1865-0937 (electronic)
Communications in Computer and Information Science
ISBN 978-981-97-0826-0   ISBN 978-981-97-0827-7 (eBook)
https://doi.org/10.1007/978-981-97-0827-7

This Springer imprint is published by the registered company Springer Nature Singapore Pte Ltd.
The registered company address is: 152 Beach Road, #21-01/04 Gateway East, Singapore 189721, Singapore

Paper in this product is recyclable.

Preface

The first International Conference on Applied Intelligence (ICAI 2023) was held during December 8–12, 2023, in Nanning, Guangxi, China. The conference was started to provide an annual forum dedicated to emerging and challenging topics in artificial intelligence, machine learning, pattern recognition, bioinformatics, and computational biology. It aimed to bring together researchers and practitioners from both academia and industry to share ideas, problems, and solutions related to the multifaceted aspects of Applied Intelligence.

This year, the conference concentrated mainly on the theories and methodologies as well as the emerging applications of Applied Intelligence. Its aim was to unify the picture of contemporary Applied Intelligence techniques as an integral concept that highlights the trends in advanced computational intelligence and bridges theoretical research with applications. Therefore, the theme for this conference was "Advanced Applied Intelligence Technology and Applications". Papers that focused on this theme were solicited, addressing theories, methodologies, and applications in science and technology.

ICAI 2023 received 228 submissions from 10 countries and regions. All papers went through a rigorous peer-review procedure and each paper received at least three review reports. Based on the review reports, the Program Committee finally selected 64 high-quality papers for presentation at ICAI 2023, and inclusion in the proceedings published by Springer: two volumes of Communications in Computer and Information Science (CCIS).

The organizers of ICAI 2023, including Eastern Institute of Technology and Guangxi Academy of Sciences, China, made an enormous effort to ensure the success of the conference. We hereby would like to thank the members of the Program Committee and the referees for their collective effort in reviewing the papers. In particular, we would like to thank all the authors for contributing their papers. Without the high-quality submissions from the authors, the success of the conference would not have been possible. Finally, we are especially grateful to the International Neural Network Society and the National Science Foundation of China for their sponsorship.

December 2023

Changan Yuan
De-Shuang Huang
Prashan Premaratne

Organization

General Chair

Changan Yuan   Guangxi Academy of Sciences, China

Steering Committee Chair

De-Shuang Huang   Eastern Institute of Technology, China

Program Committee Co-chairs

De-Shuang Huang   Eastern Institute of Technology, China
Kang-Hyun Jo   University of Ulsan, South Korea
Prashan Premaratne   University of Wollongong, Australia
Abir Hussain   Liverpool John Moores University, UK

Organizing Committee Co-chairs

Yingzhou Bi   Nanning Normal University, China
Jianbo Lu   Nanning Normal University, China

Organizing Committee Members

Xu Guilin   Nanning Normal University, China
Yuzhong Peng   Nanning Normal University, China
Xiao Qin   Nanning Normal University, China
Chao Wang   Guangxi Academy of Sciences, China
Wanxian He   Guangxi Academy of Sciences, China

Award Committee Chair

Michal Choras   Bydgoszcz University of Science and Technology,
                                       Poland

Tutorial Co-chairs

Jair Cervantes Canales   Autonomous University of Mexico State, Mexico
Yu-Dong Zhang   University of Leicester, UK

Special Issue Chair

Chandratilak De Silva Liyanage   University Brunei Darussalam, Brunei

Publication Chair

Damith Mohotti   University of New South Wales, Australia

Special Session Chair

Arturo Yee Rendon   Autonomous University of Sinaloa, Mexico

Workshop Chair

Josué Espejel Cabrera   Autonomous University of Mexico State, Mexico

International Liaison Chair

Prashan Premaratne   University of Wollongong, Australia

Publicity Co-chairs

Chun-Hou Zheng   Anhui University, China
Jair Cervantes Canales   Autonomous University of Mexico State, Mexico

Program Committee Members

Jing Chen	Suzhou University of Science and Technology, China
Chenxi Huang	Xiamen University, China
Wenzheng Bao	Xuzhou University of Technology, China
Lin Yuan	Qilu University of Technology (Shandong Academy of Sciences), China
Vasu Alagar	Concordia University, Canada
Prashan Premaratne	University of Wollongong, Australia
Chin-Chih Chang	Chung Hua University, Taiwan
Michal Choras	Bydgoszcz University of Science and Technology, Poland
Haijun Gong	Saint Louis University, USA
Daowen Qiu	Sun Yat-sen University, China
Rui Li	Montclair State University, USA
Boudhayan Bhattacharya	Brainware University, India
Jing Hu	Wuhan University of Science and Technology, China
Bo Li	Wuhan University of Science and Technology, China
Weitian Wang	Montclair State University, USA
Dingjiang Huang	East China Normal University, China
Laurent Heutte	Université de Rouen Normandie, France
Chengcai Fu	Shandong Jiaotong University, China
Fengying Ma	Qilu University of Technology, China
Wei Chen	China University of Mining and Technology, China
Song Deng	Nanjing University of Posts and Telecommunications, China
Yiran Huang	Guangxi University, China
Wei Lan	Guangxi University, China
Yuzhong Peng	Nanning Normal University, China
Qing Tian	Nanjing University of Information Sciences and Technology, China

Reviewers

Alaa Alsaig
Ammar Alsaig
Bo Li
Chin-Chih Chang
Chonglin Gu
Chunyan Liu
Dandan Zhu
Faquan Chen
Federica Uccello
Fuchun Liu
Gong Daoqing
Guokai Zhang
Hongguo Cai
Hongxuan Hua
Huan Ning
Hung-Chi Su
Jing Hu
Jingkai Yang
Jixin Sun
Jordan Murphy
Jun Li
Kaushik Chanda
Lei Wang
Liangyu Zhou
Ligang Xiao
Lihua Jiang
Lin Li
Lingyun Yu
Marek Pawlicki
Michael Yang
Minda Yao
Minglong Cheng
Nuo Yu
Odbal H.

Prashan Premaratne
Qing Ye
Qinhu Zhang
Rafal Kozik
Rongcan Chen
Rui Li
Ruizhi Fan
Shengzu Huang
Shijia Liao
Shuting Jin
Subhadip Nandi
Tong Si
Wang Zhi
Wei Lan
Wei Deng
Wrong Chang
Xiaoli Lin
Xiaoming Liu
Yang Liu
Yao-Hong Tsai
Yaqi Chen
Yi Zhao
Ying Sheng
Yongyong Chen
Yunzhe Qian
Yuquan Tong
Zhang Liang
Zhen Shen
Zhihong Zhang
Zhujun Zhang
Ziheng Duan
Zishan Xu
Ziyuan Dong
Zhongpeng Cai

Contents – Part II

Intelligent Control and Automation

Machine Learning

Natural Language Processing and Computational Linguistics

Contents – Part I

Computer Vision

Deep Learning

Intelligent Control and Automation

Advancing Short-Term Traffic Congestion Prediction: Navigating Challenges in Learning-Based Approaches

Chen Wang[1]([✉]) [ID], Travis Atkison[1] [ID], and Qiuhua Duan[2] [ID]

[1] Department of Computer Science, The University of Alabama, BOX 870290, Tuscaloosa, AL 35487-0290, USA
cwang86@ua.edu
[2] Department of Civil, Construction and Environmental Engineering, The University of Alabama, BOX 870205, Tuscaloosa, AL 35487-0205, USA

Abstract. Traffic congestion prediction has already become a significant aspect of modern transportation systems. By predicting traffic congestion, transportation planners and traffic management agencies can take steps to reduce congestion and improve traffic flow, and also inform the public about expected delays and suggest alternative routes, helping people to make more informed decisions about their travel plans. A growing body of literature has provided different methods in order to improve congestion management abilities to intelligent traffic systems, it is a trending research topic facilitating the development of transportation system prediction. This paper reviews and analyzes broadband of published articles regarding different approaches focusing on short-term real-time traffic predictions; then stresses on five core methodologies for a detailed analysis and provides a discussion on the pros and cons among different approaches with test results. In addition, this paper develops a perspective synthesis of the current status quo that could be the next steps for a more accurate, more efficient prediction. In the end, the paper yields conclusions about possible future research endeavors.

Keywords: Intelligence Traffic Systems · Real-time Modeling · Traffic Flow Prediction · Neural Networks · Deep Learning

1 Introduction

The majority of the early research focused on the relationships between highway traffic volume and vehicle occupancy times. This relationship was then cultivated into a time-series model that been used to predict a short-term traffic condition on highways. After this period, some researchers used averaging techniques on historical data to estimate traffic conditions in the same location and time period [20]. This method is data-based showing a simple linear time-invariant traffic system regardless of fluctuations in the real-time event. This estimation method cannot be called prediction since it overlooked too much information on traffic, some information would fundamentally affect the prediction results. Therefore, researchers have begun to study a variety of traffic flow prediction methods and proposed different traffic forecast strategies.

With the considerations of the above-mentioned categories, this paper mainly introduces survey work on short-term traffic flow prediction. The models used are mainly based on neural network (NN) methods, especially the Bayesian network, You Only Look Once (YOLO), Kalman filter, and deep learning.

For the remainder of this paper, Sect. 2 is a detailed discussion and analysis of various methods and algorithms. This includes proposed methods, algorithms, and their improvements as well as advantages and disadvantages. Section 3 is the synthesis of the core ideas, which provides an overall assessment of the existing works based on their commonality and differences. In addition, Sect. 4 generalizes all the potential approaches in a comprehensive fashion. This section provides the common and unique features of methods proposed in different papers and then explains accordingly in length. Section 5 yields discussions and conclusions on future pathways in traffic prediction and proposes a method that the author would use for a higher prediction efficiency and accuracy.

2 Analysis of Core Methodologies

The primary goal of this research is to find an optimal goal in accurately predicting traffic flows using neural networks and other computational tools. With regards to the most recent accomplishments in traffic flow prediction, the author recognized core methodologies that represent the seminal works and closely related to the field of study. In this section, those works will be reviewed and discussed in length for their attributes. Each method will be analyzed in two aspects: 1) key motivation of particular topics in traffic prediction area; 2) core methodologies and main seminal work in the area of traffic prediction.

2.1 Short-Term Traffic Flow Prediction

The major challenge of predicting traffic flows are sharp non-linearities. These are often the results during road condition transitions among free flow, congestion, breakdown, and recovery. Non-linearity problems are difficult to model directly due to the vast volume of variables and operation conditions, however, after careful measurements and variable selection, non-linearity can still be modeled using linear tools combined with deep learning techniques. In [19], Polson *et al.* develop an architecture that combines a linear model fitted using ℓ_1 regularization and a sequence of tanh layers. By providing an extreme case study, the proposed deep learning method yields a precise short-term traffic flow prediction.

In order to use deep learning for traffic congestion prediction under non-ideal events such as low visibility weather or special occasions, it is important toconstruct a learning architecture where the model recognizes the traffic flow condition, sorts out data irregularities, analyzes data recursiveness, and yields proper prediction result.

(a) Predictor selection problem. Proper selection of model predictors facilitates a reliable performance for the deep learning model. More than often the predictors together give an overall grasp on the basis of the model forecasting environment. Since most of the commercial traffic flow speed detectors are based on real-time Spatio-temporal (RTST) measurements, the goal is to find an algorithm to identity RTST in the collected data [21]. Amongst all options, a hierarchical linear vector autoregressive (HLVAR) model is developed such that -

$$x_{t+40}^t = Ax^t + \epsilon_t, \epsilon_t \sim N(0, V) \tag{1}$$

where A is the target sparse matrix, k is the number of previous measurements used to develop a forecast [19]. In addition, the optimal network is implemented from the SGD method [8, 14], and the hyper-parameters are found by RS. At this point, the model has everything it needs to begin the data training process.

(b) ℓ_1 regularization and trend filtering. For deep learning problems, typically a ℓ_2 norm square function is used as a loss function. ℓ_2 regularization has the advantages of preventing overfitting. Intuitively speaking, ℓ_2 is able to initialize early stopping with a growing potential of function overfitting with higher iterations. However for SGD, Polson *et al.* finds that ℓ_2 with ridge penalty could result into poor treatment of multi-modality and slow convergence. By comparison, ℓ_1 penalization outperforms ℓ_2 in better capturing the spatio-temporal relations [15], which is the characteristics of the measured traffic dataset. Furthermore, ℓ_1 works better in delivering scalable results, which is an appropriate technique for faster training convergence and hyper-parameter search time. A brief comparison between ℓ_1 and ℓ_2 is shown in Table 1. Another traffic flow modeling requirement is filtering out noisy data from measurements.

Table 1. Key features comparison between ℓ_1 and ℓ_2 regularization.

ℓ_1 regularization	ℓ_2 regularization
Penalize Σ|weights|	penalize $\Sigma(\text{weights})^2$
sparse outputs	non-sparse outputs
possibly multiple solutions	always one solution
robust to outliers	not robust to outliers

2.2 Bayesian Network Optimization

Bayesian network is another modeling efforts for traffic flows among adjacent road links in a transportation networks [13]. Bayesian network is constructed using the joint probability distribution (JPD) between the cause nodes and the effect nodes, usually cause nodes are the data for forecasting, while effect nodes are the data to be forecasted [22]. This cause-effect linage is normally facilitated by the Gaussian mixture model (GMM), whose parameters are estimated via the competitive expectation-maximization (CEM) algorithm. Taken all the nodes in the network into consideration, the goal is to output the optimal solution under the criterion of minimum mean square error (MMSE). From [13], the paper showed that the Bayesian network combined with principal component analysis (PCA) is able to outperform other short-term predictions methods for both complete and incomplete sets of data.

(a) Adjacent road conditional independent. The Bayesian network can be considered to be the combination of probability theory and graph theory, with the help of this conditional independence, the Bayesian network is good for parameter estimations and variable forecasting. By applying this into the traffic prediction realm, there are only two categories of parameters to be taken into calculations: 1) time series, which serve as timestamps in the datasets, and 2) traffic flows, which indicate the direction, speed, volume information of the targeted road links. The prediction process can be carried out by JPD using GMM [9, 22]. Several benefits of GMM are mentioned in [10], but here the focus should be on its convenience in the mathematical deduction. For the parameter estimation, a competitive expectation-maximization (CEM) algorithm for the analytical solutions of the GMM parameters can be applied [3].

(b) Traffic prediction. The traffic prediction problem is similar to the Bayesian inference problem. The main goal of inference in Bayesian networks is to estimate the values of target nodes given the values of the observed nodes. Since the parameters of GMM are used to describe the JPD in the Bayesian network. The justification criterion for the prediction result is determined by MMSE. Let (E, F) be a partitioning of the node indices of a Bayesian network into disjoint subsets and (x_E, x_F) be a partitioning of the corresponding random variables/vectors, the optimal forecasting $\hat{X}_F$ under the criterion of MMSE can be derived as -

$$\hat{x}_F = E(x_F|x_E) = \int x_F p(x_F|x_E)dx_F$$

$$= \sum_{l=1}^{M} \beta_l \int x_F G\left(x_F; \mu_{lF|E}, \Sigma_{lF|E}\right)dx_F = \sum_{l=1}^{M} \beta_l \mu_{lF|E} \qquad (2)$$

where $p(x_F|x_E)$ denotes the conditional probability density function according to Bayesian theory [2]; $G(x; \mu, \Sigma)$ denotes a a multidimensional normal density function with mean μ and covariance matrix Σ. The relationship between input and output of Bayesian network is rather concise, which is convenient for practical applications.

2.3   Stacked Autoencoders

With the explosion of traffic data collections, learning speed is of the essence in the process. In [16], Neelakandan et al. proposed a deep learning approach with stacked autoencoders (SAEs) for congestion prediction, especially under the cases when the road is fully occupied. Unlike short-term prediction methods, SAEs are desirable for generic traffic flow features. It considers the spatial and temporal correlations inherently such that the deep learning model can be constructed with multiple logistic regression layers, the model is trained in a greedy layer-wise unsupervised learning algorithm, each layer is directing to a more precise prediction leading to a better result. SAEs were firstly been used in traffic prediction and its results were beyond expectations comparing to other control groups.

(a) Autoencoders. Autoencoder attempts to reproduce the dimension of the target output. An autoencoder includes a single input layer, one hidden layer, and one output layer. In order to minimize the reconstruction error and obtain the model parameters, the size comparison between hidden layer and input layer are crucial for learning speed and prediction result. For example, when the hidden layer has the same or larger dimensions, the identity function need to be learned.

(b) Stacked autoencoders. To use the SAE model network for traffic flow prediction, Kashyap *et al.* added a standard predictor on the top layer [8], where a logistic regression layer was put on top of the network for supervised traffic flow prediction. The whole deep architecture model consists of the SAEs and the predictor comprise of traffic flow prediction. An SAE model is used to extract traffic flow features, and a logistic regression layer is applied for prediction.

(c) Training Algorithm. A greedy layer-wise unsupervised learning algorithm that pretraining the deep network layer by layer in a bottom-up fashion is proposed. After the pretraining phase, the model is fine-tuned using back-propagation (BP) algorithm, it can be applied to tune the model's parameters in a top-down direction to obtain better results at the same time.

2.4 Adaptive Kalman Filter for Stochastic Traffic Detection

Traffic management and control systems are expected to perform with certain reliability as well as abundant safety features. As a consequence of this trend, passive traffic systems such as rigid infrastructure are gradually superseded by more proactive systems. Typical features of proactive systems include dynamic surveillance, crest, and trough adjustment ability to real-time traffic flows. Proactive traffic patterns are harder to model and even harder to enforce high accuracy at all times, yet this is where deep learning techniques shine. A hybrid model consists of autoregression and state space (SS) representation with Kalman filtering (KF) technique is proposed and validated by [4]. A stochastic seasonal autoregressive integrated moving average plus generalized autoregressive conditional heteroscedasticity (SARIMA + GARCH) structure is realized with adaptive Kalman filter. The method is proven effective in 15-min forecasting and some improvement compared to conventional Kalman filter.

In [7], the formation and representation of SARIMA + GARCH structure are elaborated, then this structure is equipped with AKF in order to set up a deep learning model for the 15-min targeted traffic flow data.

(a) SARIMA + GARCH structure. In this structure, the SARIMA captures the dynamics of traffic flow levels, and the GARCH captures the dynamics of traffic flow variances. This structure is defined as -

$$SARIMA(p, d, q)(P, D, Q)_s + GARCH(u, v) \tag{3}$$

(b) $SARIMA(1, 0, 1)(0, 1, 1)_{672}$ can be used to represent traffic flow rates aggregated at a certain time interval, where the SARIMA can be decomposed into a seasonal integrated moving average operator and a short-term autoregressive moving average operator. Here, IMA and ARMA operators work in a cascade manner such that the seasonal IMA can capture the recursive pattern in traffic flows. For the GARCH process, GARCH(1, 1) is selected for simplicity.

(c) State space representation. SARIMA model can be interpreted as a cascade of a seasonal operator and a short-term operator. The SARIMA + GARCH structure can be executed in sequence using 1) the seasonal exponential smoothing operator, 2) distinct SS representations ARMA and GARCH. Since IMA is recursive as an intrinsic attribute, while the two SS models can be solved recursively, the forecast model now has everything needed to become a real-time monitoring system.

(d) Adaptive Kalman filter. As mentioned earlier, AKF can update the process variances during recursion. The adaptation mechanism uses memory for observation and state estimation errors in order to fine-tune the process variances. By converting time series into SS, the error rate is no longer regress but return in a form of accumulative result. By evaluating errors in each iteration, the memory size of AKF is updated for the next incoming error return. With a reasonable kickoff percentage, each iteration will push the system closer to the optimal operating point. Once the significant level is reached, the system can supervise traffic flow data in real-time.

2.5   Image-Based Real-Time Regression Methods

It is important for traffic prediction tools to collect enough useful resources for implementation. With the wide application of the ITS, more and more video and image processing materials are applied for detecting traffic flow. In order to conveniently process collected data, surveillance videos are typically conveniently converted into frames of images. In this case, the images are processed through classification and positioning in order for the system to extract necessary information. Among many features of classification, target detection is of most interest for researchers. YOLO, for example, is one of the popular target detection methods. [6] proposed a method that improves the limitation of YOLOv3 detection of a single-vehicle and used a first-level network architecture to detect irregular quadrilateral vehicle congestion areas. A novel SIGHTA regression network was proposed that can recognize real-time traffic conditions through video streams.

The non-maximum suppression (NMS) module is adapted to YOLO architecture for quadrilateral box outputs. The cost function is reconstructed in a sense that the misshape of images are compensated using Darknet53 as the backbone. Core methodologies of [6] are listed below.

(a) Quadrilateral object boxing. For a irregular quadrilateral, the expression of the vertices of a rectangle is identified to calculate offset between the center point and each vertex. For congestion detection, based on the appearance of the quadrilateral, four core features are used to describe the detected target: width, height, target credibility, and target type. The versatility is preserved for multi-class target detection so that the box changes while the target changes. As can be seen from (4):

$$Q = (wr,\ hr, po, pc, x0, y0, \Delta xi, \Delta yi) \tag{4}$$

where w_r and h_r are the height and width of the outer rectangle; p_o and p_c represent the credibility and the category of the target, respectively; x_0 and y_0 is the center point of the quadrilateral; Δx_i and Δy_i ($I = 1, 2, 3, 4$) denote the offset value from the vertex to the center point [6]. In the training process, after representing the quadrilateral object as above, the anchor with the largest intersection ratio with the ground truth is used to predict the center position of the target. The eight-dimensional information of the four vertices obtained by the regression is compared with the ground truth values.

(b) Loss Function. In order to calculate the location of the target, and determine the target type and the credibility of the target, In [24], Zhu *et al.* used the Darknet-based One Stage framework to process the data. In addition, the classification and regression multi-task loss were applied such that -

$$\text{Loss} = s^* + L_{reg}(y, y^*) \tag{5}$$

The loss function $L_{reg}(y, y^*)$ represents the classification error in two aspects: 1) prediction confidence, and 2) the prediction from the category. The regression loss s^* represents the weight coefficient accounting for the effects of distortion, mis-shape, positioning and tilting errors. The output target is delivered by the predicted quadrilateral from a calculation of fitting information of the inputs.

3 Synthesis of Literature Reviews

In this section, the author first specified the working condition that under investigation; then proposed a synthetic methodology from the mentioned works, finally, an overall summary regarding the synthesis will be drawn as an open thought in the field of ITS. The feasibility and reliability of this synthesis will require future implementation.

3.1 Online Prediction Features and Predictor Selection

The feasibility of traffic flow prediction depends on the data processing ability of the deep learning model. To construct a working model, the predictor selection is considered to be one of the crucial steps since this process can directly affect the outcomes of the model. The key features that constitute congestion are understood differently. However, one feature that is agreed to by all is vehicle speed. It is understandable since congestion indicates slower traffic, and traffic is the most direct information that can be extracted from the data. In [19], the vehicle speed is detected using HLVAR model with ℓ_1 regularization method. This method has advantages on non-recurrent events and faster convergence rate, which means it can be applied online with a small reaction time. In [16], the vehicle speed is described as a traffic flow rate, which is the output of the SAE model. SAE model uses logistic regression as the first layer for supervised learning, and the prediction accuracy is tied to the number of hidden layers and hidden units. Although the accuracy looked promising, the model solely relied on traffic volume as input, which is not friendly to non-recurrent events. In [13], although the vehicle speed is not heavily relied on, the Bayesian method also yields a good forecasting result by comparing traffic flow (vehicle/h) to ground truth. Luan *et al.* provides a novel way to work on traffic prediction, which is to treat the congestion on a scale of road link networks with multiple entry and egress. Thus, the vehicle speed is the foremost characteristics that can be easily recognized as the indicator of congestion.

As for the predictor selection, regressive algorithms are favored by most. This is because regressive algorithms help to control the modeling of feature variables. In addition, if properly tuned, regressive algorithms also help to detect anomalies in the data.

[11] features the HLVAR model for several reasons: 1) Vector autoregressive model can be used as sequence interval estimation and abnormal analysis, which is a perfect fit for the traffic flow prediction problem; 2) Hierarchical linear model uses data nesting such that it can independently analyze individual and group variables simultaneously; 3) HLVAR model is the integration of both, it is designed specifically to identify the Spatio-temporal relations in the traffic data [19]. In this case, collected data is expected to be fully used for a comprehensive output result with fewer residuals.

3.2 Accuracy Under Recurrent/Non-recurrent Events

Typical traffic flows are following recurrent trends determined by morning and evening peak times. During those peaks, the vehicle speed is expected to be lower and traffic volume is expected to be higher. Besides the mentioned two, other features such as flow rate between different directions and data density over time are also good indicators for researchers to properly understand and model real-world traffic scenarios. There are different triggers for traffic congestion, recurrent events, and non-recurrent events. Both lead to the same traffic pattern with the mentioned features, however, recurrent events are comparably far easier to predict since the historical data can put into use. Under this circumstance, predictions under normal peak hours are shown with better agreements with ground truths. When turning this online, the accuracy is mostly dependent on the responsive rate of each method. As for the non-recurrent events such as bad weather, large-gathered venues, or accident, the importance of historical data is vastly diminished.

　　Non-recurrent events and online monitoring are two players chasing the same goal because both require a fast reaction on what happening on the road. In this case, prediction accuracy is dependent on the versatility of the traffic flow. In [4], Guo *et al.* proposed the SARIMA + GARCH structure which specialized in highly versatile traffic flow. It is shown from the result that AKF helps the model to achieve acceptable prediction results at the compromise of losing the ability to turn online. In addition, in [19], SGD + RS is able to identify non-recurrent events with fast convergence rate. The result shows that the quick response rate of SGD outperforms the other controlled groups in the special event prediction. Not only the trend prediction is promising, but the error rate index is also the lowest amongst the rest. This proves the feasibility of the SGD + RS model to be used online for both recurrent and non-recurrent traffic flow predictions.

4   The Current State of Art

As mentioned in Sect. 3, the core methodologies analyzed are all based on time series and various traffic flow prediction methods proposed by neural networks, such as DL, SAE, BN, AKF, and YOLOv3, which analyzed traffic conditions in similar scenarios. In addition to these articles, researchers studied a variety of traffic flow prediction methods and proposed different traffic forecast strategies. Those prediction methods can be divided into four major categories, namely 1) time-series models, 2) neural-networks-based methods, 3) integration methods, and 4) simulations. The time-series models mainly include autoregressive integrated moving average (ARIMA) model and its adaptive (ARIMA-like) models. Neural-networks-based methods include artificial neural network (ANN), CNN, k-nearest neighbor (k-NN) method, Bayesian Network, Support Vector Regression (SVR), Sparse auto-encoders (SAE), long short-term memory (LSTM). The integration method is mainly a combination of various models. And the simulation method uses traffic simulation tools to predict traffic flow.

4.1   Neural-Networks-Based Methods

Neural-network-based methods are widely used in traffic network traffic prediction systems. Those methods are essentially machine learning (ML) methods that have been validated by many research experiments for their accuracy and efficiency of prediction. In [17], Olayode *et al.* proposed a new neural network (NN) method that uses hybrid method of exponential smoothing (ES) to parse traffic flow data, and then use Levenberg-Marquardt (LM) algorithm to variant training model, thus improves the generalization ability of short-term prediction.

The ANNs can be very suitable for complex data processing tasks, thus is adapted here for predicting traffic network traffic. In [12], an adaptive NN architecture is proposed. The model was used as traffic predictor during network traffic modeling and also as video generator during network design phase. In addition, [18] proposed a dynamic multi-interval traffic volume prediction model based on k-NN non-parametric regression (KNN-NPR) method. When the time series data shows fluctuations or sudden changes, the KNN-NPR model can still be independent of its prediction accuracy.

In addition, researchers have proposed network traffic prediction methods based on deep learning and spatio-temporal compressed methods. There are some general ML tools applicable for traffic prediction. For example, SVR and online SVR regression machine application of supervised statistical learning technology can be used to predict atypical conditions, such as vehicle collision, severe weather and work on short-term highway traffic flows by region and holidays [1]. Also in [25], a Bayesian network is established, which can consider the random characteristics of the total average traffic level and the variability of the traffic, then predict the traffic and update these by combining the traffic distribution rules with the Bayesian network model.

4.2   Integration Methods

There are no algorithms that can be fit into any scenarios. This statement is especially true when it comes to traffic predictions. The existence of prediction flaws could easily cause erroneous traffic administrative maneuvering, unrealistic signaling as well as increasing traffic accidents [5]. In preventing this from happening, researchers have made ways in congregating different methods together. Methods that would compensate each other are beneficial to integrate, this process has becoming the most favorite path for improving prediction accuracy. For instance, [23] proposed a traffic flow prediction aggregation method based on the hybrid of moving average (MA), ES, ARIMA and NN models. The prediction results are used as baseline for the aggregation stage NN, and the output of the trained NN is used as the final prediction. The results show that the data aggregation (DA) model can obtain more accurate predictions than any single model alone.

5 Conclusion

This paper analyzed different methodologies focusing on the real-time short-term traffic prediction methods and proposed a possible solution for future research works. Throughout this literature study, different deep learning approaches have been introduced with how to set up structures, how to process collected data, how to verify the results under different testing cases. The prediction method has to be robust enough for high versatility. Traditional history-based learning methods are no longer valid since it is not feasibility to predict non-recurrent events. This also means the method with a relatively slow learning rate will be cut loose as well. Offline network structures will need to be converted into online fashion for faster convergence and higher efficiency. Adapting existing methods for online working conditions is necessary once the learning curves can be steepened, which is achievable potentially by dimension reduction techniques. The integration of different detection methods constitutes a more informed prediction system. Such integration could be data analysis and visualization. Each of the detection methods could result from the integration of multiple theories and/or algorithms.

References

1. Agarap, A.F.M.: A neural network architecture combining gated recurrent unit (GRU) and support vector machine (SVM) for intrusion detection in network traffic data. In: Proceedings of the 2018 10th International Conference on Machine Learning and Computing, pp. 26–30 (2018)
2. Duda, R.O., Hart, P.E., Stork, D.G.: Pattern Classification, 2nd edn., vol. 58, p. 16. Wiley, New York (2001)
3. Faghih-Imani, A., Eluru, N.: A finite mixture modeling approach to examine New York city bicycle sharing system (citibike) users' destination preferences. Transportation 47(2), 529–553 (2020)
4. Guo, J., Huang, W., Williams, B.M.: Adaptive kalman filter approach for stochastic short-term traffic flow rate prediction and uncertainty quantification. Transp. Res. C Emerg. Technol. 43, 50–64 (2014)
5. Huang, D., Deng, Z., Wan, S., Mi, B., Liu, Y.: Identification and prediction of urban traffic congestion via cyber-physical link optimization. IEEE Access 6, 63268–63278 (2018)
6. Jiang, L., Wang, Y., Zhao, Y.: Real-time traffic congestion detection with sighta regression network. In: 2019 IEEE 9th International Conference on Electronics Information and Emergency Communication (ICEIEC), pp. 45–50. IEEE (2019)
7. Karthika, M.B.: Traffic flow prediction using an improved fuzzy convolutional LSTM algorithm. Turk. J. Comput. Math. Educ. (TURCOMAT) 12(10), 5541–5549 (2021)
8. Kashyap, A.A., et al.: Traffic flow prediction models–a review of deep learning techniques. Cogent Eng. 9(1), 2010510 (2022)
9. Kim, H., Ye, L.: Bayesian mixture model to estimate freeway travel time under low-frequency probe data. Appl. Sci. 12(13), 6483 (2022)
10. Li, G., Pan, Y., Yang, Z., Ma, J.: Modeling vehicle merging position selection behaviors based on a finite mixture of linear regression models. IEEE Access 7, 158445–158458 (2019)
11. Liu, J., Kang, Y., Li, H., Wang, H., Yang, X.: STGHTN: spatial-temporal gated hybrid transformer network for traffic flow forecasting. Appl. Intell. 53, 1–17 (2022)
12. Lu, S., Zhang, Q., Chen, G., Seng, D.: A combined method for short-term traffic flow prediction based on recurrent neural network. Alex. Eng. J. 60(1), 87–94 (2021)

13. Luan, S., Ke, R., Huang, Z., Ma, X.: Traffic congestion propagation inference using dynamic bayesian graph convolution network. Transp. Res. C Emerg. Technol. **135**, 103526 (2022)
14. Manikandan, S., Chinnadurai, M., Vianny, D.M.M., Sivabalaselvamani, D.: Real time traffic flow prediction and intelligent traffic control from remote location for large-scale heterogeneous networking using tensorflow. Int. J. Future Gener. Commun. Netw. **13**(1), 1006–1012 (2020)
15. Maranzano, P., Otto, P., Fassò, A.: Adaptive lasso estimation for functional hidden dynamic geostatistical model. arXiv preprint arXiv:2208.05528 (2022)
16. Neelakandan, S., Prakash, M., Bhargava, S., Mohan, K., Robert, N.R., Upadhye, S.: Optimal stacked sparse autoencoder based traffic flow prediction in intelligent transportation systems. In: Hassanien, A.E., Gupta, D., Khanna, A., Slowik, A. (eds.) Virtual and Augmented Reality for Automobile Industry: Innovation Vision and Applications. Studies in Systems, Decision and Control, vol. 412, pp. 111–127. Springer, Cham (2022). https://doi.org/10.1007/978-3-030-94102-4_6
17. Olayode, I.O., Severino, A., Campisi, T., Tartibu, L.K.: Prediction of vehicular traffic flow using levenberg-marquardt artificial neural network model: Italy road transportation system. Commun.-Sci. Lett. Univ. Zilina **24**(2), E74–E86 (2022)
18. Peng, H., et al.: Spatial temporal incidence dynamic graph neural networks for traffic flow forecasting. Inf. Sci. **521**, 277–290 (2020)
19. Polson, N.G., Sokolov, V.O.: Deep learning for short-term traffic flow prediction. Transp. Res. C Emerg. Technol. **79**, 1–17 (2017)
20. Sebai, M., Rejeb, L., Denden, M.A., Amor, Y., Baati, L., Said, L.B.: Optimal electric vehicles route planning with traffic flow prediction and real-time traffic incidents. Int. J. Electr. Comput. Eng. Res. **2**(1), 1–12 (2022)
21. Shang, P., Liu, X., Yu, C., Yan, G., Xiang, Q., Mi, X.: A new ensemble deep graph reinforcement learning network for spatio-temporal traffic volume forecasting in a freeway network. Digit. Signal Process. **123**, 103419 (2022)
22. Wang, S., Patwary, A., Huang, W.: A general framework for combining traffic flow models and bayesian network for traffic parameters estimation. Transp. Res. C Emerg. Technol. **139**, 103664 (2022)
23. Xu, X., Jin, X., Xiao, D., Ma, C., Wong, S.: A hybrid autoregressive fractionally integrated moving average and nonlinear autoregressive neural network model for short-term traffic flow prediction. J. Intell. Transp. Syst. **27**, 1–18 (2021)
24. Zhu, K., Zhang, S., Li, J., Zhou, D., Dai, H., Hu, Z.: Spatiotemporal multi-graph convolutional networks with synthetic data for traffic volume forecasting. Expert Syst. Appl. **187**, 115992 (2022)
25. Zong, F., Chen, X., Tang, J., Yu, P., Wu, T.: Analyzing traffic crash severity with combination of information entropy and bayesian network. IEEE Access **7**, 63288–63302 (2019)

Interval Observers Design for Discrete-Time Linear Systems with Uncertainties

Alexey Zhirabok[1] ⓘ, Alexander Zuev[2,4(✉)] ⓘ, Vladimir Filaretov[3,4] ⓘ, and Changan Yuan[5]

[1] Far Eastern Federal University, Vladivostok 690992, Russia
[2] Institute of Marine Technology Problems, Vladivostok 690950, Russia
avzuev@yandex.ru
[3] Institute of Automation and Process Control, Vladivostok 690950, Russia
[4] Vladivostok State University, Vladivostok 690014, Russia
[5] Guangxi Academy of Science, Nanning 530007, China

Abstract. The paper studies the problem of designing interval observers for discrete-time linear dynamic systems under the external disturbances, measurement noises, and unknown system parameters. To construct such observers, we use not the original system but its reduced-order model of the original system of minimal dimension which is insensitive to the disturbances. The observers are designed in such a way to estimate the prescribed linear function of the original system state vector. Canonical form of the observer is used that allows to simplify the design procedure. The obtained results are displayed by example of the electric servoactuator.

Keywords: Linear systems · estimation · interval observers · uncertainties · canonical form

1 Introduction

In [1], a concept of intelligent control-emergency systems was proposed using knowledge bases for a comprehensive analysis of autonomous underwater vehicles malfunctions and developing solutions to adapt to them. Knowledge bases allow obtaining formal description of malfunctions and attributes to identify them. Besides, an ontological approach to construction of these databases makes it possible to unify their structure and include domain experts in the development process who provide its qualitative content. For realization of intelligent control-emergency systems, it is necessary to promptly receive information about the correct functioning and operability of individual subsystems of the underwater vehicles. This can be done by using interval observers.

The problem to estimate the system vector of state is very important for many practical applications. The main difficulties in designing an estimator are the system

Supported by the Ministry of Science and Higher Education of the Russian Federation, Project No. FZNS-2023-0011.

complexity and uncertainty (unknown parameters, measurement noises, and external disturbances). It is known that sliding mode observers can solve this problem [2–4]; under uncertainties, however, the estimation error is not equal to zero. Recently, this problem has been successfully solved on the basis of interval observers evaluating the state of system. An advantage of such observers is that they allow to consider many types of the system uncertainties: measurement noise, external disturbances, parametric uncertainties.

Such an approach can be used to deal with significant disturbances and provide component-wise information on possible solutions. Therefore, this approach is fundamentally different from conventional techniques of robust stability analysis or control law construction for different perturbed processes. An advantage of interval observer is that it allows to take into account many types of uncertainties in the system. The main peculiarity of interval observation is that it is necessary to ensure positivity of the estimation error dynamics in addition to their stability.

For many types of dynamic systems different interval observers have been developed in many papers: for linear and non-linear continuous-time [5–15], discrete-time linear and non-linear [16, 17], for time-delay systems [18, 19], for Takagi-Sugeno fuzzy systems [20], for switched systems [21–23], and singular systems [19]. Many practical problems also successfully have been solved [24–26]. Exhaustive reviews are in [18, 27, 28].

Note that the above-mentioned papers solve the problem of estimation of full state vector. Contrary to these papers, the interval observers in the present paper are designed to estimate the prescribed linear function of the original system state vector. Such a solution may be useful in some practical applications. Our approach is closed to that based on the functional interval observers developed in [29–31] which enable estimating some linear function of the state vector. In comparison with [30, 31], we take into account system parametric uncertainties; unlike [29], our approach considers measurement noise.

The main contribution of the paper is that interval observers estimate not the state vector but its some prescribed linear function. The suggested solution is based on the reduced order model of minimal dimension which is invariant with respect to the disturbances. This allows to reduce the interval width and the dimension of observer in comparison with [11, 16, 18] and similar papers. Besides, in Sect. 2 we design interval observers insensitive to the external disturbances and parametric uncertainties. In comparison with [32] where measurement noise and external disturbances are considered, the present paper takes into account parametric uncertainties additionally. Based on the reduced order model, one can obtain more precise estimates for the full state vector. Besides, a new method to design such observers on the basis of the identification canonical form is suggested. As a result, this enables designing such observers for broadened class of dynamic systems.

2 The Main Models

Consider the following model of a linear system

$$
\begin{aligned}
x(t + 1) &= A(\mu)x(t) + Bu(t) + Q\rho(t), \\
y(t) &= Cx(t) + w(t)
\end{aligned}
\tag{1}
$$

with states $x \in R^n$, inputs $u \in R^m$, and outputs $y \in R^l$; the matrix $A(\mu)$ is of the form $A(\mu) = (A + \Delta A(\mu(t)))$, where $A \in R^{n \times n}$ is constant matrix, the matrix function $\Delta A(\mu)$ is known for a given value of μ, $\mu(t) \in \Pi \subset R^s$ is the bounded vector of parameters where Π is known; it is assumed by analogy with [18] that the values of $\mu(t)$ cannot be measured. The matrices $B \in R^{n \times l}$, $C \in R^{l \times n}$, and $Q \in R^{n \times p}$ are known constant matrices; $\rho(t) \in R^p$ is a bounded unknown function with $\|\rho(t)\| \leq \rho_*$ describing the unmatched disturbance; $w(t) \in R^l$ is a bounded unknown function with $\|w(t)\| \leq w_*$ describing the measurement noise.

The objective is to develop the method of interval observer design generating functions $\underline{v}(t)$ and $\overline{v}(t)$ under the condition $\underline{v}(t) \leq v(t) \leq \overline{v}(t)$ for all $t \geq 0$ where $v(t)$ is such that $v(t) = Mx(t)$ for known matrix M. For matrices $A^{(1)}$, $A^{(2)}$ and vectors $x^{(1)}$, $x^{(2)}$, the inequalities $A^{(1)} \leq A^{(2)}$ and $x^{(1)} \leq x^{(2)}$ one understands elementwise.

In comparison with [11, 16, 18] where the problem of estimation of full state vector is studied, the suggested solution is based on the reduced order model of minimal dimension which is invariant with respect to the disturbances. This allows to reduce the interval width and the dimension of interval observer.

Note that the problem when the parametric uncertainties are absent, that is $\Delta A(\mu(t)) = 0$ was solved in [32]. Recall briefly the main results of this solution.

The solution is based on the reduced-order model of the initial system:

$$\begin{aligned} \eta(t + 1) &= A_*\eta(t) + B_*u(t) + G_*Cx(t) + Q_*\rho(t), \\ v(t) &= C_*\eta(t) + P_*y_0(t). \end{aligned} \tag{2}$$

Here $\eta \in R^k$ is the vector of state, the matrix A_* is specified in the identification canonical form

$$A_* = \begin{pmatrix} 0 & 1 & 0 & \dots & 0 \\ 0 & 0 & 1 & \dots & 0 \\ \dots & \dots & \dots & & \dots \\ 0 & 0 & 0 & \dots & 0 \end{pmatrix}, \tag{3}$$

the matrices B_*, G_*, Q_*, P_*, and C_* of the appropriate dimensions have to be determined, $y_0(t)$ will be determined later. Since the matrix A_* is stable and nonnegative, system (2) is cooperative [18].

Remark 1. The term $G_*Cx(t)$ in the model (2) is used instead of $G_*y(t)$ due to the necessity to account measurement noise since $y(t) = Cx(t) + w(t)$. We will use the term $G_*y(t)$ in the interval observers (10) and (17).

We assume that $x_*(t) = \Phi x(t)$ for some matrix Φ satisfying the conditions [4, 34]

$$\Phi A = A_*\Phi + G_*C, \quad B_* = \Phi B, \quad Q_* = \Phi Q. \tag{4}$$

The variable $y_0(t)$ in (2) must be free of the disturbance $\rho(t)$. The method to obtain such a variable was suggested in [32]; it is based on the equation

$$(N_1 - N_2)\begin{pmatrix} Q_0 \\ C \end{pmatrix} = 0$$

for some matrices N_1 and N_2 of maximal rank. Solution of this equation produces $y_0 = N_2 y(t) = N_2 Cx$. Here Q_0 is the maximal rank matrix satisfying the condition $Q_0 Q = 0$.

The relation $v(t) = Mx(t)$ and (2) result in

$$M = C_* \Phi + P_* N_2 H = (C_* \ P_*) \begin{pmatrix} \Phi \\ N_2 C \end{pmatrix}. \tag{5}$$

This equation is solvable if

$$rank \begin{pmatrix} \Phi \\ N_2 C \end{pmatrix} = rank \begin{pmatrix} \Phi \\ N_2 C \\ M \end{pmatrix}. \tag{6}$$

The best observer generating the minimal interval width is when the model is free of the parametric uncertainties $\Delta A(\mu(t))$ and the disturbance $\rho(t)$. Such a model can be designed based on the equation

$$\left(\Phi_1 \ -G_{*1} \ ... \ -G_{*k} \right) \left(W^{(k)} \ Q^{(k)} \ \Delta^{(k)} \right) = 0, \tag{7}$$

generalizing the equation suggested in [4, 34] where

$$W^{(k)} = \begin{pmatrix} A^k \\ CA^{k-1} \\ ... \\ C \end{pmatrix},$$

$$Q^{(k)} = \begin{pmatrix} Q \ AQ \ ... \ A^{k-1}Q \\ 0 \ CQ \ ... \ CA^{k-2}Q \\ ... \ ... \ ... \ ... \\ 0 \ 0 \ ... \ 0 \end{pmatrix},$$

$$\Delta^{(k)} = \begin{pmatrix} \Delta A(\mu) \ A\Delta A(\mu) \ ... \ A^{k-1}\Delta A(\mu) \\ 0 \ C\Delta A(\mu) \ ... \ CA^{k-2}\Delta A(\mu) \\ ... \ ... \ ... \ ... \\ 0 \ 0 \ ... \ 0 \end{pmatrix},$$

G_{*i} is the i-th row of the matrix G_*. The matrices $Q^{(k)}$ and $\Delta^{(k)}$ guarantee invariance of the model with respect to the disturbance and parametric uncertainties, respectively, $W^{(k)}$ enables to design the model (2). The Eq. (7) is solvable when

$$rank \left(W^{(k)} \ Q^{(k)} \ \Delta^{(k)} \right) < n + lk. \tag{8}$$

To design the model, one has to find minimal k from (8); the matrix $(\Phi_1 \ - G_{*1} \ ... \ - G_{*k})$ is determined from (7) and the rows of the matrix Φ are found from the equations

$$\Phi_i A = \Phi_{i+1} + G_{*i} C, i = 1, ..., k-1, \ \Phi_k A = G_{*k} C \tag{9}$$

which can be obtained based on (3) and (4). Finally, the matrices B_*, C_*, and P_* are found from (5) and (4).

Based on the model (2), the interval observer is designed:

$$\underline{\eta}(t+1) = A_*\underline{\eta}(t) + B_*u(t) + G_*y(t) - |G_*|E_lw_* - |Q_*|E_p\rho_*,$$
$$\overline{\eta}(t+1) = A_*\overline{\eta}(t) + B_*u(t) + G_*y(t) + |G_*|E_lv_* + |Q_*|E_p\rho_*, \qquad (10)$$
$$\underline{\eta}(0) = \underline{\eta}_0, \overline{\eta}(0) = \overline{\eta}_0.$$

Here E_k is $(k \times 1)$-matrix: $E_k = (1\ 1\ ...\ 1)^T$, the matrix $|G_*|$ contains the absolute values of the corresponding entries of G_*.

Theorem 1. [32] When $\underline{\eta}(0) \leq \eta(0) \leq \overline{\eta}(0)$, then it follows for all $t \geq 0$.

$$\underline{\eta}(t) \leq \eta(t) \leq \overline{\eta}(t), \quad \underline{v}(t) \leq v(t) \leq \overline{v}(t), \qquad (11)$$

where

$$\underline{v}(t) = C_*\underline{\eta}(t) + P_*y_0(t), \quad \overline{v}(t) = C_*\overline{\eta}(t) + P_*y_0(t) \qquad (12)$$

if $C_* \geq 0$ and

$$\underline{v}(t) = C_*\overline{\eta}(t) + P_*y_0(t), \quad \overline{v}(t) = C_*\underline{\eta}(t) + P_*y_0(t) \qquad (13)$$

if $C_* \leq 0$.

Note that when (8) is true for some k, then $Q_* = 0$ in (10) and the interval width is minimal. If (8) is not true for all k, one has to check the condition

$$rank\left(W^{(k)}\ \Delta^{(k)}\right) < n + lk. \qquad (14)$$

If it is satisfied, then $Q_* \neq 0$, that is the disturbance affects the model, and the interval width becomes greater. It can be reduced by applying robust method based on singular value decomposition described in [32]. If (14) is not satisfied for all k, one has to use the method developed in Sect. 3.

3 Parametric Uncertainty

Assume that (14) is not satisfied for all k and $\underline{\Delta A} \leq \Delta A(\mu(t)) \leq \overline{\Delta A}$ for all $\mu(t) \in \Pi$ with some $\underline{\Delta A}, \overline{\Delta A} \in R^{n \times n}$. The interval $\left(\underline{\Delta A}, \overline{\Delta A}\right)$ can be calculated for known Π and $\Delta A(\mu)$ [18]. To simplify the procedure, assume initially that $w = 0$ and $\rho = 0$.

The model estimating the variable $v(t)$ is based on the model

$$\eta(t+1) = (A_* + \Delta A_*)\eta(t) + B_*u(t) + \left(G_* + G'\right)y(t),$$
$$v(t) = C_v\eta(t) + Py_0(t). \qquad (15)$$

The matrices ΔA_* and G' can be found as follows. It follows from (1) and (15)

$$\Phi(A + \Delta A(\mu)) = (A_* + \Delta A_*)\Phi + G_*C + G'C.$$

Since Φ satisfies (4), one obtains $\Phi\Delta A(\mu) = \Delta A_*\Phi + G'C$, or

$$\Phi\Delta A(\mu) = \begin{pmatrix} \Delta A_* & G' \end{pmatrix}\begin{pmatrix} \Phi \\ C \end{pmatrix}. \tag{16}$$

After obtaining the matrix Φ we find ΔA_* and G' from (16). Generally, the matrix G' depends on ΔA. Assume for simplicity that G' does not depend on ΔA.

Assumption. $\underline{\Delta A}_* \leq \Delta A_* \leq \overline{\Delta A}_*$ For some $\underline{\Delta A}_*$ and $\overline{\Delta A}_*$.

Given a matrix ΔA, define by analogy with [18] $\Delta A^+ = max(0, \Delta A)$ and $\Delta A^- = \Delta A^+ - \Delta A$; clearly, $\Delta A^+ \geq 0$ and $\Delta A^- \geq 0$.

Lemma 1. [19] *If $\Delta A_* \in R^{k \times k}$ is constant and $\underline{\eta} \leq \eta \leq \overline{\eta}$, then*

$$\Delta A_*^+\underline{\eta} - \Delta A_*^-\overline{\eta} \leq \Delta A_*\eta \leq \Delta A_*^+\overline{\eta} - \Delta A_*^-\underline{\eta}.$$

If $\underline{\Delta A}_* \leq \Delta A_* \leq \overline{\Delta A}_*$ for some $\underline{\Delta A}_*$, ΔA_*, $\overline{\Delta A}_*$ and $\underline{\eta} \leq \eta \leq \overline{\eta}$, then

$$\underline{\Delta A}^+\underline{\eta}^+ - \overline{\Delta A}_*^+\underline{\eta}^- - \underline{\Delta A}^-\overline{\eta}^+ + \overline{\Delta A}_*^-\overline{\eta}^- \leq \Delta A_*\eta$$
$$\leq \overline{\Delta A}_*^+\overline{\eta}^+ - \underline{\Delta A}^+\overline{\eta}^- - \overline{\Delta A}_*^-\underline{\eta}^+ + \underline{\Delta A}^-\underline{\eta}^-.$$

The interval observer is given by

$$\underline{\eta}(t+1) = A_*\underline{\eta}(t) + \left(\underline{\Delta A}^+\underline{\eta}^+ - \overline{\Delta A}_*^+\underline{\eta}^- - \underline{\Delta A}^-\overline{\eta}^+ + \overline{\Delta A}_*^-\overline{\eta}^-\right)$$
$$+ B_*u(t) + (G_* + G')y(t),$$
$$\overline{\eta}(t+1) = A_*\overline{\eta}(t) + \left(\overline{\Delta A}_*^+\overline{\eta}^+ - \underline{\Delta A}^+\overline{\eta}^- - \overline{\Delta A}_*^-\underline{\eta}^+ + \underline{\Delta A}^-\underline{\eta}^-\right) \tag{17}$$
$$+ B_*u(t) + (G_* + G')y(t),$$
$$\underline{\eta}(0) = \underline{\eta}_0, \overline{\eta}(0) = \overline{\eta}_0,$$

The estimation errors are as follows:

$$\underline{e}_*(t+1) = A_*\underline{e}_*(t) + \Delta A_*\eta(t) - (\underline{\Delta A}^+\underline{x}^+(t) - \overline{\Delta A}_*^+\underline{\eta}^-(t)$$
$$- \underline{\Delta A}^-\overline{\eta}^+(t) + \overline{\Delta A}_*^-\overline{\eta}^-(t)),$$
$$\overline{e}_*(t+1) = A_*\overline{e}_*(t) + (\overline{\Delta A}_*^+\overline{\eta}^+(t) - \underline{\Delta A}^+\overline{\eta}^-(t) - \overline{\Delta F}_*^-\underline{\eta}^+(t)$$
$$+ \underline{\Delta A}^-\underline{\eta}^-(t)) - \Delta A_*\eta(t).$$

Theorem 2. *If* $\underline{\Delta A}_* \leq \Delta A_* \leq \overline{\Delta A}_*$ *and* $\underline{\eta}(0) \leq \eta(0) \leq \overline{\eta}(0)$, *then the relations* (11) *with* (12) *and* (13) *are true for the observer* (17) *and all* $t \geq 0$.

Proof. The condition $\underline{\eta}(0) \leq \eta(0) \leq \overline{\eta}(0)$ implies $\underline{e}_x(0), \overline{e}_x(0) \geq 0$. Since $A_* \geq 0$ and $\underline{e}_x(0) \geq 0$, it follows from Lemma that $\underline{e}_x(1) \geq 0$ that is $\underline{\eta}(1) \leq \eta(1)$. It can be shown by induction that $\underline{\eta}(t) \leq \eta(t)$ is true for all $t \geq 0$. The relation $\eta(t) \leq \overline{\eta}(t)$ one can prove analogously.

If $C_* \geq 0$, it follows from (12)

$$\underline{e}_v(t) = v(t) - \underline{v}(t) = C_*\eta(t) + P_*y_0(t) - \left(C_*\underline{\eta}(t) + P_*y_0(t)\right) = C_*\underline{e}_*(t),$$
$$\overline{e}_v(t) = \overline{v}(t) - v(t) = C_*\overline{\eta}(t) + P_*y_0(t) - (C_*\eta(t) + P_*y_0(t)) = C_*\overline{e}_*(t).$$

Since $\underline{e}_x(t), \overline{e}_x(t) \geq 0$, one obtains $\underline{e}_v(t), \overline{e}_v(t) \geq 0$ which is equivalent to $\underline{v}(t) \leq v(t) \leq \overline{v}(t)$. If $C_* \leq 0$, one has from (13)

$$\underline{e}_v(t) = v(t) - \underline{v}(t) = C_*\eta(t) + P_*y_0(t) - (C_*\overline{\eta}(t) + P_*y_0(t)) = -C_*\overline{e}_*(t),$$
$$\overline{e}_v(t) = \overline{v}(t) - v_*(t) = C_*\eta(t) + P_*y_0(t) - (C_*\underline{\eta}(t) + P_*y_0(t)) = -C_*\underline{e}_*(t).$$

Taking into account $C_* \leq 0$, the relations $\underline{e}_v(t), \overline{e}_v(t) \geq 0$ can be obtained as well. Theorem has been proved. □

The case when $w \neq 0$ and $\rho \neq 0$ can be taken into account by additional addends $-|G_*|E_l w_* - |Q_*|E_p\rho_*$ and $|G_*|E_l w_* + |Q_*|E_p\rho_*$ in (17) by analogy with (10).

4 Example

Consider the control system

$$\begin{aligned}
x_1(t+1) &= \gamma_1 x_2(t) + x_1(t), \\
x_2(t+1) &= (\gamma_2 + \delta_1(t))x_2(t) + \gamma_3 x_3(t) + \rho(t), \\
x_3(t+1) &= \gamma_4 x_2(t) + (\gamma_5 + \delta_2(t))x_3(t) + \gamma_6 u(t), \\
y_1(t) &= x_1(t) + w_1(t), \\
y_2(t) &= x_3(t) + w_2(t).
\end{aligned} \tag{18}$$

Equation (18) constitute the sampled-data model of the robot electric servoactuator. The coefficients $\gamma_1 \div \gamma_6$ depend on the servoactuator parameters and the sampling time; the function $\rho(t)$ is induced by the external loading moment; the uncertainty $\delta_1(t)$ is due to change of inertia properties, $\delta_2(t)$ is due to change of active resistances.

The matrices describing the system are given by

$$A = \begin{pmatrix} 1 & \gamma_1 & 0 \\ 0 & \gamma_2 & \gamma_3 \\ 0 & \gamma_4 & \gamma_5 \end{pmatrix}, B = \begin{pmatrix} 0 \\ 0 \\ \gamma_6 \end{pmatrix}, C = \begin{pmatrix} 1 & 0 & 0 \\ 0 & 0 & 1 \end{pmatrix},$$
$$Q = \begin{pmatrix} 0 \\ 1 \\ 0 \end{pmatrix}, \Delta_1 A(t) = \begin{pmatrix} 0 & 0 & 0 \\ 0 & \delta_1(t) & 0 \\ 0 & 0 & 0 \end{pmatrix}, \Delta_2 A(t) = \begin{pmatrix} 0 & 0 & 0 \\ 0 & 0 & 0 \\ 0 & 0 & \delta_2(t) \end{pmatrix}.$$

The problem is to estimate the variables $x_1(t)$, $x_2(t)$, and $x_3(t)$ that is

$$M = \begin{pmatrix} 1 & 0 & 0 \\ 0 & 1 & 0 \\ 0 & 0 & 1 \end{pmatrix}.$$

It should be noted that to solve this problem, the approaches suggested in [11, 16, 18] and similar papers design full-order interval observer. Our approach allows reducing the observer dimension and interval width.

Since $y_1(t) = x_1(t) + w_1(t)$ and $y_2(t) = x_3(t) + w_2(t)$, then

$$\underline{x}_1(t) = y_1(t) - w_{*1}, \overline{x}_1(t) = y_1(t) + w_{*1}$$

and

$$\underline{x}_3(t) = y_2(t) - w_{*2}, \overline{x}_3(t) = y_2(t) + w_{*2}.$$

To estimate the variables $x_2(t)$, solve the Eq. (7) with $k = 1$:

$$(\Phi - G_*) \begin{pmatrix} 1 & \gamma_1 & 0 \\ 0 & \gamma_2 & \gamma_3 \\ 0 & \gamma_4 & \gamma_5 \\ 1 & 0 & 0 \\ 0 & 0 & 1 \end{pmatrix} = 0.$$

Its solution is $\Phi = (1/\gamma_1 \; -1/\gamma_2 \; 0)$ and $G_* = (1/\gamma_1 \; -\gamma_3/\gamma_2)$ that gives $B_* = 0$ and $Q_* = -1/\gamma_2$. Clearly, $C_* = -\gamma_2$, $P_* = (\gamma_2/\gamma_1 \; 0)$, $\Delta A_{*1}(t) = \delta_1(t)$, $G'_1 = (-\delta_1(t)/\gamma_2 \; 0)$; $\Delta A_{*2}(t) = 0$ and $G'_2 = 0$ according to (16).

The reduced order model is given by

$$\eta(t+1) = \delta_1(t)\eta(t) + (1/\gamma_1 - \delta_1(t)/\gamma_2)C_1 x(t) - (\gamma_3/\gamma_2)C_2 x(t) - \rho(t)/\gamma_2,$$
$$v(t) \quad = -\gamma_2 \eta(t) + (\gamma_2/\gamma_1)y_1(t).$$

The observer has the following description:

$$\eta(t+1) = \delta_1 \eta(t) + \left(1/\gamma_1 + \underline{\underline{Sg}}/\gamma_2\right)y_1(t) - (\gamma_3/\gamma_2)y_2(t)$$
$$\quad -\left(1/\gamma_1 + \overline{\delta}_1\right)w_{*1} - (\gamma_3/\gamma_2)w_{*2} - \rho_*/\gamma_2,$$
$$\overline{\eta}(t+1) = \overline{\delta}_1\overline{\eta}(t) + \left(1/\gamma_1 + \overline{\overline{Sg}}/\gamma_2\right)y_1(t) - (\gamma_3/\gamma_2)y_2(t) \qquad (19)$$
$$\quad +\left(1/\gamma_1 + \overline{\delta}_1\right)w_{*1} + (\gamma_3/\gamma_2)w_{*2} + \rho_*/\gamma_2,$$
$$\underline{v}(t) \quad = -\gamma_2 \overline{\eta}(t) + (\gamma_2/\gamma_1)y_1(t),$$
$$\overline{v}(t) \quad = -\gamma_2 \eta(t) + (\gamma_2/\gamma_1)y_1(t),$$

where

$$\underline{\underline{Sg}} = 0.5\big((1 - sign(y_1(t)))\underline{\delta}_1 + (1 + sign(y_1(t)))\overline{\delta}_1\big),$$
$$\overline{\overline{Sg}} = 0.5\big((1 - sign(y_1(t)))\overline{\delta}_1 + (1 + sign(y_1(t)))\underline{\delta}_1\big).$$

Comparing the obtained results and the results which can be obtained for this example by methods developed in [11, 16, 18] and similar papers, we may conclude that the dimension of the observer (19) is fewer than that in [11, 16, 18] and the suggested approach produces the estimations with smaller interval width since the ones for $x_1(t)$ and $x_3(t)$ do not contain the disturbance ρ_* and the uncertainties. Besides, the observer (19) does not contain the uncertainty $\delta_2(t)$.

For simulation consider the model (18) and the observer (19); the measurement noises $w_1(t)$ are $w_2(t)$ are random processes evenly distributed on $[-0.01, 0.01]$, the parametric uncertainty δ_1 is modeled as $\delta_1 = 0.03(1 + sin(10t))$. Set for simplicity $\gamma_1 = \gamma_2 = \gamma_5 = \gamma_6 = 1$, $\gamma_3 = \gamma_4 = -1$; set $\underline{\delta}_1 = 0$ and $\overline{\delta}_1 = 0.06$; $w_{*1} = w_{*2} = \rho_* = 0.01$. Figures 1 and 2 illustrate simulation results, where $\underline{v}(t)$, $\overline{v}(t)$, and $x_2(t)$ are presented for $\eta(0) = 0$, $\underline{\eta}(0) = -0.05$, and $\overline{\eta}(0) = 0.05$. In Fig. 1, the control $u(t) = 0.2$; in Fig. 2, $u(t) = 0.2sin(t/20)$.

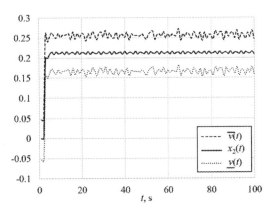

Fig. 1. Behavior of $x_2(t)$ and $\underline{v}(t)$ and $\overline{v}(t)$ with $u(t) = 0.2$

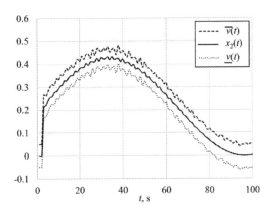

Fig. 2. Behavior of $x_2(t)$ and $\underline{v}(t)$ and $\overline{v}(t)$ with $u(t) = 0.2sin(t/20)$

5 Conclusion

The paper has studied the problem of interval observer design for linear discrete-time systems under the disturbance, measurement noise, and unknown parameters. The reduced-order model which does not depend on the disturbances or has minimal sensitivity to them and is realized in identification canonical form has been used to solve the problem. The designed interval observer has minimal dimension and estimates the prescribed linear function of the original system state vector with the reduced interval width. The limitation of the proposed approach is that it can be applied for limited class of nonlinear systems; nonlinearities should satisfy some requirements. A future research direction is the interval observer design for hybrid dynamic systems.

References

1. Timoshenko, A., Zuev, A., Mursalimov, E., Gribova, V., Inzartsev, A.: Description and fault diagnosis in autonomous underwater vehicles based on ontologies. In: 2022 International Conference on Ocean Studies (ICOS), Vladivostok, pp. 39–44 (2022). https://doi.org/10.1109/ICOS55803.2022.10033380
2. Edwards, C., Spurgeon, S., Patton, R.: Sliding mode observers for fault detection and isolation. Automatica **36**, 541–553 (2000)
3. Fridman, L., Levant, A., Davila, J.: Observation of linear systems with unknown inputs via high-order sliding-modes. Int. J. Syst. Sci. **38**, 773–791 (2007)
4. Zhirabok, A., Zuev, A., Seriyenko, O., Shumsky, A.: Fault identificaition in nonlinear dynamic systems and their sensors based on sliding mode observers. Autom. Remote Control **83**, 214–236 (2022)
5. Chebotarev, S., Efimov, D., Raissi, T., Zolghadri, A.: Interval observers for continuous-time LPV systems with performance. Automatica **58**, 82–89 (2015)
6. Degue, K., Efimov, D., Richard, J.: Interval observers for linear impulsive systems. IFAC-PapersOnLine **49**(18), 867–872 (2016)
7. Dinh, T., Mazenc, F., Niculescu, S.: Interval observer composed of observers for nonlinear systems. In: 2014 European Control Conference (ECC), Strasbourg, pp. 660–665 (2014)
8. Kolesov, N., Gruzlikov, A., Lukoyanov, E.: Using fuzzy interacting observers for fault diagnosis in systems with parametric uncertainty. In: XII-th International Symposium on Intelligent Systems, INTELS 2016, Moscow, pp. 499–504 (2016)
9. Krokavec, D., Filasova, A.: Interval observers design for systems with ostensible Metzler system matrices. Front. Aerosp. Eng. **2**, 1158718 (2023)
10. Mazenc, F., Bernard, O.: Interval observers for linear time-invariant systems with disturbances. Automatica **47**, 140–147 (2011)
11. Raissi, T., Efimov, D., Zolghadri, A.: Interval state estimation for a class of nonlinear systems. IEEE Trans. Automat. Control **57**, 260–265 (2012)
12. Zheng, G., Efimov, D., Perruquetti, W.: Interval state estimation for uncertain nonlinear systems. In: IFAC Nolcos 2013, Toulouse (2013)
13. Wang, T., Li, Y., Xiang, W.: Design of interval observer for continuous linear large-scale systems with disturbance attenuation. J. Frankl. Inst. **359**(8), 3910–3929 (2022)
14. Zhirabok, A., Zuev, A., Kim, C.: A method to design interval observers for linear time-invariant systems. J. Comput. Syst. Sci. Int. **65**(5), 485–495 (2022)
15. Zhu, F., Fu, Y., Dinh, T.: Asymptotic convergence unknown input observer design via interval observer. Automatica **147**(12), 110744 (2023)

16. Efimov, D., Raissi, T., Perruquetti, W., Zolghadri, A.: Interval observers for time-varying discrete-time systems. IEEE Trans. Autom. Control. **58**, 3218–3224 (2013)
17. Mazenc, F., Dinh, T., Niculescu, S.: Interval observers for discrete-time systems. Inter. J. Robust Nonlin. Control **24**, 2867–2890 (2014)
18. Efomov, D., Raissi, T.: Design of interval state observers for uncertain dynamical systems. Autom. Remote Control **77**, 191–225 (2015)
19. Efimov, D., Polyakov, A., Richard, J.: Interval observer design for estimation and control of time-delay descriptor systems. Eur. J. Control **23**, 26–35 (2015)
20. Li, J., Wang, Z., Shen, Y., Wang, Y.: Interval observer design for discrete-time uncertain Takagi-Sugeno fuzzy systems. IEEE Trans. Fuzzy Syst. **27**(4), 816–823 (2019)
21. Marouani, G., Dinh, T., Raissi, T., Wang, X., Messaoud, H.: Unknown input interval observers for discrete-time linear switched systems. Eur. J. Control **59**, 165–174 (2021)
22. Raissi, T.: Interval observers design for continuous and switched systems. In: GT Controle – Sorbonne Universite, 7 Octobre 2022 (2022)
23. Zammali, C., Gorp, J., Wang, Z., Raissi, T.: Sensor fault detection for switched systems using interval observer with performance. Eur. J. Control **57**, 147–156 (2020)
24. Blesa, J., Puig, V., Bolea, Y.: FDI and FTC of wind turbines using the interval observer approach and virtual actuators/sensors. Control Eng. Pract. **24**, 138–155 (2014)
25. Rotondo, D., Fernandez-Canti, R., Tornil-Sin, S.: Robust fault diagnosis of proton exchange membrane fuel cells using a Takagi-Sugeno interval observer approach. Int. J. Hydrog. Energy **41**, 2875–2886 (2016)
26. Zhang, K., Jiang, B., Yan, X., Edwards, C.: Interval sliding mode based fault accommodation for non-minimal phase LPV systems with online control application. Int. J. Control. (2019). https://doi.org/10.1080/00207179.2019.1687932
27. Khan, A., Xie, W., Zhang, L., Liu, L.: Design and applications of interval observers for uncertain dynamical systems. IET Circuits Devices Syst. **14**, 721–740 (2020)
28. Khan, A., Xie, W., Zhang, L., Liu, L.: A survey of interval observers design methods and implementation for uncertain systems. J. Frankl. Inst. **358**, 3077–3126 (2021)
29. Gu, D., Liu, L., Duan, G.: Functional interval observer for the linear systems with disturbances. IET Control Theory Appl. **12**, 2562–2568 (2018)
30. Liu, J., Xie, W., Khan, A., Zhang, L.: Finite-time functional interval observer for linear systems with uncertainties. IET Control Theory Appl. **14**, 2868–2878 (2020)
31. Meyer, L.: Robust functional interval observer for multivariable linear systems. J. Dyn. Syst. Measur. Control **141**, 094502 (2019)
32. Sergiyenko, O., et al.: Interval observers for discrete-time linear systems with uncertainties. Symmetry **14**, 2131 (2022)
33. Zhirabok, A., Shumsky, A., Pavlov, S.: Diagnosis of linear dynamic systems by the nonparametric method. Autom. Remote Control **78**, 1173–1188 (2017)
34. Zhirabok, A., Shumsky, A., Solyanik, S., Suvorov, A.: Fault detection in nonlinear systems via linear methods. Int. J. Appl. Math. Comp. Sci. **27**, 261–272 (2017)

Transformer-Based Multi-industry Electricity Demand Forecasting

Zhenli Deng[1], Fangzhao Deng[1], Meng Yang[1], Xingwu Guo[1], Jianan Si[1], Boning Yu[1], and Boxiang Feng[2(✉)]

[1] State Grid Henan Electric Power Company Economic and Technological Research Institute, Zheng Zhou, China

[2] School of Informatics, Xiamen University, Xiamen, China
boxiangfeng@stu.xmu.edu.cn

Abstract. The accuracy of electricity demand forecasting is closely related to the correctness of decision-making in the power system, ensuring stable energy supply. Stable energy supply is a necessary guarantee for socioeconomic development and normal human life. Accurate electricity demand forecasting can provide reliable guidance for electricity production and supply dispatch, improve the power system's supply quality, and ultimately enhance the security and cost-effectiveness of power grid operation, which is crucial for boosting economic and social benefits. Currently, research on electricity demand forecasting mainly focuses on the single-factor relationship between power consumption and economic growth, industrial development, etc., while neglecting the study of multiple influencing factors and considering different time dependencies.

To address this challenge, we propose a transformer-based forecasting model that utilizes transformer networks and fully connected neural networks (FC) for electricity demand forecasting in different industries within a city. The model employs the encoder part of the transformer to capture the dependencies between different influencing factors and uses FC to capture time dependencies. We evaluate our approach on electricity demand forecasting datasets from multiple cities and industries using various metrics. The experimental results demonstrate that our proposed method outperforms state-of-the-art methods in terms of accuracy and robustness. Overall, we provide a valuable framework in the field of electricity demand forecasting, which holds practical significance for stable power system operations.

Keywords: Electricity Demand Forecasting · Time Series Prediction · Transformer

1 Introduction

Electricity demand forecasting is crucial for optimizing power supply-demand structures [1]. With the evolving power industry, renewable energy growth, and unpredictable weather events, accurately predicting demand across regions and industries is essential. Recent research, shifting from traditional statistical methods to machine and deep learning models, has improved accuracy and service provision [2]. Common methods include grey system analysis [3] and regression [1], while innovative deep learning models like LSTM [4] and GRU [5] show promising results. However, these models lack interpretability. Decision trees and gradient boosting algorithms enhance accuracy by learning complex patterns within time series [6]. Feature selection and processing, including dimensionality reduction, are crucial during forecasting. This paper builds on advanced time series prediction models, capturing dependencies between influencing factors and improving prediction accuracy by analyzing historical data dependencies. This paper contributes to electricity demand forecasting in the following aspects:

1. Firstly, the transformer model can capture the dependency relationship between different positions in a sequence, achieving context awareness. We leverage the advantages of the transformer model to analyze the dependencies between different feature factors, effectively capturing the complex relationships between multiple input variables and the target variable. This is crucial for improving the performance of the model.
2. Secondly, we capture the time dependencies between different historical time series through the decoder layer composed of fully connected networks. This can potentially improve the accuracy of the prediction results.
3. Finally, we have validated our proposed method on electricity demand datasets from different cities and industries in the real world to demonstrate its effectiveness in predicting city electricity consumption.

2   Related Work

2.1   Classical Statistical Methods

In the past century, classical statistical methods dominated time series prediction, relying on experts' experience and simple relationships, resulting in lower accuracy. Methods included time series analysis, regression, exponential smoothing, and grey forecasting. Time series analysis uses historical data to model power load changes, divided into autoregressive, moving average, and integrated processes [7]. It has fast convergence but overlooks internal factors. Regression predicts future electricity levels based on historical data, offering simplicity and generalization but limited adaptability [8]. Exponential smoothing averages past sequences to predict future trends but struggles with unstable sequences and complex factors [9]. Grey forecasting suits uncertainty, with ordinary models for exponential growth and optimized models for fluctuating sequences. Advantages include simplicity, fewer parameters, and strong mathematical foundations, but they struggle with longer forecasts [10]. Classical methods require small datasets and lack adaptability to complex relationships, making them suitable for monthly predictions but challenging for practical applications involving temporal and spatial aspects [11].

2.2   Machine Learning Methods

With the development of machine learning, a series of classical algorithms have emerged. Compared to traditional statistical methods, machine learning-based time series forecasting has the advantage of powerful nonlinear fitting capabilities, resulting in higher prediction accuracy. One of the most popular time series techniques for electricity demand forecasting is Long Short-Term Memory (LSTM). Recurrent Neural Network (RNN) is a typical type of recurrent neural network that incorporates internal feedback connections and feedforward connections between processing units in different layers, enabling it to associate past information with present tasks. However, as the length of the time series increases, RNN struggles to learn long-term dependencies across distant time steps. LSTM, a special type of RNN, overcomes this limitation by incorporating three gates within the units to control internal states, thus addressing the vanishing gradient problem. As a result, it not only possesses the short-term dependency learning capability of RNN but also learns long-term dependencies. In literature [12], an algorithm for load forecasting is proposed based on the integration of LGBM and LSTM. In literature [13], the gate structure of LSTM is adjusted to reduce model parameters and improve computing speed. In literature [14], the strengths of both RNN and LSTM are combined for prediction, and an Attention mechanism is used to aggregate the prediction results of the two models, applied to small-scale monthly electricity sales datasets.

Transformer is a neural network model based on attention mechanisms, originally proposed by Google for natural language processing tasks such as machine translation, text summarization, and speech recognition. Compared to recurrent neural network models such as LSTM and GRU, which are representative of RNN and its variants, the Transformer model exhibits better parallelization and shorter training time. It performs well not only in processing long sequences but also in capturing contextual dependencies within sequences and internal dependencies between different sequences. As a result, it has found wide applications in various fields. The DehazeFormer approach proposed by Song et al. [15] modifies the Transformer model for image dehazing tasks. VideoBERT is a joint representation model based on Transformer for extracting representations from both image and language data, achieving excellent results in video content recognition datasets and serving as a fundamental architecture for multimodal fusion tasks [16]. Radford et al. proposed CLIP, a zero-shot learning method based on the ViT network, which combines language and image data and achieved promising results in various tasks [17]. Roy et al. introduced a multimodal fusion attention mechanism for extracting class labels from multimodal data using Transformer with cross-attention weights on input labels, and verified its performance on multimodal remote sensing classification tasks [18]. The relatively simple structure and outstanding performance of Transformer greatly enhance its application potential in the field of machine learning.

3 Proposed Method

This section provides a detailed description of the method proposed in this paper for predicting the electricity demand of different industries in cities. Firstly, we propose a general model framework for performing this task. Then, we analyze the different variables that influence the prediction of city electricity demand based on time series theory and select relevant covariance features. Finally, we provide a detailed description of the transformer-based model prediction framework and validate it on real-world datasets. Through this research, we aim to provide an accurate method for forecasting city electricity demand to support relevant decision-making and planning.

3.1 Overall Framework

In this article, we employ the combination of transformers and fully connected neural networks, making full use of the contextual learning ability of transformers, and considering the dependency among multiple time series. Our algorithm framework, as shown in Fig. 1, consists of two main parts. The encoder layer of the transformer is mainly responsible for capturing the spatial dependencies between different features, while the fully connected neural network primarily captures the temporal dependencies among different time series.

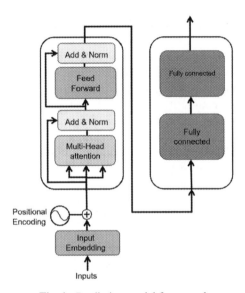

Fig. 1. Prediction model framework

Our proposed algorithm framework for electricity demand forecasting can be viewed as two stages: feature aggregation using the encoder layer of the transformer to learn the relationship between features and past electricity demand values, and prediction using the fully connected neural network layer. The encoder layer of the transformer is used to aggregate various features from the input time series data and learn their nonlinear relationship with electricity demand. The fully connected neural network layer then uses the aggregated features to predict future electricity demand values.

A high-dimensional time series regarding electricity consumption across various dimensions is as follows:

$$\{Y_t = (y_{t1}, y_{t2}, \ldots, y_{tp}), t = 1, 2, \ldots, N\} \tag{1}$$

as well as time series data for related covariates:

$$\{X_t = (x_{t1}, x_{t2}, \ldots, x_{tp}), t = 1, 2, \ldots, N\} \tag{2}$$

In this context, 't' represents the time step, and 'p' represents different dimensions of time series.

3.2 Feature Extraction Module

In this section, we analyze and address the factors affecting electricity demand from meteorological and social perspectives and construct relevant covariates.

From a social perspective, residential electricity consumption levels are generally lower during working days compared to holidays. The age distribution of the population in a given region also affects electricity demand. For instance, regions with a higher number of students during summer vacation experience a significant increase in electricity demand. Therefore, it is necessary to quantify holidays and summer vacation. Holidays are non-numeric data and need to be encoded to transform them into numerical values. For the "day of the week" data, we use one-hot encoding. For the data on "whether it is a holiday" and "festival type," we use 0–1 encoding, where 1 represents a holiday, 0 represents a working day, and 1 represents a specific festival, while the rest are represented as 0.

From a meteorological perspective, weather is the most important factor influencing electricity demand. Therefore, studying and analyzing meteorological conditions is an important step in improving the accuracy of the forecasting model.

From the mechanism of variation, it is known that temperature has the most significant impact among all meteorological factors, especially in some extreme natural environments. During cold winters and hot summers, electricity generation is significantly higher than in other seasons. Therefore, we construct three types of covariates to characterize temperature changes in the region: average temperature, maximum temperature, and minimum temperature. We also incorporate humidity information to build feature sequences that capture the meteorological impacts on electricity demand.

3.3 Time Series Prediction Module

We first introduce how to learn the dependencies of complex features.

Transformer, as the most advanced model in natural language processing, has been widely used due to its efficiency and strong contextual awareness.

In Transformer, the input to the Encoder is a sequence of text, and the output is a feature vector that represents the semantic information of the input text. The input to the Decoder is a specific token, based on which it generates a new sequence of text, and the output is a sequence of text. The Encoder is typically used for text encoding and representation learning. Therefore, we can use the encoder layers of Transformer for feature encoding and representation learning. The encoder layers primarily include four components: Positional Encoding, Multi-Head Attention, Add and Norm, and Feedforward and Add and Norm.

1. In the Positional Encoding positional encoding is performed using sine and cosine functions, as shown in the following formula:

$$PE(pos, 2i) = \sin(pos/10000^{2i/d_{mode:}}) \tag{3}$$

$$PE(pos, 2i + 1) = \cos(pos/10000^{2i/d_{mode:}}) \tag{4}$$

Here, *pos* represents the position of the feature in the entire sequence, and 'i' refers to the dimension of the feature vector. After positional encoding, we obtain an encoding array X_{pos} that is completely consistent with the input dimension. When this encoding array is added to the original feature embeddings, we obtain new feature embeddings:

$$X_{embedding} = X_{embedding} + X_{pos} \qquad (5)$$

2. The multi-head self-attention mechanism calculates the similarity between each input vector and all other input vectors, and then weights and sums them to obtain a new representation for each input vector. The mathematical expression for multi-head self-attention is as follows:

$$Multihead(Q, K, V) = Concat(head_1, head_2, \ldots, head_h)W^o \qquad (6)$$

Among them,

$$head_i = Attention(QW_i^Q, KW_i^K, VW_i^V) \qquad (7)$$

In Eq. (6), Q(Query), K(Key), V(Value) represents three vectors obtained from the input sequence through three linear mapping layers, with dimensions d_q, d_k, and d_v respectively. 'Concat' represents the concatenation function, which combines all the output results of $head_i$.

In Eq. (7), $W_i^Q \in R^{s \times d_k}$, $W_i^K \in R^{s \times d_k}$, $W_i^V \in R^{s \times d_k}$, $W_i^O \in R^{hd_v \times s}$, they respectively represent the weight matrices for the Q, K, and V vectors of the i-th 'head', and the weight matrix for the final output after dimension reduction. Here, it is mentioned that $d_k = d_v = s/h$. The computation of the attention mechanism is as follows:

$$Attention(Q, K, V) = SoftMax(\frac{QK^T}{\sqrt{d_k}})V \qquad (8)$$

In Eq. (8), $d_h = s/h$, 'SoftMax' represents an activation function, while $\sqrt{d_k}$ is used to transform the attention matrix into a standard normal distribution.

3. In the "Add and Norm" section, the input 'x' from the previous layer is connected with the output from the previous layer through residual connections.

4. In the "Feedforward and Add and Norm" section, the feature representation is obtained by passing the input through a feedforward network, which includes linear mappings and activation functions:

$$X_{hidden} = Relu(X_{hidden} * W_1 * W_2) \qquad (9)$$

In the Eq. (9), W_1 and W_2 are the weights of the two linear layers, and 'Relu' represents the activation function.

Next, we will discuss how to learn the temporal dependencies of different historical sequences.

A fully connected neural network is a multi-layer perceptron structure.

We use a 2-layer fully connected network to learn the nonlinear temporal dependencies of each time segment. In the current connection layer 'l', we have:

$$X^l = f(W^l X^{l-1} + b^l) \qquad (10)$$

In the equation, X^l represents the output of the current connection layer 'l', W^l represents the weights of the current layer, b^l represents the bias of the hidden layer, and 'f()' represents the nonlinear activation function. In this paper, we choose SoftMax as the activation function.

4 Proposed Method

4.1 Experimental Dataset

The dataset used in this paper includes the electricity demand and related covariate information for 13 industries from January 1, 2020, to January 31, 2023. The dataset is divided into a training set and a testing set in a 4:1 ratio. The training set consists of electricity demand data from January 1, 2020, to June 18, 2022, which is used for model training. The testing set includes electricity demand data from June 19, 2022, to January 31, 2023. Additionally, the dataset also includes the meteorological feature data and holiday feature data constructed in the previous section. The data has been preprocessed to eliminate outliers and missing values. We use Prophet, GBDT, and CNN-LSTM as experimental baseline methods.

4.2 Data Pre-processing

In this article, we utilize the min-max normalization method, which linearly transforms data to a specified range to eliminate dimensional impact. The commonly used ranges are [0, 1] or [−1, 1]:

$$X^* = \frac{x - x_{min}}{x_{max} - x_{min}} \tag{11}$$

In the Eq. (11), x represents the electricity demand data, while x_{max} and x_{min} represent the maximum and minimum values of the data, respectively.

Furthermore, in terms of the loss function, we use the Mean Squared Error (MSE) function to measure the average difference between the actual observed values and the predicted values. As shown in Eq. (12), Y_i represents the predicted electricity demand at the current time step, $\widehat{Y_i}$ represents the true electricity demand at the current time step, and n represents the number of training samples. Additionally, we utilize the Adam optimizer [19] to optimize the model gradients.

$$\text{MSE}_{loss} = \frac{1}{n} \sum_{i=1}^{n} (Y_i - \widehat{Y_i})^2 \tag{12}$$

4.3 Experimental Results and Analysis

The formulas for the daily average error indicator and the monthly average error indicator are as follows:

$$\text{month\_degree}_{error} = \frac{\sum_{i=1}^{n} y_{pred}(i) - \sum_{i=1}^{n} y_{true}(i)}{\sum_{i=1}^{n} y_{true}(i)} \tag{13}$$

$$\text{day\_degree}_{error} = \frac{1}{n}\sum\nolimits_{i=1}^{n}\left|\frac{y_{pred}(i) - y_{true}(i)}{y_{true}(i)}\right| \tag{14}$$

In the formulas, $y_{pred}(i)$ represents the predicted electricity demand for the i-th day, and $y_{true}(i)$ represents the true electricity demand for the i-th day.

In the experiment, we conducted electricity demand prediction tasks for different industries. Here is the specific industry breakdown, consisting of 13 industries. For the sake of readability, we use abbreviations to represent each industry. 1) Urban and rural residents' electricity demand (Ure); 2) Agriculture, forestry, animal husbandry, and fishery (Afahf); 3) Accommodation and catering industry (Aci); 4) Construction business (Cb); 5) Real estate industry (Ri); 6) Industrial sector (Is); 7) Information transmission, software, and information technology services industry (Isit); 8) Total electricity demand in society (Tes); 9) Financial business (Fb); 10) Wholesale and retail industry (Wi); 11) Leasing and business services industry (Rbi); 12) Public services and management services (Pm); 13) Transportation, warehousing, postal industry (Twp).

We selected the prediction results for each industry in August 2022 for comparison, with the monthly average error abbreviated as M-E and the daily average error as D-E. To ensure data confidentiality, we refer to the predicted city as X and the proposed prediction model based on Transformer and fully connected networks in this paper as Transformer-F:

Table 1. Error Results of Electricity Demand Forecast for 13 Industries of City X in August.

Industry	Model Error							
	Transformer-F		Prophet		GBDT		CNN-LSTM	
	M-E	D-E	M-E	D-E	M-E	D-E	M-E	D-E
Ure	**−0.18**	**0.16**	3.50	3.50	−0.38	0.38	−0.27	0.17
Afahf	−0.019	**0.02**	1.30	1.30	−0.20	0.30	**−0.01**	0.09
Aci	**−0.01**	0.01	0.18	0.19	−0.14	0.20	−0.02	**0.01**
Cb	**−0.01**	**0.01**	−0.05	0.07	−0.15	0.17	0.04	0.06
Ri	**−0.02**	0.01	0.13	0.14	−0.18	0.20	−0.02	**0.01**
Is	**0.02**	**0.02**	−0.04	0.04	0.12	0.11	0.33	0.33
Isit	**−0.02**	**0.02**	0.04	0.04	−0.06	0.08	0.02	0.03
Tes	**0.03**	0.01	0.04	0.65	−0.12	0.16	−0.04	**0.01**
Fb	−0.02	**0.01**	0.04	0.07	0.10	0.20	**0.01**	0.02
Wi	**−0.01**	0.11	0.17	0.21	−0.13	0.16	0.02	**0.04**
Rbi	−0.06	**0.01**	0.37	0.94	−0.20	0.22	**−0.01**	0.02
Pm	**0.01**	**0.01**	0.10	0.11	−0.09	0.17	−0.07	−0.1
Twp	0.02	**0.01**	−0.19	0.18	**0.01**	0.06	0.02	0.02

From Table 1, it can be seen that, compared to the comparative methods, the proposed Transformer-F prediction model in this paper has the lowest monthly average error and daily average error for 9 industries. This fully validates the effectiveness of the Transformer-F model. In contrast, the Prophet model performs the worst, indicating that the Prophet model may have limitations in predicting long-term trends. Additionally, the Prophet model typically requires the original data to have certain seasonal variations. If the training set lacks noticeable seasonal patterns, the Prophet model may struggle to effectively model the data.

Furthermore, the GBDT model has higher error results compared to the CNN-LSTM model and the proposed model in this paper. This is because tree-based models are generally not suitable for high-dimensional sparse data and are sensitive to parameter values, requiring careful tuning.

As shown in Fig. 2, we also presented the fitting performance of various models for the overall societal electricity demand in August. The overall societal electricity demand is defined as the sum of daily electricity demands across 13 industries. The curves of different colors in the graph represent the predicted values of different models.

From Fig. 2, it can be observed that our proposed model shows a good fit to the real curve, and the performance of the CNN-LSTM model is also considerable. However, the prediction results of the GBDT model are slightly worse compared to our proposed method and CNN-LSTM.

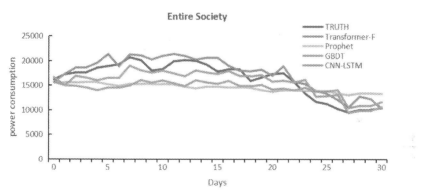

Fig. 2. Comparative results of different models in predicting the electricity demand of entire society in August.

In addition, our proposed method also outperforms the comparative methods in the segmented 13 industries.

As shown in Fig. 3 below, the curve fitting of the model for the real estate industry in August 2022 is very close to the actual situation on the ground. The GBDT method also performs well in some industries, but its performance is not as good as CNN-LSTM. Compared to the other three methods, the Prophet method performs relatively poorly across all industries:

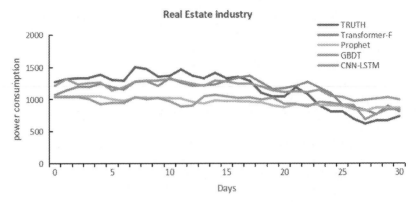

Fig. 3. Comparative results of different models in predicting the electricity demand of real estate industry in August.

Overall, the prediction results, as evidenced by comparing different numerical indicators and examining the fitting of different models to the real curves, demonstrate the effectiveness of our proposed model in electricity demand forecasting.

5   Conclusion

Translation: In this paper, we utilized time series statistical analysis methods to analyze historical electricity demand data. We established a multi-category electricity prediction model. We validated the effectiveness of our proposed method using a real-city electricity demand dataset. By comparing the prediction errors of different models across various industries, we demonstrated that our proposed method outperforms others in terms of accuracy.

Acknowledgement. This work is supported by the Research Funds from State Grid Henan (SGHAYJ00NNJS2310068);

References

1. Amjady, N.: Short-term hourly load forecasting using time-series modeling with peak load estimation capability. IEEE Trans. Power Syst. **16**(3), 498–505 (2001)
2. Li, J., Jiao, R., Wang, S., et al.: An ensemble load forecasting model based on online error updating. Proc. CSEE **43**(4), 1402–1412 (2023)
3. Vähäkyla, P., Hakonen, E., Léman, P.: Short-term forecasting of grid load using box-jenkins techniques. Int. J. Electr. Power Energy Syst. **2**(1), 29–34 (1980)
4. Jing, O., Lü, Y., Kang, Y., et al.: Short-term load forecasting method for integrated energy system based on ALIF-LSTM and multi-task learning. Acta Energiae Solaris Sinica **43**(9), 499–507 (2022)
5. Deng, D., Li, J., Zhang, Z., et al.: Short-term electric load forecasting based on EEMD-GRU-MLR. Power Syst. Technol. **44**(2), 593–602 (2020)

6. Ahmad, A.S., et al.: A review on applications of ANN and SVM for building electrical energy consumption forecasting. Renew. Sustain. Energy Rev. **33**, 102–109 (2014)
7. Moghram, I., Rahman, S.: Analysis and evaluation of five short term load forecasting techniques. IEEE Trans. Power Syst. **4**(4), 1487–1491 (1989)
8. Amral, N., Ozveren, C.S., King, D.: Short term load forecasting using Multiple Linear Regression. In: International Universities Power Engineering Conference. IEEE (2007)
9. Pang, Y.: Research on market share prediction of highway passenger transport based on exponential smoothing method. Am. J. Traffic Transp. Eng. **8**(2) (2023)
10. Kumar, T.S., Rao, K.V., Balaji, M., et al.: Online monitoring of crack depth in fiber reinforced composite beams using optimization Grey model GM (1, N). Eng. Fract. Mech. **271** (2022)
11. Shumway, R.H., Stoffer, D.S., Shumway, R.H., Stoffer, D.S.: Arima models. Time Series Analysis and Its Applications: With R Examples, pp. 75–163 (2017)
12. Torres, J.F., Martínez-Álvarez, F., Troncoso, A.: A deep LSTM network for the Spanish electricity consumption forecasting. Neural Comput. Appl. **34**(13), 10533–10545 (2022)
13. Greff, K., Srivastava, R.K., Koutnik, J., et al.: LSTM: a search space odyssey. IEEE Trans. Neural Netw. Learn. Syst. **28**(10), 2222–2232 (2017)
14. Cui, Q., Sun, M., Na, M., et al.: Regional electricity sales forecasting research based on big data application service platform. In: 2020 IEEE 3rd International Conference on Electronics and Communication Engineering (ICECE), pp. 229–233. IEEE, Xi'an (2020)
15. Song, Y., He, Z., Qian, H., Du, X.: Vision Transformers for Single Image Dehazing. arXiv 2022, arXiv:2204.03883
16. Sun, C., Myers, A., Vondrick, C., Murphy, K., Schmid, C.: VideoBERT: a joint model for video and language representation learning, arXiv preprint arXiv:1904.01766 (2019)
17. Radford, A., et al.: Learning transferable visual models from natural language supervision, arXiv preprint arXiv:2103.00020 (2021)
18. Roy, S.K., Deria, A., Hong, D., Rasti, B., Plaza, A., Chanussot, J.: Multimodal fusion transformer for remote sensing image classification, arXiv preprint arXiv:2203.16952 (2022)
19. Kingma, D.P., Ba, J.L.: Adam: a method for stochastic optimization. In: International Conference on Learning Representations, San Diego, CA, USA, pp. 1–15, May 2015

Method of Visual Formation Control for Large Group of AUV in Environment with Obstacles

Filaretov Vladimir[1,2] (ID), Yukhimets Dmitry[2,3(✉)] (ID), Oseledets Vitaly[2] (ID),
and Changan Yuan[4] (ID)

[1] Sevastopol State University, Sevastopol, Russia
[2] Institute of Automation and Control Processes FEB RAS, Vladivostok, Russia
undim@iacp.dvo.ru
[3] Institute of Marine Technology Problems FEB RAS, Vladivostok, Russia
[4] Guangxi Academy of Science, Nanning 530007, China

Abstract. The paper proposes a modification of the method for formation control of the group of autonomous uninhabited underwater vehicles (AUVs) in the unknown environment containing obstacles for large group of AUV. AUVs move in the "leader-follower" mode in the given formation. The AUV-leader has information about the mission, moves to the target and defines the motion trajectory to safely avoid the detected obstacles. AUV-followers follow the leader, in accordance with the place given to them in the formation. For this movement, information about the current position of the AUV-leader is used. In the basic proposed method, the followers receive this information via hydroacoustic communication channels. Obstacles and the distance to them are determined via onboard rangefinders. The low bandwidth of hydroacoustic channels and large delays in data transmission do not provide safe and accurate movement of group members when they are close to an obstacle or to another AUV. To solve this problem, using onboard video cameras of AUV-followers and technical vision to determine the position of the leader, on which a special light beacons are installed, is proposed. This approach makes possible eliminating delays in the receiving of information by the followers and ensure the safe movement of the AUV group when using high-precision control systems. The main difficulty of using visual information in underwater environment is the limited visibility distance. To consider this limitation, some AUV-followers can act as leaders for other followers. This will allow to form groups from a large number of AUV. At the same time, a control system with a predictive model is used to ensure high accuracy of controlling the movement of the AUV inside the formation when bypassing obstacles. The effectiveness of the proposed method is confirmed by the results of mathematical simulation.

Keywords: Formation control · Underwater vehicle · Path planning · Obstacle avoidance · Computer vision

1 Introduction

Currently, using autonomous underwater vehicles (AUVs) is one of the most promising approach for the studying and development of the World Ocean. AUVs are capable of performing a wide range of tasks related to survey and prospecting and geological exploration, oceanographic research.

A certain disadvantage of using AUVs is the limited operation time during the mission, which is conditioned by a limited energy reserve of onboard batteries. This disadvantage complicates the performance of operations for the survey of sufficiently large areas. To eliminate this shortcoming, many researchers suggest using groups of AUVs cooperatively performing the mission [1].

One of the main strategies for using groups of AUVs is the "leader-follower" mode, in which one of the AUVs is appointed as a leader who has complete information about the given mission and defines the movement trajectory of the group, and other AUVs are considered as followers, receiving information from the leader about its current position and correcting their movement, taking into account their prescribed position in the formation.

An important condition for the implementation of this strategy is the availability of communication channels between all AUVs of the group, which allows exchanging information about the current position of the AUV-leader and each member of the group. This is especially important when a group of AUVs performs a joint mission in an environment containing obstacles. When avoiding obstacles, the trajectories of the leader and followers can change unpredictably. In this case, AUVs must move in such a way as to avoid these obstacles at a safe distance, while excluding the possibility of collision with other members of the group.

One of the approaches to ensure the safe movement of robots is using of a special schedule in which the robot collision is not allowed [2, 3]. But using of such a method is permissible only in cases where the working environment is known. At the same time, an additional problem for AUVs is large errors in determining their coordinates in the absolute system, which requires using of additional mechanisms for matching the indications of their navigation systems and thereby leads to a significant complication of the implementation of AUV group control systems [4–6].

Methods for controlling a group of robots based on algorithms that are intended for a swarm of robots are described in the works [7, 8]. These algorithms are hardly applicable for controlling a group of AUVs, because swarm control implies a frequent exchange of information between group members, which can be difficult in the underwater environment. Also, the control of a group of robots in an unknown environment is considered in [9], where collisions with obstacles are prevented by changing the type of a given formation or the speed of movement of a group of robots so the group avoids the obstacle at the safe distance.

In all the considered examples implementation of the correction of the robots group trajectories occurs due to the constant exchange of information with each other, containing data on the current position of group members, which makes possible to coordinate the trajectories of the robots. However, this approach is not always applicable to AUVs, because the possible data transmission via hydroacoustic channels is carried out with large delays and has low bandwidth. The method without the necessity for a large exchange of information between the group members is presented in [10]. The disadvantage of the proposed approach is using of hydroacoustic communication channels, which do not ensure the safe and accurate movement of a group of AUVs, when they are located at a small distance from each other.

At the same time, a method for generating the trajectories of a group of AUVs during their movement in the "leader-followers" mode in an unknown environment with obstacles that does not require the additional data transmission between the AUVs of the group to coordinate their trajectories when avoiding obstacles was proposed in [11]. This method assumes that all AUVs of the group must know their coordinates in the absolute coordinate system quite accurately, which requires expensive hydroacoustic navigation systems, as well as the use of methods for coordinating their readings.

A method for solving the problem of the necessity of using hydroacoustic systems on AUV-followers, for the coordinated movement of the AUVs group based on visual information about the position and orientation of the leader, obtained from the on-board cameras of the followers was proposed in [12]. When using this method for implementing the movement of the AUVs group in an environment with obstacles, a problem arises that, when avoiding an obstacle, AUV-followers can cover the AUV-leader from other AUVs moving behind them, which can lead to incorrect and unsafe movement of these AUVs relative to other AUVs of the group.

This article solves the problem of developing such a method of group movement that would allow the formation of the AUVs group in which the followers do not need expensive sonar systems to follow the leader and use high-speed communication channels to coordinate the trajectories of these AUVs in the avoiding obstacles process. At the same time, the approach described in [12] will be used as the basic method that implements the "leader-follower" strategy for the AUV group, and the method [11] will be used to implement the method for generating safe obstacle avoidance trajectories in the group, which will be modified taking into account the features of the method [13].

2   Problem Formulation

A group of AUVs, consisting of AUV-leader, AUV-followers and AUV which are followers for other AUV-followers considers in this article. The AUV-leader has all the information about the mission and forms its trajectory in such a way as to ensure the accomplishment of this mission. Several light beacons are installed on AUV-leader board, which can be observed by AUV-followers through their onboard video cameras. Using this video information, they form data about the position and orientation of the AUV-leader relatively to the follower. This information AUV-followers use to follow the leader and keep given them place in the formation (Fig. 1). At the same time, AUV-followers are leaders for other AUV-followers.

AUV-followers do not have information about their current position in the absolute coordinate frame (ACF). In this case, the position of the AUV-follower in the CF relative the AUV-leader will be determined by the following expression:

$$X_F^L = -(R_L^F)^T X_L^F, \tag{1}$$

where X_F^L is the coordinate vector of the AUV-follower position in the CF of the AUV-leader; X_L^F is the coordinate vector of the AUV-leader position in the CF of the AUV-follower; R_L^F is the is the orientation matrix of the AUV-leader in the CF of the AUV-follower.

The desired position of the AUV-follower in the formation is calculated in the CF relative to the AUV-leader, which makes possible to form the desired formation of the AUV group, regardless of their position and orientation in the ACF during the mission.

To determine the position and orientation of the AUV-leader relative to the followers, these followers are equipped with video cameras that observe light beacons on the video image, the location on the leader's body and the position of which is known relative to the leader's center of mass. The position of the beacons in the CF of the AUV-leader is known and is set by the coordinates $B = (B_1, B_2, \ldots, B_k)$, where $B_i = (x_{bi}, y_{bi}, z_{bi})^T$ is the coordinates of the i-th beacon in the CF of the AUV-leader; k is the number of beacons on the leader AUV. The beacons have such characteristics that they can be identified in the image received from the video camera.

The main difficulty of using visual information in underwater environment is the limited visibility distance. To consider this limitation, some AUV-followers can act as leaders for other followers. This will allow to form groups from a large number of AUV.

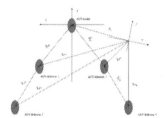

Fig. 1. AUV group formation

From the image received from the video camera of the follower, the vector $P = (P_1, P_2, \ldots, P_k)$ is formed, where $P_i = (p_{xi}, p_{yi})$ are the pixel coordinates of the i-th beacon.

The expressions binding the pixel coordinates of the beacons and the position and orientation of the leader in the camera CF:

$$P_i = F_{B2P}\left(U, B_i, X_L^C, A_L^C\right),$$
$$X_{Bi}^C(P_i) = R_L^C\left(A_L^C\right)B_i + X_L^C, \, i = \left(\overline{1, k}\right), \tag{2}$$

where $X_{Bi}^C \in R^3$ is the coordinate vector of the i-th beacon in the CF of the AUV-follower; $X_L^C \in R^3$ is the coordinate vector AUV-leader in follower's camera CF; R_L^C is the transformation matrix AUV-leader relatively to CF of follower's camera; $A_L^C \in R^3$ is the vector of orientation angles of the AUV-leader relatively to the axes of the camera's CF.

The system of Eqs. (2) contains 6 unknown variables, at least 3 beacons are required for their definite determination. Equations (2) are significantly non-linear and cannot be solved analytically, the determination of the values A_L^C and X_L^C is possible using numerical optimization methods [14].

The coordinate vector of the AUV-leader in CF of follower can be determined by equation:

$$X_L^F = R_C^F X_L^C + X_C^F, \tag{3}$$

where $X_C^F \in R^3$ is the camera's position coordinate vector of i-th follower in CF of follower, $R_C^F \in R^{3 \times 3}$ is the orientation matrix of onboard camera in CF of follower.

To provide the ability of the AUVs to determine surrounding obstacles, they are equipped with rangefinders (sonars), which form information about the distance to the obstacle in the direction of these rangefinders. During the operation of rangefinders and the movement of the AUV, a vector $D = (d_1, d_2, \ldots, d_n)$ is formed, where d_i is the distance to the obstacle determined by the i-th rangefinder. This vector is used for calculating the AUV target point.

The problem of forming such trajectories of AUV-followers moving behind the AUV-leader as part of a group in an environment containing obstacles is solving in this paper. These trajectories must satisfy the following requirements:

- The formation of the trajectories of the AUVs of group should occur independently of each other and do not require data exchange between the AUVs of group via hydroacoustic communication channels. At the same time, AUV-followers form their trajectories using information about the position of the AUV-leader received from their onboard video cameras and readings from onboard rangefinders.
- The trajectories of the AUVs of group pass at a safe distance from the detected obstacles.
- During movement and avoiding obstacles, the possibility of collision between members of the group should be excluded.
- In the process of avoiding obstacles, keeping a given formation is not required.
- Due to limited visibility, another AUV-follower can act as a leader.

3   Trajectories Formation Method

Due to the lack of communication channels between the AUV-followers, the problem of coordinating the trajectories of the group members movement in the process of avoiding obstacles arises. In this case, the desired position of the AUV-followers is determined relatively to the position of the AUV-leader, and all followers can determine the position of the leader relatively to themselves. This fact can be used to guarantee the safe movement of followers when avoiding obstacles without directly coordinating their trajectories.

The method described in [13] will be used as a base method, which will be modified for the case when the position determination of the leader is based on video information, and the followers do not have information about their coordinates in the ACF.

This method is illustrated in Fig. 2. This figure shows the AUV-followers moving in a given formation behind the AUV-leader and being at their prescribed positions in the formation.

The black dots on Fig. 2 indicate the desired positions of the AUV-followers in the formation, and the dotted lines indicate the trajectories of the adjusted displacement of the AUV-followers within the formation when avoiding the detected obstacles.

When the onboard rangefinders detect obstacles located at a distance less than safe, the desired position of the AUV-follower inside the formation begins to shift along the dotted lines towards the trajectory along which the AUV-leader passed. Since the AUV-leader has already found a safe trajectory, shifting to this trajectory, the AUV-followers will also provide themselves with a safe passage. At the same time, the trajectories of AUV-followers movement within the formation are chosen to ensure their safe movement relatively to each other.

Setting the trajectories of AUV-followers movement inside the formation is as follows. The indicated motion trajectories are segments of straight lines in the CF of the AUV-leader. The starting point of the i-th AUV-followers trajectory corresponds to its desired position in the formation with coordinates $\tilde{X}_{Fi} = (\tilde{x}_{fi}, \tilde{y}_{fi})$, and the end point lies on the axis $\tilde{x}$ in CF of AUV-leader and has coordinates $\tilde{X}_{Fi}^0 = (\tilde{x}_{fi} \pm D_a/2, \ 0)$. The value D_a is chosen so that when different AUVs move along these given trajectories inside the formation, the distances between them would always be no less than the safe distance $D_{\min}$. The choice of sign when determining the point $\tilde{X}_{Fi}^0$ depends on the location of the AUV-follower relatively to the axis $\tilde{x}$ (sign "+" indicates that the AUV is located to the left of the $\tilde{x}$, sign "−" indicates that the AUV is located to the right of the $\tilde{x}$). Value $D_a > D_{\min}$ depends on the number of columns, on the features of the formation and the number of AUV-followers.

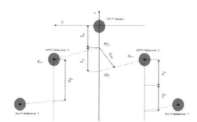

Fig. 2. AUV-followers movement trajectories when avoiding obstacles

The equation of the displacement trajectory of the i-th AUV-follower inside the formation, passing through the points $\tilde{X}_{Fi}$ and $\tilde{X}_{Fi}^0$ in CF of AUV-leader in horizontal plane [15]:

$$\pm \frac{D_a}{2}(\tilde{y} - \tilde{y}_{Fi}) + \tilde{y}_{Fi}(\tilde{x} - \tilde{x}_{Fi}) = 0, \tag{4}$$

As information about the position of the AUV-leader is obtained by AUV-follower using video information from the onboard video camera that observes the leader's light beacons, it becomes possible that during the process of displacement, AUV-follower will cover the visibility area of the beacons for other group members.

To eliminate this situation, it is proposed to carry out an additional displacement in the vertical plane along the axis (Fig. 3). The trajectory of such displacement passes through points $\tilde{X}_{Fi}^{\tilde{z}} = (0, \tilde{y}_{fi})$ and $\tilde{X}_{Fi}^{\tilde{z}0} = (\pm D_b, \ 0)$, where $\tilde{X}_{Fi}^{\tilde{z}}$ – is the initial position of the i-th AUV-follower in the absence of obstacles, $\tilde{X}_{Fi}^{\tilde{z}0}$ – is the point located on the axis $\tilde{z}$, that determines the position of the i-th follower when moving behind the AUV-leader in the plane $\tilde{x}\tilde{z}$ of the CF of the AUV-leader. The choice of the sign when determining the point $\tilde{X}_{Fi}^{z0}$ depends on the desired displacement of the AUV-follower relative to the axis $\tilde{z}$ when avoiding the detected obstacle.

The formation of the displacement trajectory along the axis $\tilde{z}$ occurs similarly to Eq. (4):

$$\pm\frac{D_b}{2}(\tilde{y} - \tilde{y}_{Fi}) + \tilde{y}_{Fi}\tilde{z} = 0, \tag{5}$$

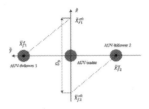

Fig. 3. Changing the depth of AUV-followers during displacement

The value D_b depends on the characteristics of the video cameras installed on the AUV-followers, the characteristics of the beacons of the AUV-leader and the type of formation. D_b is chosen for each follower in such a way that any displacement along the given trajectories within the formation does not cover the beacon's visibility area for other members of the group.

In the process of moving behind the leader, the followers determine the presence and proximity of obstacles to the trajectories using onboard rangefinders. If this distance is less than a safe distance $D_{\min}$, the program point of this follower, which set its position in the formation, is shifted along the trajectory described by Eqs. (4), (5). When several rangefinders are triggered at once, the distance to all obstacles is estimated and the nearest one is selected. If the distance to obstacles is more than safe, then AUV-followers continue to move in the place, gave them in the formation.

To determine the required shift of the AUV-follower program point, it is necessary to determine the obstacle point closest to the follower, which must be avoided. As the shift of the AUV-follower is set in the AUV-leader's CF, the coordinates of the extreme point of the obstacle are respectively translated into the leader's CF. This calculation is carried out as follows:

$$X_{dj} = X_F^L + R_L R_F^T \begin{bmatrix} d_j\cos(\alpha_j) \\ d_j\sin(\alpha_j) \end{bmatrix}, j = \overline{(1, n)}, \tag{6}$$

where X_{dj} are the coordinates of detected point of obstacle in ACF, fixed by j-th rangefinder of AUV-follower; ψ_f is the follower heading angle; α_j is the angle of orientation of j-th rangefinder relatively to longitudinal axis of the AUV-follower; d_j is the distance to the obstacle point, determined by the j-th range finder.

Next, it is necessary to determine where the program point, which sets the position of the AUV-follower in the formation, should move to ensure its safe avoidance of the detected obstacle.

For this, first need to determine which point on the obstacle is closest to the AUV-follower trajectory. As the AUV-follower does not know in advance the trajectory of its movement, because it is determined by the movement of the AUV-leader, will assume that the predicted trajectory of the AUV-follower is a straight line parallel to the axis $\tilde{x}$. Therefore, the proximity of the detected points X_{dj} to the specified trajectory can be calculated by the formula:

$$\delta_j = S(\tilde{y}_{dj})(\tilde{y}_{dj} - \tilde{y}_F), j = (\overline{1, m}),$$

$$S(\tilde{y}_{dj}) = \begin{cases} 1, & if \ \tilde{y}_{dj} \geq 0, \\ -1, & if \ \tilde{y}_{dj} < 0 \end{cases} \tag{7}$$

The multiplier $S(\tilde{y}_{dj})$ is necessary to take into account on which side of the slave an obstacle is detected.

The closest point to the trajectory of movement X_{dc} is considered to be the point X_{dj} for which the value will have a minimum value.

The new program position of the point in the CF of the AUV-leader for the AUV-follower can be calculated by the expression:

$$\tilde{y}_F^* = \begin{cases} y_{dc} - sign(y_{dc}) \cdot D_{\min}, & if \ \min(\delta_j) < D_{\min}, \ j = (\overline{1, m}) \\ \tilde{y}_F, & if \ \min(\delta_j) \geq D_{\min}, \ j = (\overline{1, m}) \end{cases},$$

$$\tilde{x}_F^* = \tilde{x}_F \pm \frac{D_a(\tilde{y}_F - \tilde{y}_F^*)}{2\tilde{y}_F},$$

$$\tilde{z}_F^* = \pm \frac{D_b(\tilde{y}_F^* - \tilde{y}_F)}{2\tilde{y}_F}. \tag{8}$$

where $\tilde{x}_F^*, \tilde{y}_F^*, \tilde{z}_F^*$ – coordinates of the new position of the target point $\tilde{X}_F^*$ AUV-follower during obstacle avoidance in formation.

If obstacles are detected or there is a lot of noise in the data generated by the sonar, the position of the target point may suddenly change, which will lead to the generation of incorrect control signals. To eliminate this situation, it is possible to use a low-pass filter [16] to smooth the trajectories:

$$\hat{X}_F^*(k) = \hat{X}_F^*(k - 1) + \beta(\tilde{X}_F^*(k) - \hat{X}_F^*(k - 1)), \tag{9}$$

where $\tilde{X}_F^*$ is the desired position of the AUV-follower at the current step of the system operation. $0 \leq \beta \leq 1$ is the smoothing coefficient.

Thus, the proposed method for generating AUV trajectories during their movement in a group makes it possible to dispense with the use of acoustic communication channels to coordinate the AUV trajectories when avoiding detected obstacles.

4   AUV Control System

AUV is a complex object described by a system of nonlinear equations. At the same time, in the process of moving in a group and bypassing obstacles, the ability to move along complex spatial trajectories is required. Thus, it is necessary to obtain a control system that ensures the movement of the AUV in space with sufficient accuracy.

4.1 Mathematical Model of AUV

Mathematical model of spatial movement of AUV has following view [14]:

$$M\dot{v} + (C(v) + D(v))v + g(\eta) = \tau,$$
$$\dot{\eta} = J(\eta)v,$$

(10)

where $M = M_R + M_A \in R^{6\times 6}$; $M_R \in R^{6\times 6}$ is an AUV inertia matrix; $M_A \in R^{6\times 6}$ is a matrix of added mass and moment inertia; $C(M, v) \in R^{6\times 6}$ is a matrix of Coriolis and centripetal forces and torques; $D(d_1, d_2, v) \in R^{6\times 6}$ is a matrix of hydrodynamic forces and moments; $g(\eta) \in R^6$ is a vector of hydrostatic forces and torques; $\eta = [x_a, y_a, z_a, \varphi_a, \theta_a, \psi_a]^T \in R^6$ is a vector of the AUV position and orientation in an ACF; $J(\eta)$ is a transition matrix from body-fixed CF (BCF) to ACF; $\tau = [\tau_x, \tau_y, \tau_z, M_x, M_y, M_z]^T \in R^6$ is a vector of propulsion forces and moments in the AUV BCF; $v = [v_x, v_y, v_z, \omega_x, \omega_y, \omega_z]^T \in R^6$ is a vector of linear and angular velocities in the AUV BCF.

The matrix of Coriolis and centripetal forces and moments described by expressions [17]:

$$C(M, v) = C_{RB}(M_R, v) + C_A(M_A, v),$$

(11)

where $C_{RB}(M_R, v) = \begin{bmatrix} 0_{3\times 3} & -S(M_{11}v_1 + M_{12}v_2) \\ -S(M_{11}v_1 + M_{12}v_2) & -S(M_{21}v_1 + M_{22}v_2) \end{bmatrix}$, $M_R = \begin{bmatrix} M_{11} & M_{12} \\ M_{21} & M_{22} \end{bmatrix}$, $M_{ij} \in R^{3\times 3}$, $i, j = \overline{(1, 2)}$;

$C_A(M_A, v) = \begin{bmatrix} 0_{3\times 3} & -S(A_{11}v_1 + A_{12}v_2) \\ -S(A_{11}v_1 + A_{12}v_2) & -S(A_{21}v_1 + A_{22}v_2) \end{bmatrix}$, $M_A = \begin{bmatrix} A_{11} & A_{12} \\ A_{21} & A_{22} \end{bmatrix}$, $A_{ij} \in R^{3\times 3}$, $i, j = \overline{(1, 2)}$; $v_1 = \begin{bmatrix} v_x & v_y & v_z \end{bmatrix}^T$; $v_2 = \begin{bmatrix} \omega_x & \omega_y & \omega_z \end{bmatrix}^T$, and operator S(.) described

by expression: $S(\lambda) = \begin{bmatrix} 0 & -\lambda_3 & \lambda_2 \\ \lambda_3 & 0 & -\lambda_1 \\ -\lambda_2 & \lambda_1 & 0 \end{bmatrix}$, $\lambda = \begin{bmatrix} \lambda_1 & \lambda_2 & \lambda_3 \end{bmatrix}^T \in R^3$ is the parameter of operator S.

The elements of diagonal matrix $D(d_1, d_2, v)$ are described following expression:

$$D_{ii} = d_{1i} + d_{2i}|v_i|, i = \overline{(1, 6)},$$

(12)

where $d_{1i}, d_{2i}, i = \overline{(1, 6)}$ are hydrodynamic coefficients respective linear and quadratic dependances of hydrodynamic forces and torques from AUV velocities along all degree of freedom.

The vector of hydrostatic forces and torques has the following form [14]:

$$g(\eta) = \begin{bmatrix} (W - B)\sin\theta \\ -(W - B)\cos\theta\sin\varphi \\ -(W - B)\cos\theta\cos\varphi \\ -B_y\cos\theta\cos\varphi + B_z\cos\theta\sin\varphi \\ B_z\sin\theta + B_x\cos\theta\cos\varphi \\ -B_x\cos\theta\sin\varphi - B_y\sin\theta \end{bmatrix},$$

(13)

where W_a is the gravity force; B_a is the buoyancy force; $B_x = W_a x_G - B_a x_B$, $B_y = W_a y_G - B_a y_B$, $B_z = W_a z_G - B_a z_B$; x_G, y_G, z_G are coordinates of the center of gravity (CG) in the AUV BCF; x_B, y_B, z_B are coordinates of the center of buoyancy (CB) in the AUV BCF.

4.2 Model Predictive Control for AUV

The essence of the control approach with a predictive model is in the formation of current control actions based on the analysis of the predicted AUV response to various sequences of their changes. This makes it possible to improve the adaptive and robust properties of the developed control systems (CS). To build an CS with a predictive model, it is rational to use a discrete model of AUV dynamics [18]:

$$x_{i+1} = f(x_i, u_i) + w_i, i = 0, 1, 2 \ldots$$
$$y_i = H x_i + v_i \qquad (14)$$

where the vectors $x_i \in E^n$, $u_i \in E^m$, $y_i \in E^r$ represent the current state of the object, the control actions and the observation vector, i is the current step of the system, f is a known nonlinear vector function, H is the matrix measurements. Such a model is random in nature, since it is impossible to determine in advance the values of external disturbances w_i and measurement errors v_i.

Eliminating unknown random components, a predictive model of the form is selected:

$$x_{i+1} = f(x_i, u_i), i = 0, 1, 2 \ldots$$
$$y_i = H x_i \qquad (15)$$

Model (15) is initialized at the initial cycle by the current state of the control object and allows one to approximately predict its dynamics. The final sequence of vectors $x_{i+j}, j = 1, \ldots, P$ - calculated according to the system (15), is called the forecast of the movement of the object on the prediction horizon P.

To quantify the quality of control, the following functional should be specified [19]:

$$J = J(\bar{x}, \bar{u});$$
$$\bar{x} = (x_{i+1}, x_{i+2} \ldots x_{i+P}) \in E^{nP}, \qquad (16)$$
$$\bar{u} = (u_i, u_{i+1} \ldots u_{i+P-1}) \in E^{mP}.$$

The quality of AUV motion control depends on the method of forming control actions on the forecast horizon and the performance of the onboard computer system.

Consider the solution of the optimal control problem based on predictive model (15). The quality of the control process is determined by the functional (16). The behavior of system (15) on cycles $i = 1, 2, \ldots, P$, uniquely depends on the choice of the vector $\bar{u}$. Considering that there is a functional dependence $\bar{x} = f(\bar{u})$, we can assume that $J = J(\bar{x}, \bar{u}) = J(\bar{u})$. Thus, the constrained optimization problem is formulated as follows:

$$J = J(\bar{u}) \to \begin{matrix} min \\ u \in \Omega \subset E^{mP} \end{matrix}, \qquad (17)$$

where $\Omega = \{\overline{u} \in E^{mP} : u_{i+j-1} \in U, j = 1, 2, \ldots, P\}$ – admissible set of finite sequences of m- dimensional vectors. $J(\overline{u})$ is a function of mP arguments.

Thus, the control scheme for a specific optimization problem (17) takes the following form:

1. The state vector y_i is measured;
2. Optimization problem (17) is solved for predictive model (15) with initial conditions $\overline{x} = x_i$. The extremum of the functional $J(u_i, u_{i+1} \ldots u_{i+P-1})$ is calculated on the admissible set of values Ω.
3. From the generated optimal sequence $u_i^*, u_{i+1}^* \ldots u_{i+P-1}^*$ the first vector is used as a control action at the next cycle of the system.
4. For the next measure, operations 1–3 are repeated. When controlling the AUV movement based on a predictive model, the control quality functional (17) on the forecast horizon is chosen as:

$$J = \rho_1 \sum_{j=1}^{P} e_{i+j}^2 + \rho_2 \sum_{j=1}^{P} (u_{i+j} - u_{i+j-1})^2, \qquad (18)$$

where e_{i+j} is the system output error, ρ_1 is the contribution of the cost of changing the error to the final functional J, ρ_2 is the contribution of the cost of changing the control signal to the final functional J. This form allows minimizing not only the system output error, but also abrupt changes in the control impact on the device.

The structure of the CS with a predictive model is shown in Fig. 4. The setting signal r_i and the current estimate y_i^* are fed to the input of the optimization block. This block generates sequences of control actions $\widetilde{u}_i$ applied to the predictive model of the control object, and receives a prediction of the object behavior $\widetilde{y}_i$ for P cycles ahead. The value of the functional $J(\overline{y}, \overline{u})$ is calculated for each sequence. The found optimal value u_i^* is the input of the control object, which is also affected by external disturbances w_i. The state of the object x_i changes, measurements of y_i are made with unknown noises v_i, and the state of the object is specified by the observer.

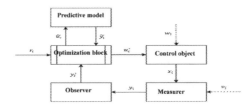

Fig. 4. AUV CS structure

5   Simulation Results

To test the effectiveness of the method, mathematical modeling of the movement of a group of underwater vehicles consisting of five robots (1 leader, 2 followers, 2 followers by followers) moving in a triangular formation was carried out. Each AUV-follower is equipped with a video camera with a resolution of 512×512 pixels, has 4 rangefinders on board, located in front of the AUV. Four beacons are installed at the stern of the leader AUV. The simulation was carried out in the CoppeliaSim environment (Fig. 5). The AUV-leader moves in a straight line, passing through a narrow passage between two obstacles, and the AUV-followers, being unable to pass the obstacles while keeping formation, move beyond the trajectory of the leader.

AUV CS was performed in MATLAB Simulink. The AUV with following parameters is considered in this simulation:

$$m_a = 325kg, J_{xx} = 225kg \cdot m^2, J_{yy} = 175kg \cdot m^2, J_{zz} = 215kg \cdot m^2,$$

$$Y_c = 0.05m, \lambda_{ijmin} = 40kg\,(i,j = 1, 2, 3), \lambda_{ijmax} = 300kg\,(i,j = 1, 2, 3),$$

$$\lambda_{ijmin} = 20kg \cdot m^2\,(i,j = 4, 5, 6), \lambda_{ijmax} = 225kg \cdot m^2\,(i,j = 4, 5, 6),$$

$$d_{1min} = \text{Ns/m}, d_{1max} = 50\text{Ns/m}, d_{2min} = 75kg \cdot m^{-2}, d_{2max} = 125kg \cdot m^{-2}.$$

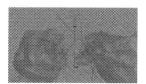

Fig. 5. The movement of the AUV group in an unknown environment

Figures 6 and 7 show the processes of changing the coordinates of the AUV group during their movement while avoiding obstacles. Figure 6 shows that in the process of movement, AUV-follower detect obstacles in their path and avoid them along the computed safe trajectories. At the same time, the AUV-followers trajectories follow the shape of obstacles, and the distances between the AUVs of the group are always greater than the specified safe distance (Figs. 9, 10, 11). Figure 8 shows that the distance between

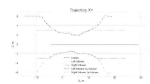

Fig. 6. Movement trajectories of the AUV group in the horizontal plane when avoiding obstacles

the AUV-followers and obstacles does not exceed 2 m. After passing the obstacles, the AUV-followers return to their prescribed position in the formation. At the same time, AUV-followers are also shifted in the vertical plane (Fig. 7), which provides them with continuous tracking of the leader AUV with the help of onboard video cameras, even when these AUVs line up one after another in the process of avoiding obstacles.

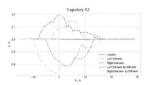

Fig. 7. Movement trajectories of the AUV group in the vertical plane when avoiding obstacles

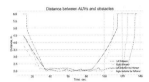

Fig. 8. Distances between AUV-followers and detectable obstacles

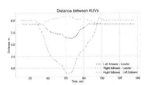

Fig. 9. Distances between AUVs while moving

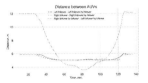

Fig. 10. Distances between AUVs while moving

Fig. 11. Minimal distance between AUVs

6 Conclusions

The paper presents a method for formation control of the AUVs group in the "leader-follower" mode in an environment containing unknown obstacles. A feature of the proposed method is to ensure the safe movement of these AUVs in conditions when there is no data transmission between the AUVs of the group via hydroacoustic communication channels to coordinate their movement trajectories. This is achieved by using information from the onboard video cameras of the followers to determine the position and orientation of the leader AUV relatively to these followers, and by presetting the movement trajectories of each follower AUV within the formation. A feature of the proposed method is the setting of such a displacement of the followers within the formation, which excludes the loss of beacons of the AUV-leader in the visibility area of the cameras of the followers.

Acknowledgement. This work is supported by Russian Science Foundation (grant 22-19-00392).

References

1. Das, B., Subudhi, B., Pati, B.: Cooperative formation control of autonomous underwater vehicles: an overview. Int. J. Autom. Comput. **13**(3), 199–225 (2016)
2. Spensieri, D., Carlson, J.S., Ekstedt, F., Bohlin, R.: An iterative approach for collision free routing and sheduling in multirobot stations. IEEE Trans. Autom. Sci. Eng. **13**(2), 950–962 (2015)
3. Langerwisch, M., Wagner, B.: Dynamic path planning for coordinated motion of multiple mobile robots. In: Proceedings of the 14th International IEEE Conference on Intelligent Transportation Systems (ITSC), pp. 1989–1994 (2011)
4. Vo, C., Harrison, J.F., Lien, J.: Behavior-based motion planning for group control. In: Proceedings of the IEEE/RSJ International Conference on Intelligent Robots and Systems, pp. 3768–3773 (2009)
5. Eustice, R., Whitcomb, L., Singh, H., Grund, M.: Experimental results in synchronous-clock one-way-travel-time acoustic navigation for autonomous underwater vehicles. In: Proceedings of the 2007 IEEE International Conference on Robotics and Automation, Roma, pp. 4257–4264 (2007)
6. Cario, G., et al.: Clock synchronization and ranging estimation for control and cooperation of multiple UUVs. In: Proceedings of the MTS/IEEE International Conference OCEANS 2016, Shanghai, China, 10–13 April 2016
7. Filaretov, V., Subudhi, B., Yukhimets, D., Mursalimov, E.: The method of matching the navigation systems of AUV-leader and AUV-followers moving in formation. In: Proceedings of the 2019 International Russian Automation Conference (RusAutoCon), Sochi, Russia, September 2019

8. Yong, L., Yu, L., Yipei, G., Kejie, C.: Cooperative path planning of robot swarm based on ACO. In: Proceedings of the IEEE 2nd Information Technology, Networking, Electronics and Automation Control Conference (ITNEC), pp. 1428–1432 (2017)

9. Lui, S., Sun, D., Zhu, C.: Coordinated motion planning for multiple mobile robots along designed paths with formation requirement. IEEE/ASME Trans. Mechatron. 16(6), 1021–1031 (2011)

10. Figueiredo, A.B., Ferreira, B.M., Matos, A.C.: Vision-based localization and positioning of an AUV. In: Proceedings of the OCEANS 2016, Shanghai, pp. 1–6 (2016)

11. Reyes, L.A., Tanner, H.G.: Flocking, formation control, and path following for group of mobile robots. IEEE Trans. Control Syst. Technol. 23(4), 1268–1282 (2015)

12. Filaretov, V.F., Yukhimets, D.A.: Control Method of AUV-follower on the base of visual information about AUV-leader position. Underw. Investig. Robot. 4, 29–35 (2022)

13. Filaretov, V.F., Yukhimets, D.A.: The path planning method for AUV group moving in environment with obstacles. Mechatron. Autom. Control 21(6), 356–364 (2020)

14. Strutz, T.: Data Fitting and Uncertainty (A Practical Introduction to Weighted Least Squares and Beyond), 2nd edn., Springer, Wiesbaden (2016)

15. Korn, T., Korn, G.: Mathematic Handbook. Nauka, Moscow (1973)

16. Rablinear, L., Goud, B.: Theory and Application of Digital Signal Processing. Mir, Moscow (1978)

17. Fossen, T.: Handbook of Marine Craft Hydrodynamics and Motion Control, Willey, Chichester (2011)

18. Siek., Yu.L., Borisov, A.N.: Motion control of an autonomous uninhabited underwater vehicle based on a predictive model. Morskoy vestnik 4(76), 98–101 (2020)

19. Orlovsky, I.A.: Using a predictive neurocontroller to control an electromechanical object. Sci. Pract. Donetsk Natl. Tech. Univ. Ser. Electr. Eng. Energy 1(14), 200–205 (2013)

A Broader Study of Spectral Missing in Multi-spectral Vehicle Re-identification

Tianying Yan[1,2] ⓘ, Changhai Wang[3], Changan Yuan[4], and De-Shuang Huang[1](✉) ⓘ

[1] Eastern Institute of Technology, Ningbo 315200, China
dshuang@eitech.edu.cn
[2] Shanghai Jiao Tong University, Shanghai 201100, China
[3] Guangxi Communications Design Group Co., Ltd., Nanning 530000, China
[4] Guangxi Academy of Science, Nanning 530000, China

Abstract. Recent advances in vehicle re-identification based on complex illuminance rely on multi-spectral data: visible (VI) spectra, near-infrared (NIR), and thermal infrared (TIR) spectra. However, in various applications, the spectra are always partially missing in terms of the mismatch between the training data assumptions and the actual spectral types in the test data. Spectral missing becomes an open-world challenge for testing the robustness of models. However, only a few models have spectral missing accuracy tests. In this work, criteria for constant independent spectral missing, constant Siamese spectral missing, random individual spectral missing, random Siamese spectral missing, and random individual & Siamese spectral missing are established. Extensive experiments have been conducted on the proposed criteria to evaluate the accuracy difficulties of state-of-the-art one-stream learning and multi-stream learning in spectral missing. The result shows that the most advanced multi-stream learning performed better than the one-stream learning models. In some cases, the performance of multi-stream learning is even worse than that of one-stream learning methods in Siamese spectral missing. In the benchmark test, due to the complementarity of TIR and VI and the redundancy of NIR and VI, VI + TIR Siamese spectra (1.6%) have a more moderate loss of accuracy than VI + NIR Siamese spectra (2.8%). In the end, the accuracy of all models tends to favor the VI that is more relevant to open-world scenarios. This work validates the value of benchmarking to better represent the spectral missing diversity seen in open-world practice and to guide future research.

Keywords: Spectral Missing · Vehicle Re-identification · One-stream Learning · Multi-stream Learning

1 Introduction

Training vehicle re-identification models for complex illuminance scenes typically requires large amounts of multi-spectral data [1]. The generalization of vehicle re-identification models depends largely on the complementation of multi-spectral data, which is reflected in low-illuminance, conventional illuminance, and high-illuminance

© The Author(s), under exclusive license to Springer Nature Singapore Pte Ltd. 2024
D.-S. Huang et al. (Eds.): ICAI 2023, CCIS 2015, pp. 51–63, 2024.
https://doi.org/10.1007/978-981-97-0827-7_5

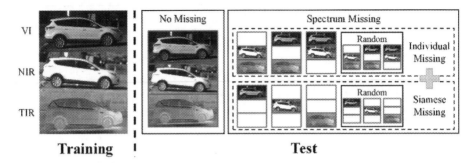

Fig. 1. Spectral Missing Protocol.

scenes [2, 3]. The multi-spectral data sets are presented in the form of near-infrared (NIR)/thermal infrared (TIR) spectra for low illuminance, visible (VI) spectra for conventional illuminance, and TIR spectra for high illuminance. The existing vehicle re-identification models focus on multi-spectral data sets in conventional and abnormal (low and high) illuminance scenes. While spectral missing may not come at a heavy cost, as a cross combination of multi-spectral groups, correct vehicle re-identification in spectral missing cases is critical.

In some cases, such as under conventional illuminance, the existing models can obtain rich information based on the existing VI. However, under abnormal illuminance, existing models are difficult to mine enough information in VI filled with light noise [4]. The introduction of other spectra that are not sensitive to illuminance is a new means of solving the difficulty, which is satisfying with challenging. One of the challenges is spectral missing, especially VI missing. In the case of VI missing, NIR and TIR cannot provide color information, which is a significant feature of model dependence in conventional illuminance. In the case of abnormal illuminance missing by NIR and TIR, the existing models fall into a visual fault caused by abnormal noise [3, 5]. In other words, existing vehicle re-identification models for complex illuminance scenes are challenged in disposing of spectral missing expertise.

From another perspective, spectral missing is a robust research topic for models. The robustness is essentially the tolerance of the models to data changes, which is desired as (1) The models have high accuracy in the raw data; (2) Small biases in the data have a small effect on the model outputs; (3) Large biases in the data with no catastrophic impact on the model outputs [6–8]. Deep learning-based models can be trusted for robustness to small deviations from existing data [9]. However, recent studies have pointed out that deep learning-based models, as special cases with large data biases, perform poorly when spectral data is missing [10–12]. In essence, spectral missing is a special case for raw data bias in multi-spectral models. However, spectral missing is an extremely significant open-world problem: the detection viability of existing models in mixed data. Specifically, the open world exists in the form of pure data (VI, VI + NIR, VI + TIR, and VI + NIR + TIR), and mixed data (no missing and spectral missing). Although a benchmark for conventional noise interference for pure data has been established, a benchmark for spectral missing noise for pure data is in the bursting stage. Models designed for pure data may have significant effects in spectral missing noise, and further may not perform well in mixed data.

To fill the gap of spectral missing in multi-spectral vehicle re-identification, a broader study of spectral missing in multi-spectral vehicle re-identification is proposed (Fig. 1). Spectral missing covers individual missing, Siamese missing as exemplified by triplets of (VI, NIR, TIR). The missing criteria contain constant missing and random missing. Existing state-of-the-art re-identification models are rigorously tested in the proposed criteria. This work would help the community to understand the robust differences of models in spectral missing and contribute to the advancement of open-world applications.

In summary, the contributions of this work are itemized as follows:

- Establish a broader study benchmark for the spectral missing in multi-spectral vehicle re-identification, based on individual missing and Siamese missing of the triplet VI, NIR, and TIR, with accompanying constant and random criteria.
- Evaluate the performance of current re-identification models, including one-stream learning for individual spectra, as well as multi-stream learning tailored to multiple spectra.
- The result shows that the most advanced multi-stream learning is superior to one-stream learning, and the accuracy gain of all re-identification models is lost.
- The result also shows that the accuracy of all models is related to the extent of spectral missing proposed, validating the diversity of problem representations and the value of benchmarks for future research.

2 Related Work

This work involves two types of models: one-stream learning and multi-stream learning. One-stream learning is trained separately for each independent spectral instance and combines each independent spectral feature information for testing, while multi-stream learning is trained and tested directly for the whole spectral instance (triplet instance).

2.1 One-stream Learning

One-stream learning is a method based on deep learning, which is originally applied to the feature extraction of the visible spectral data [13]. The success has inspired the extension of one-stream learning to be applied to feature extraction of other spectral types [14]. One-stream learning in re-identification includes two subdivisions of person and vehicle, due to non-rigid and rigid structure [15, 16]. Fortunately, the concept of technical solutions to address the landing of two types of re-identification applications is mutually applicable, such as video re-identification [17, 18], unsupervised re-identification for handling pseudo label [19–21], and domain adaptive re-identification [21, 22]. To further improve the performance of one-stream learning, previous works have proposed methods to combine local and global information [23, 24]. The methods of combining local information and global information can promote one-stream learning to understand the structure of the image and improve the accuracy of re-identification. In addition, some works have focused on designing new loss functions for networks that attempt to constrain feature distributions in metric spaces [24, 25].

To solve the re-recognition dilemma caused by complex illuminance in the open world, multi-spectral images are introduced to reduce the illuminance effect. The existing work improves the accuracy of re-identification by fusing spectral images of different spectra [26]. Specifically, images from different spectra can be treated as separate domains and one-stream learning processes feature at different spectra separately, ultimately fusing head features. In practical applications, one-stream learning for multi-spectral images does not take into account the complementary information inside the multi-spectral, which limits applicability in abnormal illuminance, which creates a need for the exploration of re-identification solutions based on multi-stream learning.

2.2   Multi-stream Learning

Multi-stream learning reads multi-spectral data directly and uses a multi-branch backbone to process fusion head features. Li et al. proposed the first multi-spectral vehicle re-identification dataset, RGBNT100 [27]. Subsequently, Zheng et al. extended the data protocol and proposed a multi-spectral vehicle re-identification dataset MSVR310 for the open world [5]. Li et al. propose a three-stream network with multi-spectral feature fusion to handle multi-spectral variations to ensure feature discriminability [27]. Wang et al. considered spectral-specific information and proposed to fuse other spectral information through multiple graduations [28]. Zheng et al. proposed a fusion strategy of one-branch local information and multi-branch global information [8]. At the same time, Zheng et al. proposed the measurement loss of spectral differences and identity differences to enhance network characterization [5]. However, the existing multi-stream learning considers the complementary relationship of multi-spectral data, ignoring the decoupling operation of multi-spectral missing. In the case of no spectral missing, the multi-spectral complementary information introduced by multi-stream learning improves the recognition accuracy. However, in the case of spectral missing, the robustness of multi-stream learning may be weakened compared with one-stream learning.

3   Proposed Benchmark

In this section, a broader study of spectral missing in multi-spectral vehicle re-identification benchmarks is presented, which includes the data source, missing criterion, and experimental setup.

3.1   Data Source

Since only two publicly available (VI, NIR, TIR) image datasets are available for evaluating multi-spectral vehicle re-identification benchmarks, RGBNT100 and MSVR310 can be used as baseline data sources. Although RGBNT100 covers the abnormal illuminance scene, which contains a total of 17,250 image triples of 100 vehicles. RGBNT100 contains a large number of similar frames in the image triples. MSVR310 extends the RGBNT100 data protocol, and the training set contains 155 vehicles with a total of 1032 image triples as the training set. The test set consists of the remaining 155 vehicles with 1055 image triples. The query set contains 52 vehicles with a total of 591 image triples, which are randomly selected from the test set. Considering the open-world application, MSVR310 is chosen, which has little repetitive information on a frame-by-frame basis, along with view changes (VC), partly occlusion (PO), and various resolutions (VR) practical challenges.

3.2 Missing Criterion

Considering the future application of multi-spectral vehicle re-identification in the open world, the spectral missing criteria are defined as constant missing and random missing (Fig. 1).

- Constant missing. Fixed missing specified types of triplet images, including individual missing and Siamese missing.
- Random missing. Random missing of triplet images, including individual missing, Siamese missing, and two kinds of mixed missing.

The pseudo-code is shown in Algorithm 1.

3.3 Experimental Setup

- Baseline. This work uses one-stream networks ABD [29], StrongBaseline [30], OSNet [31], and AGW [32], multi-stream networks HAMNet [27], PFNet [8], IEEE [28] and CCNet [5] as baselines. This work follows the baseline manual of image preprocessing, parameter optimization, and loss design.
- Evaluation. In this work, Cumulative Matching Characteristics (CMC) and Mean Average Precision (mAP) are used as evaluation indicators. The CMC score reflects the retrieval accuracy, which is reported in this work as Rank 1 (R1), Rank 5 (R5), and Rank 10 (R10) scores. mAP measures the average of all queries that reflect the average accuracy of the Recall (the area under the precision-recall curve).
- Platform. This work experiment environment is built in a Docker container of the high-performance computing platform. The hardware configuration is an eight-core CPU, 24G memory, 4096M shared memory, and GeForce RTX 3090 GPU. The GPU acceleration package version is CUDA 11.1. The deep learning package is Pytorch 1.8 implemented in Python 3.8.

Algorithm 1. Spectral Missing.

Input: multi-spectral data and spectral missing criterion

Output: modified data with spectral missing criterion

1   **switch spectral_missing_criterion**
       // constant missing
2   case **miss_VI**
3   set data[:, 0] to 0 // miss VI
4   case **miss_NIR**
5   set data[:, 1] to 0 // miss NIR
6   case **miss_TIR**
7   set data[:, 2] to 0 // miss TIR
8   case **retain_VI**
9   set data[:, 1:] to 0 // retain VI, miss NIR, TIR
10   case **retain_NIR**
11   set data[:, [0, 2]] to 0 // retain NIR, miss VI and TIR
12   case **retain_TIR**
13   set data[:, :2] to 0 // retain TIR, miss VI and NIR
       // random missing
14   case **random_loss_1**
15   randomly choose one spectral type data and set it to 0
16   case **random_loss_2**
17   randomly choose two spectral type data and set them to 0
18   case **random_loss_3**
19   randomly miss at least one spectral type data and set them to 0, while retaining at least one spectral type data
20   **end switch**
21   **return modified_data**

4   Experiments and Analysis

In this work, constant individual spectral missing, constant Siamese spectral missing, and random spectral missing are analyzed successively.

4.1   Constant Individual Spectral Missing

As shown in Table 1, in the case of spectral missing as VI, the best performing models are CCNet (31.4), ABD (24.9), and OSNet (23.0) according to mAP, while HAM-Net (20.5) is slightly behind in mAP. However, HAMNet performs well on the R1 (45.5), R5 (60.7), and R10 (66.7). In the case of spectral missing as NIR, the best-performing models are CCNet (34.6), AGW (27.7), and ABD (26.4). In the case of TIR spectral missing, the best-performing models are CCNet (34.0), AGW (30.0), and ABD (25.4). In the one-stream models, the spectral missing case has own outstanding place, which comes from the independence of the one-stream models in feature extraction. The multi-stream models are similar to the one-stream models, however, the performance of PFNet and IEEE for multi-spectral person re-identification is slightly lower than most one-stream models, which may be related to the design of non-rigid structures for the person. However, no single model performs best in all spectral missing cases, and existing models vary in performance for new spectral missing requirements, tailored to own independent needs.

4.2 Constant Siamese Spectral Missing

As shown in Table 2, the performance of all models decreases when the spectral data is reserved only for VI, however, CCNet (mAP: 30.7) performs best, followed by ABD (mAP: 23.7) and AGW (mAP: 21.8). When the spectral data is only NIR, the performance order of the models does not change, and the mAP values are 26.3, 21.8 and 20.5. When only TIR is retained in the spectral data, the performance of the models declined further, however, CCNet still performs best on various indicators (mAP: 19.6, R1: 35.7, R5: 53.5, R10: 61.9). The one-stream models perform well in all spectral missing cases, especially ABD. AGW and OSNet also perform well under different spectral missing cases, but not as well as ABD. The performance of the multi-stream models is relatively uneven in all spectral missing cases, which may be related to the application requirements that do not consider spectral missing cases. Relatively speaking, individual one-stream models are as stable as most multi-stream models.

4.3 Random Spectral Missing

As shown in Table 3, the performance of all models decreases when the random spectral is missing, however, the robustness of CCNet, AGW, and ABD is better than that of other models. The one-stream model performs well in all cases of random spectral missing, especially ABD. The multi-stream model has the lowest CCNet loss (mean of mAP loss: 7.0) in all spectral missing cases. The random accuracy loss is shown in Fig. 2.

As shown in Fig. 2, this work compares multiple types of spectral missing cases. Figure 2(a) and Fig. 2(b) show that missing and retention of VI have the highest impact on model performance (4.2% and 5.8%). However, the missing and retention of NIR and TIR have an impact on model performance related to the degree of complementary information. In Fig. 2(a), since NIR and VI are highly correlated in image texture, the missing NIR has a smaller effect (1.6%), while TIR has a larger negative effect (2.8%). Then, in Fig. 2(b), the independent retention of NIR (7.5%) and TIR (10.6%) confirms the importance of rich image textures for the models. At the same time, the trends of R1, R5, and R10 confirm that in the case of independent spectral data, the existing models rely more on texture information, while in the case of multi-spectral data, the existing models tend to complement the information. Figure 2(c) shows that individual spectral missing (2.9%) is less harmful than Siamese spectral missing (8.0%).

Table 1. State-of-the-art Models for Vehicle Re-ID on MSVR310 with Individual Missing (in%).

Model		No Missing				Individual Missing											
						VI				NIR				TIR			
		mAP	R1	R5	R10	mAP	R1	R5	R10	mAP	R1	R5	R10	mAP	R1	R5	R10
One-Stream	ABD [29]	28.7	44.8	66.2	73.1	24.9	45.1	59.5	64.2	26.4	46.9	62.6	68.5	25.4	36.2	58.1	62.1
	StrongBaseline [30]	23.0	49.3	65.9	72.2	19.9	45.0	57.9	63.2	21.9	40.2	52.7	57.4	20.6	37.9	52.1	58.6
	OSNet [31]	27.7	48.1	61.7	67.9	23.0	44.3	59.6	65.0	26.0	51.5	67.0	72.5	22.8	35.5	54.1	67.8
	AGW [32]	29.0	48.1	61.5	68.1	24.7	45.3	59.8	66.3	27.7	42.9	57.2	63.2	30.0	23.1	36.1	45.2
Multi-Stream	HAMNet [27]	25.4	39.2	50.6	56.4	20.5	45.5	60.7	66.7	23.9	42.8	54.8	61.8	21.3	25.5	22.9	20.3
	PFNet [8]	23.4	40.4	54.6	61.1	18.7	38.2	52.3	58.9	21.5	41.2	51.6	65.2	19.9	38.5	54.6	60.8
	IEEE [28]	20.1	33.7	46.5	54.0	16.7	33.1	49.3	55.6	19.1	45.6	63.8	71.3	17.0	43.1	60.8	68.1
	CCNet [5]	36.4	55.2	72.4	79.7	31.4	51.6	68.9	76.6	34.6	52.8	68.7	75.5	34.0	53.6	70.2	76.3

Table 2.   State-of-the-art Models for Vehicle Re-ID on MSVR310 with Siamese Missing (in%).

Model		No Missing				Siamese Missing												
						w/VI				w/NIR				w/TIR				
		mAP	R1	R5	R10	mAP	R1	R5	R10	mAP	R1	R5	R10	mAP	R1	R5	R10	
One-Stream	ABD [29]	28.7	44.8	66.2	73.1	23.7	40.7	51.6	59.6	21.8	41.5	58.4	65.3	19.2	37.2	53.1	59.9	
	StrongBaseline [30]	23.0	49.3	65.9	72.2	17.4	31.6	48.3	53.4	16.4	37.1	52.8	56.7	13.2	34.4	54.1	62.3	
	OSNet [31]	27.7	48.1	61.7	67.9	20.1	44.7	59.1	65.6	19.1	39.2	56.3	60.2	17.1	36.3	52.7	58.4	
	AGW [32]	29.0	48.1	61.5	68.1	21.8	38.9	58.8	68.2	20.5	40.8	57.5	63.1	17.3	39.0	51.0	57.8	
Multi-Stream	HAMNet [27]	25.4	39.2	50.6	56.4	19.6	38.1	54.4	60.5	18.2	38.4	54.1	59.2	16.4	30.0	43.4	48.5	
	PFNet [8]	23.4	40.4	54.6	61.1	17.2	37.8	51.7	57.5	16.3	36.5	50.4	55.9	14.0	38.0	58.0	66.3	
	IEEE [28]	20.1	33.7	46.5	54.0	16.4	30.0	43.4	48.5	14.8	35.8	52.4	60.3	12.4	34.5	52.6	59.6	
	CCNet [5]	36.4	55.2	72.4	79.7	30.7	49.4	65.5	73.3	26.3	45.5	67.3	73.1	19.6	35.7	53.5	61.9	

Table 3. State-of-the-art Models for Vehicle Re-ID on MSVR310 with Random Missing (in%).

| Model | | No Missing | | | | Random Missing | | | | | | | | | | | |
| | | | | | | Individual | | | | Siamese | | | | Individual + Siamese | | | |
		mAP	R1	R5	R10	mAP	R1	R5	R10	mAP	R1	R5	R10	mAP	R1	R5	R10
One-Stream	ABD [29]	28.7	44.8	66.2	73.1	25.6	42.7	60.1	64.9	21.6	39.8	54.4	61.6	23.6	41.3	57.3	63.3
	StrongBaseline [30]	23.0	49.3	65.9	72.2	20.8	41.0	54.2	59.7	15.7	34.4	51.7	57.5	18.3	37.7	53.0	58.6
	OSNet [31]	27.7	48.1	61.7	67.9	23.9	43.8	60.2	68.4	18.8	40.1	56.0	61.4	21.4	42.0	58.1	64.9
	AGW [32]	29.0	48.1	61.5	68.1	27.5	37.1	51.0	58.2	19.9	39.6	55.8	63.0	23.7	38.4	53.4	60.6
Multi-Stream	HAMNet [27]	25.4	39.2	50.6	56.4	21.9	37.9	46.1	49.6	18.1	35.5	50.6	56.1	20.0	36.7	48.4	52.9
	PFNet [8]	23.4	40.4	54.6	61.1	20.0	39.3	52.8	61.6	15.8	37.4	53.4	59.9	17.9	38.4	53.1	60.8
	IEEE [28]	20.1	33.7	46.5	54.0	17.6	40.6	58.0	65.0	14.5	33.4	49.5	56.1	16.1	37.0	53.8	60.6
	CCNet [5]	36.4	55.2	72.4	79.7	33.3	52.7	69.3	76.1	25.5	43.5	62.1	69.4	29.4	48.1	65.7	72.8

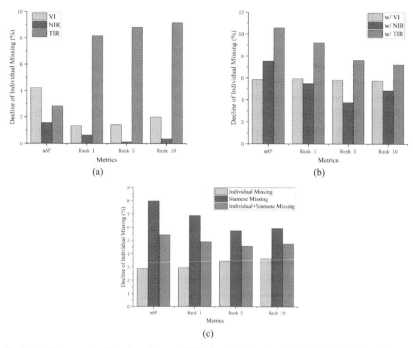

Fig. 2. Model Loss under Random Spectral Missing. (a) Random Individual Missing, (b) Random Siamese Missing, and (c) Two Kinds of Mixed Missing.

In summary, VI contributes the most to the overall performance, VI matches the open-world scene, and the vehicle re-identification scene is mostly normal illuminance. In multi-spectral fusion, when the spectra are missing to NIR and TIR, the performance indexes are reduced, indicating that the complementary information of NIR and TIR has a positive effect on the performance.

5   Conclusion

In this work, a broader study of spectral missing in multi-spectral vehicle re-identification is formally introduced, with benchmarks covering multiple spectral missing cases. One-stream learning and multi-stream learning are extensively analyzed and evaluated for effect gain in spectral missing. The result shows that state-of-the-art multi-stream learning outperforms one-stream learning, while spectral missing degrades the performance of open-world applications. In addition, compared with multi-stream learning, partial one-stream learning performs better in individual spectral missing due to feature mining capability. In fact, with Siamese spectra (VI + TIR) preserved, the gain of complementary information is higher than that of redundant Siamese spectra (VI + NIR). In addition, the accuracy of all models correlates with the proposed correlation with spectral missing, validating the diversity of the proposed benchmarks in terms of complex spectral missing representations and value to guide future research. Together, this work would help the community understand which models work best in spectral missing practices and help drive further advancements that would bring benefits to open-world applications faster.

Acknowledgment. This work is supported by the High-Performance Computing (HPC) Center at the Eastern Institute for Advanced Study (EIAS).

Funding. This work is supported in part by STI 2030-Major Projects, under Grant No. 2021ZD0200403, and partly supported by grants from the National Science Foundation of China, Nos. 62333018, 62372255, U22A2039, 62073231, and 62372318, and supported by the Key Project of Science and Technology of Guangxi (Grant No. 2021AB20147), Guangxi Natural Science Foundation (Grant Nos. 2022JJD170019, 2021JJA170204, and 2021JJA170199) and Guangxi Science and Technology Base and Talents Special Project (Grant Nos. 2021AC19354 and 2021AC19394) and by Guangxi Key Lab of Human-machine Interaction and Intelligent Decision, Guangxi Academy Sciences, and supported by Key Research and Development (Digital Twin) Program of Ningbo City under Grant Nos.2023Z219 and 2023Z226, and supported by the China Postdoctoral Science Foundation under Grant No. 2023M733400.

References

1. Kamenou, E., Del Rincón, J.M., Miller, P., Devlin-Hill, P.: A meta-learning approach for domain generalisation across visual modalities in vehicle re-identification, pp. 385–393 (2023)
2. Guo, J., Zhang, X., Liu, Z., Wang, Y.: Generative and attentive fusion for multi-spectral vehicle re-identification. In: Proceedings of 2022 7th International Conference on Intelligent Computing and Signal Processing (ICSP), pp. 1565–1572 (2022)
3. He, Q., Lu, Z., Wang, Z., Hu, H.: Graph-based progressive fusion network for multi-modality vehicle re-identification. In: IEEE Transactions on Intelligent Transportation Systems, pp. 1–17 (2023)
4. H.B. A., P.M.M. M., V. U., M.P. R., Enhanced vehicle re-identification for ITS: a feature fusion approach using deep learning. In: Proceedings of 2022 IEEE International Conference on Electronics, Computing and Communication Technologies (CONECCT), pp. 1–6 (2022)
5. Zheng, A., Zhu, X., Ma, Z., Li, C., Tang, J., Ma, J.: Cross-directional consistency network with adaptive layer normalization for multi-spectral vehicle re-identification and a high-quality benchmark. Inf. Fusion 101901 (2023)
6. Whang, S.E., Roh, Y., Song, H., Lee, J.: Data collection and quality challenges in deep learning: a data-centric AI perspective. VLDB J. **32**, 791–813 (2023)
7. Subbaswamy, A., Adams, R., Saria, S.: Evaluating model robustness and stability to dataset shift. In: Proceedings of the 24th International Conference on Artificial Intelligence and Statistics, PMLR, pp. 2611–2619 (2021)
8. Zheng, A., Wang, Z., Chen, Z., Li, C., Tang, J.: Robust multi-modality person re-identification, pp. 3529–3537 (2021)
9. Wang, Q., et al.: Viewpoint adaptation learning with cross-view distance metric for robust vehicle re-identification. Inf. Sci. **564**, 71–84 (2021)
10. L. J., W. J., H. N., Z. Q., H. J.: Revisiting modality-specific feature compensation for visible-infrared person re-identification. IEEE Trans. Circ. Syst. Video Tech. 7226–7240 (2022)
11. Jiang, K., Zhang, T., Liu, X., Qian, B., Zhang, Y., Wu, F.: Cross-modality transformer for visible-infrared person re-identification. In: Avidan, S., Brostow, G., Cissé, M., Farinella, G.M., Hassner, T. (eds.) Computer Vision – ECCV 2022, ECCV 2022, LNCS, vol. 13674, pp. 480–496. Springer, Cham (2022). https://doi.org/10.1007/978-3-031-19781-9_28

12. Zhang, Q., Lai, C., Liu, J., Huang, N., Han, J.: Fmcnet: feature-level modality compensation for visible-infrared person re-identification. In: Proceedings of the IEEE/CVF Conference on Computer Vision and Pattern Recognition (CVPR), pp. 7349–7358 (2022)

13. Huang, N., Liu, J., Miao, Y., Zhang, Q., Han, J.: Deep learning for visible-infrared cross-modality person re-identification: a comprehensive review. Inf. Fusion **91**, 396–411 (2023)

14. Zhang, X.: Deep learning-based multi-focus image fusion: a survey and a comparative study. IEEE Tran. Pattern Anal. Mach. Intell. **44**, 4819–4838 (2022)

15. Li, Y., Zhang, T., Duan, L., Xu, C.: A unified generative adversarial framework for image generation and person re-identification, pp. 163–172 (2018)

16. Jiang, M., et al.: Robust vehicle re-identification via rigid structure prior. In: Proceedings of the IEEE/CVF Conference on Computer Vision and Pattern Recognition (CVPR) Workshops, pp. 4026–4033 (2021)

17. Ye, M., Shen, J., Lin, G., Xiang, T., Shao, L., Hoi, S.C.: Deep learning for person re-identification: a survey and outlook. IEEE Trans. Pattern Anal. Mach. Intell. **44**(6), 2872–2893 (2022)

18. Wu, D., et al.: Deep learning-based methods for person re-identification: a comprehensive review. Neuro Comput. **337**, 354–371 (2019)

19. Bai, Y., Wang, C., Lou, Y., Liu, J., Duan, L.Y.: Hierarchical connectivity-centered clustering for unsupervised domain adaptation on person re-identification. IEEE Trans. Image Process. **30**, 6715–6729 (2021)

20. Cho, Y., Kim, W.J., Hong, S., Yoon, S.: Part-based pseudo label refinement for unsupervised person re-identification. In: Proceedings of the IEEE/CVF Conference on Computer Vision and Pattern Recognition (CVPR), pp. 7308–7318 (2022)

21. Song, L., et al.: Unsupervised domain adaptive re-identification: theory and practice. Pattern Recogn. **102**, 107173 (2020)

22. Zheng, K., Liu, W., He, L., Mei, T., Luo, J., Zha, Z.: Group-aware label transfer for domain adaptive person re-identification. In: Proceedings of the IEEE/CVF Conference on Computer Vision and Pattern Recognition (CVPR), pp. 5310–5319 (2021)

23. He, B., Li, J., Zhao, Y., Tian, Y.: Part-regularized near-duplicate vehicle re-identification. In: Proceedings of the IEEE/CVF Conference on Computer Vision and Pattern Recognition (CVPR), pp. 3997–4005 (2019)

24. Bai, Y., Liu, J., Lou, Y., Wang, C., Duan, L.Y.: Disentangled feature learning network and a comprehensive benchmark for vehicle re-identification. IEEE Trans. Pattern Anal. Mach. Intell. **44**, 6854–6871 (2022)

25. Zhang, Q., Lai, J., Xie, X.: Learning modal-invariant angular metric by cyclic projection network for VIS-NIR person re-identification. IEEE Trans. Image Process. **30**, 8019–8033 (2021)

26. Basaran, E., Gökmen, M., Kamasak, M.E.: An efficient framework for visible–infrared cross modality person re-identification. Signal Process. Image Commun. **87**, 115933 (2020)

27. Li, H., Li, C., Zhu, X., Zheng, A., Luo, B.: Multi-spectral vehicle re-identification: a challenge. In: Proceedings of the AAAI Conference on Artificial Intelligence, pp. 11345–11353 (2020)

28. Wang, Z., Li, C., Zheng, A., He, R., Tang, J.: Interact, embed, and enlarge: Boosting modality-specific representations for multi-modal person re-identification. In: Proceedings of the AAAI Conference on Artificial Intelligence, pp. 2633–2641 (2022)

TimesNet Model for Multi-industry Electricity Demand Forecasting

Fangzhao Deng[1], Hujun Li[1], Zhenli Deng[1], Jianan Si[1], Boning Yu[1], Xingwu Guo[1], and Yizhou Zeng[2(✉)]

[1] State Grid Henan Electric Power Company Economic and Technological Research Institute, Zheng Zhou, China
[2] School of Informatics, Xiamen University, Xiamen, China
zengyizhou@stu.xmu.edu.cn

Abstract. Accurately predicting electricity demand is of crucial importance for optimizing power resource allocation, improving grid operation safety and providing significant economic benefits. In recent years, deep learning models have been widely employed in time series analysis tasks, such as recurrent neural networks (RNNs), gradient boosting decision trees (GBDT), and transformer networks. However, these methods only focus on a single time dimension. For a specific periodic process, the temporal variations at each time point within the process are not only related to neighboring moments but also highly correlated with neighboring periods, exhibiting both intraperiod and interperiod temporal changes. In this study, we propose the TimesNet temporal forecasting model and conduct experiments using electricity consumption data from the A city, combined with a series of covariates such as temperature and holidays, to train the model and predict electricity consumption. We compare the performance of TimesNet with other methods, and the results demonstrate that the TimesNet model outperforms the other models. Overall, the hybrid model we propose provides a valuable framework for accurately predicting electricity demand and holds practical significance for managing and operating power grids in urban areas.

Keywords: Timesnet · Electricity Demand Forecasting · Time Series Forecasting

1 Introduction

Electricity demand prediction is a critical task for urban energy management, infrastructure planning, and cost control [1]. It provides important insights for planners to make informed decisions on the construction and expansion of power generation facilities to meet future electricity needs [2].Accurate demand prediction facilitating the efficient utilization of renewable energy through proper energy conversion and storage which contributes to the accelerated development of clean energy, reduces reliance on traditional fossil fuels, and promotes the transformation and adoption of sustainable energy sources [3].

Nowadays, data-driven deep learning models based on machine learning frameworks have gradually become the mainstream for electricity demand prediction [4]. LSTM [5] is a type of Recurrent Neural Network (RNN) [6] that can capture long-term dependencies and nonlinear relationships in time series data. Gradient Boosting Decision Trees (GBDT) [7] which uses decision trees as base learners and combines their predictions by training new trees at each step to adapt to the residuals between the current prediction and the true value. To extract temporal dependencies and capture long-term dependencies in the context of time series analysis, some methods have utilized the local properties of one-dimensional convolutional kernels [8]. However, due to the local nature of one-dimensional convolutional kernels, they can only model the changes between adjacent time points and therefore still struggle to capture long-term dependencies. Recently, attention mechanisms [9] have been widely applied in Transformer models for sequence modeling [10]. In the domain of time series analysis, many Transformer-based models incorporate attention mechanisms or their variants to capture pairwise temporal dependencies between time points [11].

These methods only consider a single time dimension, whereas realworld electricity demand time series data often exhibit multi-process superposition, such as annual cyclic variations due to seasons and holidays. This intrinsic property of multiple cycles makes the time series fluctuations highly complex. For a specific periodic process, the temporal variations at each time point within the process are not only related to neighboring moments but also highly correlated with neighboring periods, exhibiting both intraperiod and interperiod temporal changes. The intraperiod changes correspond to short-term processes within a cycle, while the interperiod changes reflect long-term trends across consecutive cycles.

Based on the observations above, the multi-period property naturally inspires a modular design approach, which involves designing a specific module to capture the temporal variations dominated by a particular period. This modular design approach allows for the decomposition of complex temporal changes into multiple individual time transformations, which facilitates time series prediction. By folding one-dimensional time series based on multiple periods, we can obtain multiple two-dimensional tensors. The columns and rows of each two-dimensional tensor respectively reflect the temporal variations within a period and across periods, resulting in a two-dimensional representation of temporal changes.

2 Proposed Method

2.1 Overall Framwork

Predicting electricity demand is of great importance for the planning, operation, and management of energy markets and power systems. Electricity demand prediction is closely related to time series forecasting, as electricity demand data often exhibit patterns in a time series format, with regular monthly and yearly cycles. It aims to forecast the amount of electricity demand for a future period (typically days or months) by analyzing the time series patterns in historical electricity demand data.

The main approaches for time series analysis include statistical models, regression models, and machine learning models. These methods analyze historical data to learn patterns and trends in electricity consumption and utilize this information to predict future electricity demand. Therefore, time series forecasting can be employed for electricity demand prediction, which is a key component of the overall process.

This study is based on a dataset of the electricity consumption across the entire A city, covering the data from the past three years (2020–2023).

Additionally, the dataset includes meteorological information such as the daily maximum and minimum temperatures, wind speed, and atmospheric pressure for the city. This dataset provides a high-dimensional feature sequence for each electricity demand, and integrating this multidimensional data is crucial for constructing accurate and reliable electricity demand prediction models. It can provide insights into the patterns and trends of the electricity consumption for model predictions.

The objective of this study is to establish and analyze appropriate statistical and artificial intelligence methods for forecasting a set of target data sequences, with a focus on the accuracy of each dimension's target prediction. In this report, the prediction model also takes into consideration the different electricity demands across various industries. The electricity consumption in different industries exhibits different patterns and time cycles. This study aims to predict the electricity consumption for different industries, targeting specific patterns, and can effectively improve the accuracy of the predictions.

2.2   Transform 1d-Variations into 2d-Variations

In this paper, we propose the TimesNet model, which decomposes complex temporal variations into different periods through a modularized structure. By transforming the original one-dimensional time series into a two-dimensional space, we achieve a unified modeling of within-period and between-period variations. The original 1D time series structure can only capture changes between adjacent time points. To address this limitation, we explore the use of a two-dimensional structure to represent temporal variations, which can explicitly capture both within-period and between-period changes. This provides greater expressive power and facilitates subsequent representation learning.

The conversion of a one-dimensional time series into a two-dimensional representation can be summarized by the following equations:

$$A = Avg\left(Amp\left(FFT\left(X_{\{1d\}}\right)\right)\right) \tag{1}$$

$$f_1, \ldots, f_k = argTopk_{\left\{f_*=1,\ldots,\frac{T}{2}\right\}} \tag{2}$$

$$p_i \in \frac{T}{f_i}, i \in \{1, \cdots, k\} \tag{3}$$

Specifically, for a time series of length T containing C recorded variables, the 1D time series can be represented as X1D $\in$ RC*T. To extract the periodicity, we first perform the Fast Fourier Transform (FFT) [12] on the 1D time series and calculate the amplitude values (Amp), and then take the average (Avg) across the C dimensions, resulting in A $\in$ RT, which represents the amplitudes corresponding to each frequency.

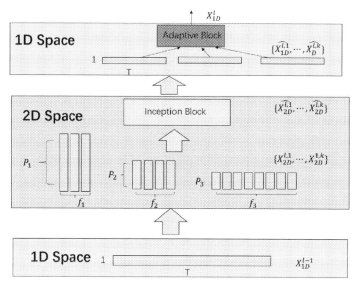

Fig. 1. A example to illustrate dimension transformation in time series

As shown in Fig. 1, for all frequencies $\{f_1,\cdots,f_n\}$, considering the sparsity of frequency domain and to avoid the noises brought by meaningless high frequencies [13], we only select the top k frequencies $\{f_1,\cdots,f_k\}$ with the highest amplitudes $\{A_{f\,1},\cdots,A_{fk}\}$ are selected, where k is a tunable hyperparameter. The length of each period is obtained by dividing the length T of the recorded variables by the corresponding frequency. Considering the conjugate domain of frequencies, we only consider the range $\{1,\cdots,\frac{2}{T}\}$. The above process can be simplified using the Period function.

In summary, the process can be expressed using the Period function as follows:

$$A, \{f_1, \cdots, f_k\}, \{p_1, \cdots, p_k\} = Period(X_{1d}) \tag{4}$$

For each frequency f_i, the corresponding period length is denoted as p_i. The one-dimensional time series X_{1D} of dimension T is padded using the Padding function to a length T' that is divisible by p_i, where $T' = f_i \cdot p_i$.

Since the one-dimensional time series X_{1D} is in $R^{T,C}$, we can reshape it using the Reshape function to obtain a two-dimensional time series X_{2D} in $R^{fi,pi,C}$. The above process can be summarized using the following equation:

$$X_{2d} = \text{Reshape}(\text{Padding}(X_{1d})) \tag{5}$$

It is important to note that for a given frequency f_i, each row of the two-dimensional time series $X_2{}^i{}_D$ corresponds to the temporal variations within a period, and each column corresponds to the temporal variations between periods. Therefore, the 2D temporal variations can be easily processed using 2D convolutional kernels.

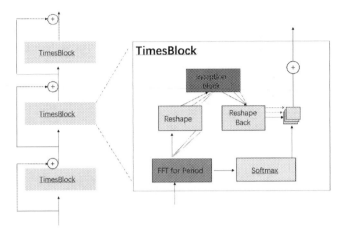

Fig. 2. Overall architecture of TimesNet.

2.3 TimesBlock

As shown in Fig. 2, We organize the TimesBlock in a residual manner, following the approach introduced by He, Kaiming and Zhang [14]. The input to the TimesBlock is a one-dimensional time series X_{1D} belonging to $R^{T,C}$. The TimesBlock receives the input from the previous layer and adds the output to the input of the previous layer. Specifically, for layer l, we have

$$X_{1D}^{l} = \text{TimesBlock}\left(X_{1D}^{l-1}\right) + X_{1D}^{l-1}.$$

For the i-th TimesBlock, its actions mainly involve capturing 2D temporal variations and adaptively aggregating representations from different periods.

2.4 Capture 2D Temporal Transformation Features

Based on the analysis from the previous section, we deconstruct k periods $\{p_1, \cdots, p_k\}$ using the Period function and obtain the two-dimensional time series X_{2D} corresponding to each period i, where $X_{2D} \in R^{fi,pi,C}$.

The Inception module [15] is a special module used in convolutional neural networks, designed to extract feature representations at different scales. It consists of a series of parallel convolutional and pooling layers, and constructs a deep network by stacking these layers. The core idea of the Inception module is to process the input feature maps in parallel using different convolutional kernels and pooling layers, and then concatenate their outputs to form the final output of the module. This approach aims to capture feature information at multiple scales, leading to richer and more diverse feature representations.

For a given frequency f_i, we utilize the Inception module with two-dimensional convolutional kernels to capture features from the two-dimensional time series, resulting in $X\hat{}_{2D}^{i} \in R^{fi,pi,d\text{model}}$. In other words, the Inception module performs the transformation from the C dimension to the d_{model} dimension for the two-dimensional time series.

Finally, we need to convert the two-dimensional time series back to a one-dimensional time series. Specifically, we use the *Reshape'* function to unravel the two-dimensional time series $X\hat{~}_2{}^i{}_D \in R^{fi,pi,d\text{model}}$ into a one-dimensional time series $X_1{}^i{}_D \in R^{1,fi\cdot pi,d\text{model}}$, and then use the Trunc function to reduce the dimensionality from $f_i \cdot p_i$ to the initial dimension T, resulting in $X\hat{~}_1{}^i{}_D \in R^{T,d\text{model}}$. The above process can be described using the following equation:

$$A, \{f_1, \cdots, f_k\}, \{p_1, \cdots, p_k\} = Period(X_{1d}) \tag{6}$$

$$X_{2d}^i = Reshape(Padding(X_{1d})), i \in \{1, \cdots, k\} \tag{7}$$

$$\widehat{X_{2d}^i} = Inception\left(X_{2d}^i\right), i \in \{1, \cdots, k\} \tag{8}$$

$$\widehat{X_{1d}^i} = Trunc\left(Reshape'\left(\widehat{X_{2d}^i}\right)\right), i \in \{1, \cdots, k\} \tag{9}$$

In summary, by transforming the one-dimensional time series into a twodimensional representation, we can conveniently utilize the two-dimensional convolutional kernels in the Inception module to capture the relationships between temporal variations within and between periods. The number of periods can be controlled by the hyperparameter k. Importantly, for different frequencies, the parameters of the Inception module are shared, which helps reduce the parameter count.

Additionally, the TimesBlock is designed using a residual approach, allowing it to capture information from different network layers. This helps in capturing hierarchical information and enables the model to learn representations at multiple scales.

Overall, this approach enables efficient extraction of temporal patterns and relationships in the time series data.

2.5 Adaptive Aggregation

Finally, inspired by Auto-Correlation [16], we merge and pass the one-dimensional time series representations $\{X\hat{~}_1{}^i{}_D, \cdots, X\hat{~}_1{}^k{}_D\}$ obtained from k different frequencies to the next layer. We apply the softmax operation to the amplitudes $A^l{}_f{}^{-1}1, \cdots, A^l{}_f{}^{-k} 1$ corresponding to different frequencies to obtain the corresponding weights $\alpha_1, \cdots, \alpha_k$. Finally, we perform a weighted summation to obtain the fused one-dimensional time series output that incorporates the information from all k frequencies.

This process can be represented using the following equations:

$$\alpha_1, \cdots, \alpha_k = Softmax\left(A_{f_1}^{l-1}, \cdots, A_{f_k}^{l-1}\right) \tag{10}$$

$$X_{1D}^l = \sum_{i=1}^{k} \alpha_i \widehat{X_{1D}^{l,i}} \tag{11}$$

As the differences within and between periods involve highly structured 2D tensors, the TimesBlock can fully capture the multi-scale temporal variations. Therefore, TimesNet achieves more effective representation learning compared to directly learning from the 1D time series data.

3 Experiment

This study utilizes a dataset that includes electricity demand data from January 1, 2020, to January 1, 2023, across ten industry sectors including (1) Agriculture, Forestry, Animal Husbandry, and Fishery (2) Industry (3) Construction Industry (4) Transportation, Warehousing, and Postal Services (5) Information Transmission, Software, and Information Technology Services (6) Wholesale and Retail Industry (7) Accommodation and Catering Services (8) Financial Industry (9) Real Estate Industry (10) Rental and Business Services. Specifically, the dataset is divided into a training set and a test set in 4:1 ratio. The electricity demand data from January 1, 2020, to June 17, 2022, is used for training, while the data from June 18, 2020, to January 1, 2023, is used for testing. The training data consists of 900 observations, and the prediction data consists of 225 observations. This dataset provides valuable resources for developing and testing machine learning models that can capture the complex dynamics of electricity demand patterns in urban areas.

The dataset includes the following features: 1) Time-related features: year, month, date, day of the week, and the position of the month, week, and day within the year (e.g., the nth day of the year, the nth week of the year). Whether it is a holiday: This variable serves as a covariate to account for activities in the residential and tertiary sectors, including weekends and all statutory holidays. Whether it is summer vacation: Used to capture the peak electricity demand. 2) Temperature-related features: average temperature, minimum temperature, and maximum temperature. 3) Weather-related features: u10, v10, sp, tp. Here, u10 and v10 represent the east-west and north-south wind speeds at a height of 10 m, SP represents atmospheric surface pressure, and TP represents total precipitation. 4) Maximum load: This variable is provided in the electricity demand dataset and is the most important covariate.

3.1 Data Pre-processing

Data pre-processing plays a crucial role in machine learning and data analysis. It refers to the process of cleaning, transforming, and organizing raw data to improve its quality and suitability for subsequent modeling and analysis tasks.

First, we can remove any missing or duplicate values from the data. Then, for non-numeric attributes like whether it is a statutory holiday or summer vacation, we can encode them as 0–1 variables. For example, a statutory holiday or summer vacation can be represented as 1, while a nonstatutory holiday or non-summer vacation can be represented as 0. After that, we can normalize the data. The purpose of normalization is to eliminate unit restrictions on the data and transform it into a dimensionless value, enabling the comparison and weighting of indicators with different units or scales.

One commonly used method is Min-Max normalization, which scales the values of a feature to a fixed range, typically between 0 and 1. To normalize a feature, we first determine the minimum and maximum values of that feature in the dataset. Then, for each value in the feature, we subtract the minimum value and divide it by the difference between the maximum and minimum values to obtain the final result. This maps all feature values to the range of [0, 1]. To revert the normalized feature values back to their original scale, we can perform the inverse process, known as Min-Max inverse normalization. This involves multiplying each normalized value by the range of the original feature and then adding the minimum value. This operation restores the original distribution of the feature values while retaining the scale of the normalized values.

3.2 Experimental Baseline Methods

This article uses the following methods as comparison methods to demonstrate their superiority over the main method, Timesnet.

Prophet is an open-source time series forecasting method that was released by Facebook in 2017. It aims to provide a simple and flexible approach to time series analysis and forecasting, suitable for various business scenarios. The Prophet method emphasizes interpretability and explainability of the forecasted results. It provides the ability to model and explain time series patterns such as trends, seasonality, and holidays. The Prophet method is highly flexible and can handle time series with irregular sampling frequencies, missing data, outliers, and anomalies. In the context of electricity demand forecasting, Prophet can capture complex patterns of electricity demand at different time scales, such as daily, weekly, and yearly trends.

GBDT (Gradient Boosting Decision Trees) is an ensemble learning method that progressively constructs a powerful predictive model by using decision trees as base learners through gradient boosting techniques. GBDT can combine multiple decision tree models to fully utilize their complementarity and achieve more accurate predictions. GBDT automatically selects the most relevant features for modeling based on their importance in decision tree models, eliminating the need for manual feature selection. In the context of electricity demand forecasting, GBDT can learn the non-linear relationship between input features and the target variable (i.e., electricity demand).

The CNN-LSTM algorithm is a hybrid model that combines Convolutional Neural Network (CNN) and Long Short-Term Memory (LSTM) for processing time series data. The main idea of the CNN-LSTM algorithm is to use CNN layers to extract features from the time series data and then input the extracted features into LSTM layers for sequence modeling and prediction. In the context of electricity demand forecasting, CNN-LSTM can capture both spatial and temporal features in time series data, automatically extracting features for electricity demand prediction.

3.3 Structure of Network and Parameter Settings

The input sequence length is 30, the predicted sequence length is 1, the value k for frequency selection is 5, the batch size is set to 32, the learning rate is set to 0.001, the number of training epochs is set to 50, and the dropout ratio is set to 0.1. The model has two Convolutional layers connected by an activation layer (GELU) and uses Inception modules. Each Inception module has 6 2D-convolutional-kernels, which the hidden layer dimension is set to 2048.

In our experiments, we used the mean squared error (MSE) function as the loss function for the regression prediction task. It is defined by formula 9, where x represents the predicted power demand value by the model at the current time step, y represents the actual value of power demand at the current time step, and n represents the number of training samples.

$$LOSS_{MSE} = \frac{\sum_{i=1}^{n}(x-y)^2}{n} \qquad (12)$$

3.4 Experimental Results and Analysis

The evaluation metrics for this experiment are daily average error and monthly average error. By comparing these two metrics, we can understand the performance of the model at different time scales. Generally, due to the finer granularity of daily predictions compared to monthly predictions, the daily average error may be relatively larger. On the other hand, the monthly average error can better reflect the overall performance of the model over a longer time span. Furthermore, for the daily average error, each term in the summation is taken as an absolute value, eliminating the possibility of positive and negative errors canceling each other out. Compared to the monthly average error, the daily average error provides a stricter characterization of the model's performance. The formulas for calculating the daily average error and monthly average error are as follows:

$$Error_{month} = \frac{\sum_{i=1}^{n} y_{pred}(i) - \sum_{i=1}^{n} y_{act}(i)}{\sum_{i=1}^{n} y_{act}(i)} \qquad (13)$$

$$Error_{day} = \frac{1}{n} \sum_{i=1}^{n} |\frac{y_{pred}(i) - y_{act}(i)}{y_{act}(i)}| \qquad (14)$$

Due to the differences in electricity consumption patterns and trends across industries, we trained multiple independent models for different industries. We plotted the predictions of the TimesNet method along with other baseline methods (Prophet, GBDT, CNN_LSTM) and the actual values.

Taking the industry sector as an example, as shown in Fig. 3, the timesnet-model's predicted results closely match the distribution of real data. All the models we trained have captured data features related to seasonality, trends, and other aspects to some extent. This approach not only provides more granular and accurate estimates of power consumption but also helps identify potential areas for energy saving and efficiency improvement.

We used the Prophet, GBDT, CNN_LSTM, and TimesNet methods to make predictions for eleven different industries separately. We computed the errors between the predicted values and the actual values and calculated the daily average error and monthly average error by taking the average of all the errors. The final results are presented in the table below to demonstrate the effectiveness of each method.

Overall, the experimental results demonstrate the effectiveness of the proposed TimesNet model in electricity demand forecasting, outperforming traditional methods such as Prophet, Gbdt, and cnn_lstm in terms of accuracy and reliability. In summary, we proposed a TimesNet model for electricity demand forecasting, which achieved good performance in terms of daily average error and monthly average error. This study provides insights for future electricity demand forecasting research and has practical implications for energy management across various industries (Table 1).

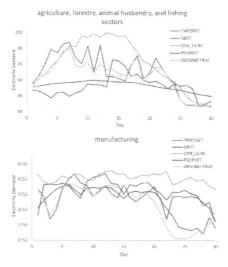

Fig. 3. Comparative results of electricity demand predictions for different models in the agriculture, forestry, animal husbandry, and fishing industry and manufacturing industry

Table 1. Electricity demand prediction error results of different models in 10 industries in A-City

Industry	Evaluation Metrics (Model-Error)							
	TimesNet		GBDT		CNN_LSTM		Prophet	
	ME	DE	ME	DE	ME	DE	ME	DE
1	**0.062**	**0.114**	−0.111	0.333	0.132	0.308	−0.566	0.574
2	**0.017**	**0.084**	0.043	0.100	0.069	0.128	0.041	0.899
3	**-0.054**	**0.081**	−0.074	0.128	−0.096	0.096	−0.058	0.089
4	0.077	**0.120**	**0.069**	0.139	0.099	0.135	0.171	0.168
5	−0.039	0.055	0.043	0.081	0.041	0.062	0.040	**0.045**
6	**−0.013**	**0.131**	−0.014	0.157	−0.043	0.143	0.124	0.185
7	−0.073	**0.184**	0.103	0.185	**−0.056**	0.238	0.188	0.268
8	**0.040**	**0.117**	0.055	0.136	−0.091	0.118	0.047	0.162
9	**0.041**	**0.135**	−0.175	0.217	-0.045	0.137	0.131	0.148
10	**0.033**	**0.129**	−0.039	0.142	0.042	0.156	0.094	0.146

4 Conclusion

In this paper, we proposed a TimesNet model to captures intra-cycle and inter-cycle information using two-dimensional convolution. We evaluated our model on a real dataset of electricity demand in A city and achieved good results with low prediction errors in different industries. Our experiments also demonstrated the effectiveness of our proposed model compared to other baseline methods.Overall, the TimesNet model we proposed shows great potential in forecasting the electricity demand for different industries in urban areas. We believe further research can improve its performance and broaden its applicability.

5 Future Work

As for future research directions, we believe there are several areas worth exploring further:

– Explore alternative network architectures such as attention mechanisms and transformer networks to capture more complex relationships between industries and regions.
– Investigate the transferability of the proposed model to other citiesor regions and adapt it to different datasets and scenarios.
– Explore the robustness of the model in the presence of data gaps ormissing values.

Acknowledgments. This work is supported by the Research Funds from State Grid Henan (SGHAYJ00NNJS2310068);

References

1. Amjady, N.: Short-term hourly load forecasting using time-series modeling with peak load estimation capability. IEEE Trans. Power Syst. **16**(3), 498–505 (2001)
2. Al-Alawi, S.M., Islam, S.M.: Principles of electricity demand forecasting. I. Methodologies. Power Eng. J. **10**(3), 139–143 (1996)
3. Gebremeskel, D.H., Ahlgren, E.O., Beyene, G.B.: Long-term evolution of energy and electricity demand forecasting: the case of Ethiopia. Energy Strategy Rev. **36**, 100671 (2021)
4. Gonzalez-Briones, A., et al.: Machine learning models for electricity consumption forecasting: a review. In: 2019 2nd International Conference on Computer Applications & Information Security (ICCAIS), pp. 1–6. IEEE (2019)
5. Shi, X., et al.: Convolutional LSTM network: a machine learning approach for precipitation nowcasting. In: Advances in Neural Information Processing Systems, vol. 28 (2015)
6. Medsker, L.R., Jain. L.C.: Recurrent neural networks. Des. Appl. **5**(64–67), 2 (2001)
7. Ke, G., et al.: Lightgbm: a highly efficient gradient boosting decision tree. In: Advances in Neural Information Processing Systems, vol. 30 (2017)
8. He, Y., Zhao, J.: Temporal convolutional networks for anomaly detection in time series. In: Journal of Physics: Conference Series, vol. 1213. No. 4, IOP Publishing, p. 042050 (2019)
9. Vaswani, A., et al.: Attention is all you need. In: Advances in Neural Information Processing Systems, vol. 30 (2017)
10. Dosovitskiy, A., et al.: An image is worth 16 x 16 words: transformers for image recognition at scale. In: arXiv preprint arXiv:2010.11929 (2020)
11. Kitaev, N., Kaiser, Ł., Levskaya, A.: Reformer: the efficient transformer. In: arXiv preprint arXiv:2001.04451 (2020)
12. Brigham, E.O.: The Fast Fourier Transform and its Applications. Prentice-Hall, Inc. Hoboken (1988)

13. Zhou, T., et al.: Fedformer: frequency enhanced decomposed transformer for long-term series forecasting. In: International Conference on Machine Learning, PMLR. 2022, pp. 27268–27286 (2022)
14. He, K., et al.: Deep residual learning for image recognition. In: Proceedings of the IEEE Conference on Computer Vision and Pattern Recognition, pp. 770–778 (2016)
15. Szegedy, C., et al. Going deeper with convolutions. In: Proceedings of the 2015 IEEE Conference on Computer Vision and Pattern Recognition (CVPR), pp. 1–9 (2015)
16. Haixu, W., et al.: Autoformer: decomposition transformers with autocorrelation for long-term series forecasting. Adv. Neural. Inf. Process. Syst. **34**, 22419–22430 (2021)

Design and Utilization of an Auto-Visual-Inspection Composite System for Suspension Cables with Fast Flaw Identification

Donglong Meng[1,2], Xiaolin Wang[2(✉)], Di Lu[2(✉)], Jianhui Li[2], Di Gan[2], and Huien Shi[2]

[1] School of Mechanical Engineering, Guangxi University, Daxue East Road, Nanning 530000, Guangxi, China
[2] Guangxi Academy of Sciences, Institute of Advanced Equipment and Manufacturing, Daling Road, Nanning 530000, Guangxi, China
`wangxiaolin@gxas.cn, dilu.gxas@gmail.com`

Abstract. The traditional manual inspection of bridge cables has drawbacks such as low detection efficiency and compromised safety. This paper presents an automated visual inspection system for suspension cables, which incorporates a dual-engine wheeled robot with four U-shaped wheels and an offline artificial intelligence (AI)-based apparent flaw identification method. With its capability to overcome significant obstacles, the robot enables a safe, robust, and efficient on-site collection of images highlighting cable flaws. The climbing ability of the robot and its AI-based auto-inspection of cable flaws were experimentally evaluated through trials under various conditions. Indoor experimental results demonstrated the robot's carrying capacity of up to 6.8 kg, as well as its crossing capability over the obstacle of 6.3-mm in height on the cable's surface. Furthermore, field trials conducted on suspension cables of arch bridges strongly evidence the effectiveness of the proposed robot, and the utilization of YOLOv7 demonstrates the rapid, autonomous, and accurate identification of flaw features.

Keywords: Cable-climbing Robot · Cable Inspection · Obstacle Crossing · Image Identification · Flaw Identification

1 Introduction

The cable-supported bridges, such as the cable-stayed bridge, and suspension bridge, represent the epitome of long-span bridge design worldwide. More than 90 percent of these bridges have spans ranging from 200 m to 2000 m and beyond [1]. As pivotal components of these bridges, the cables are usually protected by outer polythene (PE) sheaths. However, due to ultraviolet radiation, wind and rain erosion, harmful gas corrosion, and tensile stress [2–4], the PE sheaths inevitably age, and cracks may thus develop, compromising the bridge's load-bearing capacity. Moreover, man-made damages to the cables' PE sheaths often occur and go unnoticed during the tumultuous construction process, and there is no timely repair on the site, which stands as the primary cause for the most severe sheath flaws in the cables.

Human inspection is mostly common and conventional for finding apparent flaws of cables [5], which is always limited by their scope and suffers from time-consuming and labor-intensive procedures, low efficiency and accuracy, high costs, and safety concerns [6]. Therefore, unmanned cable inspection robots are anticipated to completely replace manual inspections in this field by significantly enhancing the efficiency and accuracy of flaw identification. Being designed for nondestructive testing (NDT) of cables, cable inspection robots always follow the cable naturally during operations. Wheeled robots are one kind of them. Through solid contact between their rolling wheels and the cable, wheeled robots can smoothly move along the cable, facilitating the detection of any noteworthy flaws. For instance, Mahdavi et al. designed a cable climbing robot that incorporated a wheel attachment structure [7–10]. Its main frame comprised three beams arranged in a triangular formation with each featuring two wheels attachment structures. Similarly, H.M Kim designed a wheeled climbing robot that demonstrated adaptability to different cable diameters, albeit with a relatively complex and heavy structure [11, 12]. Dharmawan et al. introduced a wheel-legged wall-climbing robot [13], while Koh et al. devised a dual-wheel miniature climbing robot [14]. Xu et al. proposed a truss-type wheeled robot, enabling nondestructive evaluation (NDE) of the cable's inner metal materials and capturing apparent flaws in the PE sheath [15, 16]. Later, they further developed a cable climbing robot for repair work, which was capable of grinding, cleaning the PE sheath, and applying a protective coating or winding protective tape onto cables [17]. Wang et al. designed a novel wheel-clamping type inspection robot for bridge stay cables with an in-house control system, which made the clamping process easy and the overall control straightforward [18].

Despite the existence of multiple cable inspection robots, there remains a need to enhance their performance and address practical concerns like lowering weight, simplifying installation procedures, and automating the flaw identification. In this paper, a comprehensive auto-inspection system was developed by integrating an onsite wheeled robot with an offline image processing platform for unsupervised identification of visible flaws on cables. The robot was designed with an improved driving system comprising two sets of combinations of brushless direct current (BLDC) motors and U-shaped wheels, along with two passively U-shaped wheels. This dual-engine configuration not only enhanced the robot's carrying capacity but also provided traction redundancy in case of single actuator failure. The robot carried 4K cameras to capture the cable surface flaws. After completing an inspection trip, the recorded video was processed using a pretrained image-identification platform. This platform implemented a frame-wise YOLOv7-based filtering method to capture the images of interest, i.e., the flaws on PE sheaths. By combining the onsite filming capabilities enabled by the designed robot with the automatic video post-processing technique, we successfully conducted inspections of suspension cables for three arch bridges. This approach significantly improved the efficiency of onsite data collection and flaw detection of the PE sheaths, showing its remarkable advantages in practical cable inspection tasks.

The remaining sections of this paper are organized as follows (Fig. 1): Sect. 2 outlines the system construction of the robot and the image identification platform. Section 3 describes the indoor experiments conducted using the developed robots. Section 4 presents the results obtained from field trials. Finally, Sect. 5 presents the conclusions derived from the paper.

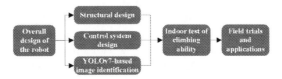

Fig. 1. The flowchart of this work.

2 System Overview

2.1 System Structure of the Designed Cable Inspection Robot

The robot that was designed as a wireless mobile platform. It comprised two components, namely the driving unit and the passive unit (Fig. 2), which were connected to each other, forming a structure encasing the inspected cable. The driving unit featured a pair of synchronized engines, each consisting of a BLDC motor and a V-shaped wheel. Enough traction provided by the driving wheels was necessary to surmount the cable's various inclinations and overcome significant obstacles. So, the surface of the driving wheels was enhanced with replaceable rubber belts for higher roughness. On the contrary, the two wheels of the passive unit moved passively. The harder and smoother surface these passive wheels had, the lower the adverse friction would be. Therefore, the passive wheels were made from nylon and machined to the U-shaped same as the driving wheels. A tension spring was integrated between the upper and lower passive wheels, ensuring high enough contact pressure between the wheels and the cable surface. Once a reliable contact was established between the wheels and the cables, the driving wheels could generate great traction to push the robot smoothly moving along the cables.

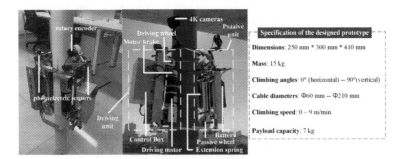

Fig. 2. Structure of the cable inspection robot

2.2 Onboard Electronics

The onboard control system of the designed robot (Fig. 3) was responsible for basic commands such as moving forward/backward, adjusting climbing speed, stopping, and broadcasting measured states, among others. Movement was achieved through the synchronized control of two identical 24V BLDC motors. The micro control unit (MCU), an Arduino Mega 2560, sent the same pulse width modulation (PWM) and logical signals to two BLDC motor drivers. The PWM signals were set to regulate motor speed while the logical signals determined the spinning direction of the motors. All components were powered by a 16.75-Ah 24V Lipo battery. To meet different voltage requirements of the onboard electronics, a power management module was designed, providing 5V, 9V, 12V, and 24V power sources. To find and locate cable flaws, four 4K motion cameras and a rotary encoder were employed. The 4K cameras continuously filmed the surface condition of cables during the inspection. A small measuring wheel, fixed coaxially to the measuring axis of the encoder, was in contact with the cable's surface and rotated correspondingly as the robot moved. Subsequently, the encoder sent pulse signals to the MCU to count the rotation number of the measuring wheel. By factoring in the wheel's diameter, the travel distance from the starting point could be determined. Moreover, photoelectric sensors were integrated to avoid collisions with the bulkheads/anchors at the cable ends. As the robot approached either end of a cable within the preset safe range, the photoelectric sensors would notify the MCU to stop motion. All command signals and state measurements were wirelessly transmitted via a pair of telemetry radios between the robot and the ground control station (GCS) that ran a self-developed software. However, the high-framerate, high-definition video stream was stored directly on the cameras during inspections. Upon completion of an inspection task, the video was retrieved and utilized for identifying flaws in the image post-processing stage.

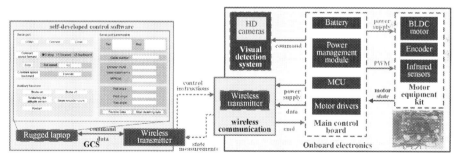

Fig. 3. Control system layout including the onboard electronics and the self-developed GCS.

2.3 YOLOv7-Based Image Identification

Screening flaws from inspection videos of cables is a laborious task. To enhance screening efficiency and mitigate the complexity and errors of manual image processing, YOLOv7 [19] was utilized to achieve rapid and autonomous flaw identification. YOLOv7 possesses the following key features:

1. Model Re-parameterization: YOLOv7 proposes a strategic re-parameterization model that applies to layers in different networks, incorporating the concept of gradient propagation path.
2. Dynamic Label Assignment: YOLOv7 introduces a novel label assignment method that employs a coarse-to-fine approach in guided label assignment. During model training with multiple output layers, dynamic targets are assigned to the outputs of different branches.
3. Extended and Compound Scaling: YOLOv7 presents "extend" and "compound scaling" methods for real-time object detection, effectively utilizing parameters and computation.
4. Efficiency: The YOLOv7-based approach significantly reduces parameters and computation by approximately 40% compared to state-of-the-art real-time object detectors. This results in faster inference speed and higher detection accuracy.

These characteristics establish YOLOv7 as a preferred image recognition and segmentation model currently in use. Thus, we employ this model to identify cable flaws in the retrieved videos (Fig. 4). Initially, the input image was resized to 640 × 640 dimensions and fed into the subsequent backbone network. The head layer network generated three feature maps of varying sizes. In this paper, a representative flaw database comprised of PE sheaths of bridge cables was employed, with training conducted for five distinct categories. Finally, utilizing Bounding Box Decoding and Non-Maximum Suppression (NMS), the predicted offset values of the network were used to decode the output bounding box coordinates. This process yielded the absolute coordinates on the feature maps, as well as the position and size of the bounding box in the original image. Consequently, possible flaws in the PE sheath of the image frame could be identified.

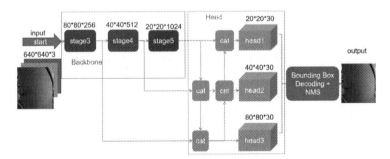

Fig. 4. Flaw identification framework

3 Indoor Evaluation of Climbing Ability

To evaluate the robot's performance in lab, a set of tests were conducted on a vertical metal pipe and an inclined metal pipe with diameters of 110 mm. These pipes served as simulations for the stay and suspension cables found in bridges. As in Fig. 5, the basic functional prototype equipped no cameras, encoder, and photoelectric sensors in all indoor experiments to access the payload capacity and its mobility.

Fig. 5. The prototype for indoor tests had no cameras, encoder, and photoelectric sensors.

3.1 Climbing Performance Under Various Payload Conditions

It was imperative for the prototype to possess sufficient load capacity to carry devices. To evaluate the allowable maximum load, the prototype underwent climb tests with different additional weights (0 kg, 3.5 kg, and 6.8 kg loads) on the vertical pipe. The climbing speed was fixed at 3.75 m/min. Figure 6 illustrates the prototype carrying 6.8 kg rubber blocks. The output power of the prototype was measured under different weight conditions, as shown in Fig. 7. The blue, red, and green curves represent the system's power under no load, 3.5 kg, and 6.8 kg, with average values of 46.55 W, 50.53 W, and 53.80 W, respectively. It is evident that the prototype necessitated higher output power as the load increased. These tests verified the prototype''s load capacity was about 7 kg. Subsequently, an additional test was conducted to determine the prototype's maximum vertical climbing speed under the maximum load. As the black curves in Fig. 7 shows, a max speed of around 9 m/min with a power output of approximate 93.53 W was confirmed. Given the battery capacity, it can be inferred that the prototype had a maximum travel range of 1.675 km while continuously climbing vertically at 3.75 m/min with its maximum load.

Fig. 6. The prototype carried extra loads of 6.8 kg during the vertical climb tests

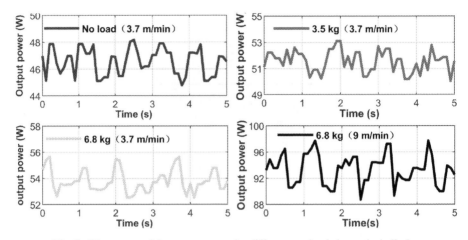

Fig. 7. The power of the prototype under different payloads in vertical climbs.

3.2 Obstacle Crossing Performances of Both Driving and Passive Units

The observable flaws found on cable surfaces are commonly protrusions or cracks, often resulting from long-term exposure to harsh environmental conditions. These flaws hinder the smooth motion of the robot along cables. Therefore, the ability to overcome obstacles was crucial for the robot. To evaluate the prototype's obstacle-crossing capability indoors, nuts of different heights (3.2 mm, 4.9 mm, and 6.3 mm) were placed on the vertical pipe, as depicted in Fig. 8, to simulate protuberances on PE sheaths of cables. Given that the height of most apparent flaws is less than 5 mm, the simulated obstacles were appropriate for assessing the performance of the designed robot in field conditions. During the experiments, the prototype was tested to climb over the set obstacles under no load and full load conditions respectively at 3.75 m/min. In different tests, the prototype intentionally ran over the obstacles with its driving wheels and passive wheels separately to see the distinct obstacle-crossing capabilities of the driving and passive units.

Obstacle Crossing of the Driving Unit
Vertical upward climb was the most power-consuming working situation. Because the actuators had to overcome the entire weight of the system. Therefore, an experiment for evaluating the obstacle crossing performance of the driving unit of the prototype under zero and full (6.8 kg) payload was firstly conducted. The vertical climbing speed of prototype was set to 3.75 m/min. As in Fig. 9, the nuts with heights of 3.2 mm, 4.9 mm, and 6.3 mm were arranged in sequence from the low to the high right in the path of the driving wheels.

The fluctuation of the prototype's output power during the experiment was documented in Fig. 10, illustrating the dynamic impact of the obstacle on the system. The blue and red curves depict the variation in output power under no load and full load conditions, respectively. In both scenarios, the driving power experienced a significant surge as the driving wheels encountered the obstacles. In the figure, the first two peaks

Fig. 8. The set obstacles with various heights of (a) 3.2 mm, (b) 4.9 mm, and (c) 6.3 mm that simulated (d) the protrusion on a real cable's surface (often smaller than 5 mm in height).

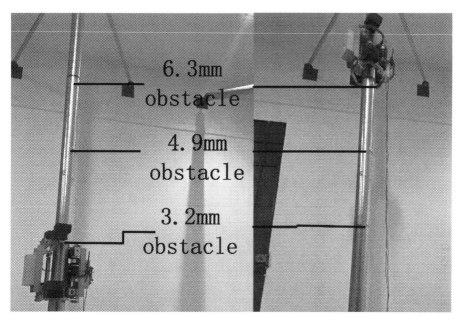

Fig. 9. The driving unit of the prototype encountered three set of obstacles with respective heights of 3.2 mm, 4.9 mm, and 6.3 mm sequential as it moved from the bottom to the top.

were the instances when the upper and lower driving wheels ran over the 3.2-mm nuts, respectively. Similarly, the third and fourth peaks indicated the impact of the two driving wheels against the 4.9-mm nuts while the last two peaks were caused by the strike of the 6.3-mm nuts. Notably, there were variations in the peaks corresponding to different obstacles and wheels. The surge in power increased with obstacle height. And it was intriguing that the upper driving wheel always required a higher output torque than the lower one when they went over the same obstacles, meaning the fact the upper wheel generated greater traction than the lower. This disparity was closely correlated to the mass distribution of the system.

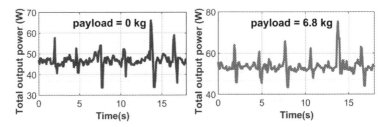

Fig. 10. The output power history of the prototype during the driving wheels crossing three different layers of obstacles at 3.75 m/min.

Obstacle Crossing of the Passive Unit

Compared with the driving wheels, the passive wheels are more likely to get stuck in cracks or protrusion since they are dragged passively. So, the obstacle crossing ability of the passive unit was evaluated by setting the three layers of nuts with different heights in the path of the passive wheels when the prototype climbs vertically at 3.75 m/min. However, the prototype could only get over the 3.2-mm nuts under zero load successfully (Fig. 11) but failed passing over the 4.9-mm and 6.3-mm nuts. Figure 12 shows the output power of the system getting over the 3.2-mm obstacles. As in the figure, two power surge representing the impact of the 3.2-mm nuts on the upper and lower passive wheels happened. The result shows that the power output for the system dealing with the 3.2-mm obstacle in this case was greater than the case where the driving wheels ran over the same obstacle in the previous experiment. It was because the connection between the driving and passive units was not ideally rigid. The low rigidity of the connecting portions led to inadequate force transmission from the driving unit to the passive wheels, resulting in an increased motor output for overcoming the obstacle.

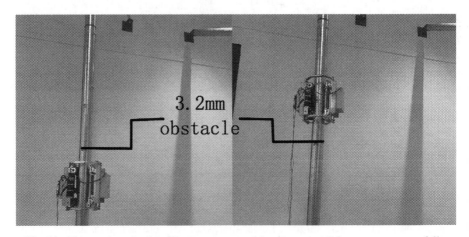

Fig. 11. The passive wheels of the prototype could only go over 3.2-mm nuts successfully.

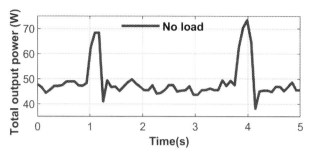

Fig. 12. The system output power during its passive wheels crossing the 3.2-mm obstacles.

4 Field Trials and Applications

The fully operational robot was deployed for cable inspection missions on three arch bridges (Fig. 13): Nanning Bridge, Pumiao Bridge, and Ximing Bridge, located in Nanning, Guangxi, China. Specially, Nanning bridge possesses asymmetric inclined arch ribs and suspension cables with helical fillets. The main parameters pertaining to the cables of these bridges are detailed in Table 1. To validate the feasibility of the designed robot and the effectiveness of the image post-processing for automatic flaw identification, field operations incorporated the integration of rotary encoders, photoelectric sensors, and high-definition cameras, as outlined in Sect. 2. The robot had total weight of 15 kg, and its wireless communication was validated over 500 m.

Table 1. Cable specifications of Nanning Bridge, Pumiao Bridge, and Ximing Bridge

Bridge	Cable Length (m)		Diameter (mm)	The Slope of Cables (deg)	
	Longest	Shortest		Minimum	Maximum
Nanning Bridge	57.6	18.5	80	15	75
Pumiao Bridge	29.4	6.5	190	90	90
Ximing Bridge	28	9	140	90	90

In practice, cable surface was frequently covered with dust and subject to unexpected surface flaws such as dents, cracks, and protrusions. The presence of dust and obstacles as well as shock-absorbing lines like the helical fillets on the cables of Nanning bridge posed a significant threat to the safe and effective operation of the robot. Dust might compromise wheel traction while obstacles might get the robot stuck. Impressively, our prototype succeeded in these demanding field operations, demonstrating the system's utility and proving its feasibility and efficiency in harsh conditions.

Massive video data was collected by the onboard 4K cameras during the operation of the robot. To capture a comprehensive view of each cable, the four cameras were placed at 90-° intervals around the cable, allowing for simultaneous filming from four perspectives. As for Nanning Bridge, all 104 suspension cables underwent inspection,

Fig. 13. Field inspection for suspension cables of (a) Nanning Bridge, (b) Ximing Bridge, and (c) Pumiao Bridge were done by the application of the designed robot.

resulting in the collection of 416 raw high-definition video files with a total size of over 523 G. The proposed flaw identification algorithm was applied to automatically detect flaws on the PE sheaths of the cables based on the raw video data, and then efficiently categorized the identified flaws into three distinct groups, as illustrated in Fig. 14. During the comprehensive inspections of all three bridges, 361 flaws were finally identified using the proposed auto-identification method. It was significantly faster than manual screening, and the accuracy of the auto-identification results was manually rechecked and reached 92%. The feasibility, effectiveness, and efficiency of the proposed auto-inspection have been strongly validated in these field trials, standing as a practical example for unmanned cable inspection as well.

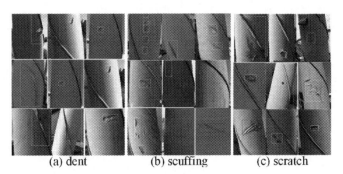

(a) dent   (b) scuffing   (c) scratch

Fig. 14. Typical flaws were found on the PE sheaths of the inspected cables automatically by the proposed flaw identification algorithm.

5 Conclusion

The damage to the PE sheath of cable is a serious threat to the structural safety of bridge. Therefore, a wheeled cable inspection robot with an improved driving system was developed for unmanned cable inspection. With its enhanced obstacle-crossing ability, the robot could handle poor surface conditions of cables and automatically inspect bridge cables. The application of the robot could reduce human engagement into dangerous job. Besides, an YOLOv7-based auto-identification method was implemented to accelerate the screening of cable flaws.

Indoor and field tests were conducted to evaluate the climbing and obstacle-crossing performance of the robot. The impact of cable inclination angle and payload on the robot's obstacle-climbing capability was also analyzed. Characterization of the robot's features was experimentally done, including payload capacity of 6.8 kg, adaptability to 60-to-210-mm diameters and any inclinations, and obstacle-crossing ability over 6.3-mm height. The YOLOv7-based flaw identification algorithm was testified using inspection video data of real cables and an accuracy over 90% was validated. This work provides an engineering reference for the auto-inspection and analysis of bridge cable.

In future, the presented auto-identification method and feature extraction algorithm will be integrated into the onboard system of the robot for real-time detection.

References

1. Gimsing, N.J., Georgakis, C.T.: Cable Supported Bridges: Concept and Design, 3rd edn. Wiley, New jersey (2011)
2. Hoang, Vu, Kiyomiya, O., An, T.: Experimental and numerical study of lateral cable rupture in cable-stayed bridges: case study. J. Bridge Eng. **23**(6), 05018004 (2018)
3. Mehrabi, A.B.: Performance of cable-stayed bridges: evaluation methods, observations, and a rehabilitation case. J. Perform. Constr. Facil. **30**(1), 4014007 (2016)
4. Zhou, P., Zhou, G., Zhu, Z., et al.: A review of non-destructive flaw detection methods for steel wire ropes. Appl. Sci. **9**(13), 2771 (2019)
5. Liu, S., Sun, Y., Jiang, X., et al.: A review of wire rope detection methods, sensors and signal processing technique. J. Nondestr. Eval. **39**(4), 1–18 (2020)
6. Yun, H.B., Kim, S.H., Wu, L., et al.: Development of inspection robots for bridge cables. Sci. World J. **2013**, 1–17 (2013)
7. Mahdavi, S., Noohi, E., and Ahmadabadi, M. N.: Basic movements of a nonholonomic wheel-based pole climbing robot. In: 2007 IEEE/ASME International Conference on Advanced Intelligent Mechatronics, pp. 1–16. IEEE, Zurich, Switzerland (2007)
8. Baghani, A., Ahmadabadi, M.N., Harati, A.: Kinematics modeling of a wheel-based pole climbing robot (UT-PCR). In: Proceedings of the 2005 IEEE International Conference on Robotics and Automation, pp. 2099–2104. IEEE, Barcelona, Spain (2005)
9. Mahdavi, S., Noohi, E., Ahmadabadi, M. N.: Path planning of the Nonholonomic pole climbing robot UT-PCR. In: 2006 IEEE International Conference on Robotics and Biomimetics, pp. 1517–1522. IEEE, Kunming, China (2006)
10. Ahmadabadi, M. N., Moradi, H., Sadeghi, A., et al.: The evolution of UT pole climbing robots. In: 2010 1st International Conference on Applied Robotics for the Power Industry, pp. 1–6. IEEE, Montréal, Canada (2010)

11. Kim, H.M., Cho, K.H., Liu, F., et al: Development of cable climbing robotic system for inspection of suspension bridge. In: Proceedings of the 28th International Symposium on Automation and Robotics in Construction (ISARC), pp.1422–1423. IEEE, Korea (2011)

12. Kim, H.M., Cho, K.H., Jin, Y.H., et al.: Development of cable climbing robot for maintenance of suspension bridges. In: 2012 IEEE International Conference on Automation Science and Engineering (CASE), pp. 602–607. IEEE, Seoul, Korea (2012)

13. Dharmawan, A.G., Xavier, P., Hariri, H.H., et al.: Design, modeling, and experimentation of a bio-inspired miniature climbing robot with bilayer dry adhesives. J. Mech. Robot. **11**(2), 1–12 (2019)

14. Koh, D. C. Y., Dharmawan, A. G., Hariri, H. H., et al.: Design and analysis of a miniature two-wheg climbing robot with robust internal and external transitioning capabilities. In: 2019 International Conference on Robotics and Automation (ICRA), pp. 9740–9746. IEEE, Montreal, QC, Canada (2019)

15. Xu, F., Wang, X.: Design and experiments on a new wheel-based cable climbing robot. In: 2008 IEEE/ASME International Conference on Advanced Intelligent Mechatronics, pp. 418–423. IEEE, Xi'an, China (2008)

16. Xu, F., Wang, X., Wang, L.: Cable inspection robot for cable-stayed bridges: design, analysis, and application. J. Field Robot. **28**(3), 441–459 (2011)

17. Xu, F., Dai, S., Jiang, Q., et al.: Developing a climbing robot for repairing cables of cable-stayed bridges. Autom. Constr. **129**, 103807 (2021)

18. Wang, Y., Li, X., Yang, M., et al.: Design and implementation of the wheel-clamping stay cable inspection robot. Adv. Mech. Eng. **15**(8), 1–12 (2023)

19. Wang, C. Y., Bochkovskiy, A., Liao, H. Y. M.: YOLOv7: trainable bag-of-freebies sets new state-of-the-art for real-time object detectors. In: Proceedings of the IEEE/CVF Conference on Computer Vision and Pattern Recognition (CVPR), pp. 7464–7475. IEEE, Vancouver, Canada (2023)

A Critical Review of Multi Criteria Decision Analysis Method for Decision Making and Prediction in Big Data Healthcare Applications

Ammar Alsaig[1(✉)], Alaa Alsaig[2], and Vasu Alagar[3]

[1] Umm Al-Qura University, Makkah, Saudi Arabia
aasaig@uqu.edu.sa
[2] Jeddah University, Jeddah, Saudi Arabia
aalsaig@uj.edu.sa
[3] Concordia University, Montreal, Canada
vangalur.alagar@concordia.ca

Abstract. Multi-criteria decision making (MCR) has a rich history, employed since the early 1950s to mathematically model decision problems and identify optimal solutions from a set of alternatives. Recently, MCR has found applications in biomedical engineering and healthcare big data. The healthcare domain, characterized by a multi-stakeholder perspective involving patients, doctors, medical device manufacturers, and insurers, presents complex decision-making challenges. With diverse criteria and extensive healthcare records, it becomes challenging to make fair decisions. This paper outlines the requirements for a fair decision-making algorithm, highlighting the limitations of MCR in meeting fairness criteria. An algorithmic framework incorporating fairness criteria is presented, employing a similarity-based approach for decision-making on Electronic Health Records (EHR) to predict prospective cohort groups for a target patient.

Keywords: Multi-criteria decision making · biomedical engineering · healthcare big data · patient-centric healthcare · fairness requirements · fair algorithm for co-hort prediction

1 Introduction

Multi-criteria decision making (MCR) [1], also called Multi-criteria decision analysis (MCDA), is a structured method for evaluating options with conflicting criteria and selecting the best solution. Early research across various domains, including Economics, Mathematical Systems, and government policy formulation, has contributed to a rich set of methods for analysis and decision making.

The MCR problem involves alternatives, criteria, and normalized weights. Below are a few examples.

© The Author(s), under exclusive license to Springer Nature Singapore Pte Ltd. 2024
D.-S. Huang et al. (Eds.): ICAI 2023, CCIS 2015, pp. 89–100, 2024.
https://doi.org/10.1007/978-981-97-0827-7_8

Example 1

1. An individual wants to buy a car from a finite set of car models. Some criteria include price, fuel type, efficiency, safety, reliability, and size. The decision criteria are of heterogeneous types, and the respective weights assigned are 5, 3, 5, 5, 5, 4.
2. A company may want to evaluate different investment opportunities. Criteria include return on investment, risk level, liquidity, and market potential for profit. The respective weights for these criteria may be 5, 3, 4, 7.
3. After surgery, a patient wants to move to a senior residence. Criteria include transportation, personal care support, medication management, nursing support, social support, and technology safety. The respective weights assigned to the criteria are 4, 5, 6, 5, 4, 5.

These examples vary in complexity, requiring individual decisions to collective decisions with conflicting goals. MCR problem solving in patient-centric healthcare is more challenging, requiring fairness to both the patient and service-providing group. Recent papers [1–3] highlight challenges in selecting fair criteria and assigning weights in biomedical engineering and healthcare decision making. In this paper, we propose a fair algorithm to address these challenges. The paper is organized as follows. Section 2 provides a brief concise review of recent papers in MCR applied to healthcare. Section 3 explains the patient-centered health model, its emerging importance, and highlights the inadequacy of MCR in current healthcare applications. Section 4 lists fairness characteristics needed for decision-making algorithms in patient-centered care, comparing MCR with our developed algorithm and demonstrating MCR's failure in meeting fairness criteria. Section 5 presents a skeleton of fair algorithms and semantic scoring functions used to compute similarity measures. Section 6 concludes the paper by listing problems solved using our fair algorithm and others currently under investigation.

2 A Brief Review of MCR Applied in Healthcare Domain

After reviewing the most recent publications that address the use of MCR in health- care domain, we have classified them into the following four categories.

General Survey: Papers in this category provide a broad survey of published works on MCR concepts and methods, and which of them are most commonly used for evaluation of new medicines, healthcare services, and treatment options in healthcare.

Safety-critical Aspects: Papers that focus on specific problems in healthcare domain where safety criterion is paramount. Example problems include the selection of medical devices and procedures, and deciding the best healthcare infrastructure location.

Treatment Options: Papers in this category discuss the MCR methods currently used to determine treatment options, and self-evaluate the adequacy of these methods.

Weight and Criteria Selection: Papers in this category discuss the important issue of criteria selection and weight assignment in problems that involve many stakeholders, and they in turn have very specific interests and concerns on health technology assessment should be done.

Category I: The paper [4] presents extensive statistics on the popularity of MCR methods across various subject areas. The engineering domain has the highest number of MCR-based publications, while agricultural and biological sciences have the fewest. Surprisingly, the healthcare domain is absent from the "Subject Areas." The paper lists 20 different MCR methods, and among them, AHP, DEA, TOPSIS, FUZZY AHP, ELECTRE, PROMETHEE, VIKOR, and FUZZY TOPSIS are identified as the most commonly used MACDM methods in healthcare assessments. Comparisons of five methods (TOPSIS, VIKOR, PROMETHEE, MULTIMOORA, and COPRAS) on an environmental problem in a hydropower plant in Latvia lead to conclusions about their differing results and priority selections.

Category II: Four papers address the MCR approach in healthcare safety-critical scenarios. In [5], the first problem involves selecting the "infrastructure location" for health- care facilities, senior homes, and long-term therapy, integrating Analytical Hierarchy Process (AHP) and Evidential Reasoning (ER). Criteria include safety, size, total cost, accessibility, design, risks, and population profile. Decision-making for medical device manufacturers is explored in [6]. The approach, lacking an explanation for the choice of the geometric mean, determines weights through a geometric mean formula on expert weight vectors. In [7], the MCR framework selects sustainable suppliers in the health- care industry using AHP in the MARCOS method, comparing it with AHP, VIKOR, and ELECTRE.

Category III: MCR methods are increasingly applied to determine the best treatment for diseases from a set of alternatives [3, 8]. In [3], fuzzy MCR methods are proposed for cancer treatment options, including brain cancer, skin cancer, and prostate cancer, using fuzzy PROMETHEE and VIKOR methods. Challenges include the necessity of fuzzy methods, the difficulty of choosing weights even with expert input, and the need for interdisciplinary expert compromise. In [8] 72 papers on MCR methods in COVID-19 analysis are surveyed. It reveals concerns about combining fuzzy sets with MCR, computational complexities, and result validation challenges.

Category IV: The central issue in MCR modeling is on selection of criteria and weights, and it requires humans (experts in groups). The 3-tier model [9] is a good model to start with. However its usefulness depends on (1) the number of criteria and analysis on "how complete" this list is, and (2) determining whether all attributes for each criteria have been identified to its "atomic level". Depending on these two and the choice of weights "MCR models" differ. The paper [1] explain a group-based experimental activity for choosing the criteria and for quantifying the weights. However, they do not provide a procedure for experts in a group activity in selecting, agreeing, and quantifying the criteria and weights.

In summary, existing multi-criteria decision making methods in the field of health-care has raised many open problems. In Sect. 3 we introduce the importance of patient-centered healthcare, explain is multi-dimensional perspective by identifying the stakeholder, criteria for delivering healthcare services, and how the weights should be determined at attribute level for each criteria.

3 Shared Decision Making - Patient-Centered Healthcare

In this section we explore *Patient-centered healthcare* (PCH) paradigm, an active area of current research and practice [10–15], to emphasize "multi-criteria decision making" is central to the success of patient care. We review the current work on it, and propose our new model and analysis approach to solving criteria-based shared decision making.

According to the paper [11], PCH paradigm was introduced in 1988. It defined it as follows:

PCH is "care that is respectful of and responsive to individual patient preferences, needs, and values" and that ensures "that patient values guide all clinical decisions".

This paper also emphasizes that "for some decisions there is clearly only one path, and patient preferences play little or no role". Patient-doctor relationship should be "trustworthy" to make patients aware where choices exist and where there is only one path. In situations where many options are available, PCH considers patient's stand point, context, and input during the entire spectrum of decision making process [10]. It not only encourages but requires the enhancement of patient's awareness on the treatment choices, collaborative and shared decision making between patients, families, care- givers, and healthcare providers. That is, it goes beyond just physical well-being to include emotional, social, ethical, and financial aspects during the different contexts of treatment selection and treatment delivery. Patients are given full control over sharing their clinical and treatment data with the level of privacy they choose, and in particular to bring in their choice of support group to share that information. Clinical support is set at the level of fully informing patients on their status, making them knowledgeable to understand the consequences of different options, and leaving the final decision to the patient or patient's delegated authority. Based on the National Research Council report *Crossing the quality Chasm* (referred in [11], PCH rests on 8 principles listed below. 1. Respect for patient's values, preferences, and expressed needs, 2. Coordination and integration of care, 3. Improved patient awareness, communication, and knowledge, 4. Emotional support, 5. Enhancing comfort levels (at senior homes, ICU, Long term Re- habilitation), 6. Involvement of Social Support Groups, 7. Continuation of care (Post surgery/treatment), 8. Improved Access to Service.

So, PCH models are expected to include entities who fulfill these principles and include "trustworthiness" factors that include the four components *safety*, *security*, *reliability*, and *availability*. Although survey papers [12, 15] mention trusted relationship must be part of PCH model, no model exists to include all of the above it in the current PPC model [13]. A sufficiently complete model is necessary to suggest the services, and constraints provided by the entities in the model fulfill the 8 principles and trustworthiness. From such a model we can identify the criteria for shared decisions and evaluate them in determining the best service option. With this goal in mind, in Sect. 3.1 we propose a new PCH model.

3.1 Proposed PCH Model

We need a patient-centered healthcare model in which the health determinants comprehensively and collectively cover the 8 principles stated earlier. We propose such a model in Fig. 1. This model is a rich refinement of the health model proposed in [16] for *mHealth* (Mobile Health). We explain below the significant elements of healthcare determinants within each ring, and motive how well it covers the 8 PCC principles stated earlier.

In Fig. 1, a ring model with six layers represents various health determinants. Beginning with the innermost "Patient" layer, subsequent layers detail physical and

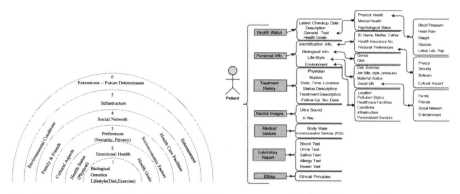

Fig. 1. Patient-centered Health Care Model   **Fig. 2.** Patient Model

emotional status, health goals, cultural aspects, preferences, and socioeconomic factors [16]. Expanding further, layers include social networks, family & friends, healthcare facilities, environmental conditions, and infrastructure support [16]. Each layer encompasses specific determinants, such as "Health Care" for healthcare providers, with patient-defined data sharing policies. The model provides a high-level conceptual view, while detailed models, illustrated in Fig. 2 for the "Patient" [16] and Fig. 3 for a "Health Care Facility," can be developed for each determinant, offering a comprehensive representation of the healthcare ecosystem [16].

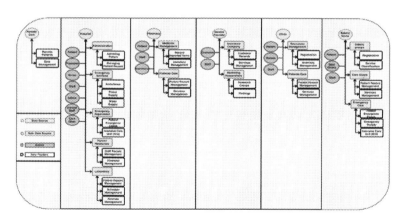

Fig. 3. Health Sources Model

3.2 Coverage and Comparison

The only paper that gives a conceptual PCH model is [13]. Their model is given in Fig. 2 (Page 21) of this paper. This model has essentially three levels. At the top, "Provider Characteristics" is associated with "Hospital" and "Physician, Nurse". At the next level

"Patient Centered Care" is associated with "Patient" and "Cost". At the lowest level, the box includes "Health related quality of life/Functional status/Psychological wellbeing". They have not given any explanation on how this diagram reflects PCC principles. In our opinion. It does not reflect any of the 8 stated principles.

As opposed to this we justify below how our conceptual model, along with a family of models for the determinants in it, add clarity and comprehensively fulfills the 8 principles of PCH.

1. *Principle-1*: "Respect for patient's values, preferences, and expressed needs" - Fulfilled in ring 1, ring 2 and ring 3.
2. *Principle-2*:"Coordination and integration of care" - Fulfilled by Family & Friends (ring 4), Healthcare Facilities (ring 3), and Social Network (ring 4).
3. *Principle-3*: "Improve patient awareness, communication, and knowledge" - Fulfilled by Healthcare (ring 3), Social Network (ring 4), Infrastructure (ring 5), Socioeconomic (ring 3).
4. *Principle-4*: "Emotional support" - Fulfilled by Emotional Health (ring 2), Cultural, Socioeconomic (ring 3), Family, Social Network (ring 4), Infrastructure, Entertainment (ring 5)
5. *Principle-5*: "Enhancing comfort levels" - Fulfilled by Preferences (ring 3), Healthcare Facilities (ring 4), all determinants (ring 5)
6. *Principle-6*: "Involvement Social Support Groups"-Fulfilled by all determinants (ring 4)
7. *Principle-7*: "Continuation of care (Post surgery/treatment)" - Fulfilled by Infrastructure, Environmental Conditions (ring 5), Healthcare Facilities (ring 3: Senior Home included)
8. *Principle-8*: "Improved Access to Service" - Fulfilled by Infrastructure (ring 5), Healthcare Facilities (ring 3)

In addition to the coverage of PCC principles, we motivate below how "models of interactions" between determinants can be developed and how such models are formal enough to guide us identify the criteria and suggest mathematically sound methods for estimating weights for preferences.

3.3   Criteria, Preferences, and Decision Quality Measurement

All reviewed papers in Sect. 2 emphasize the necessity of a collaborative approach that values and respects patients' needs and wishes for identifying criteria and weights. However, a systematic method for this purpose is lacking, and existing approaches are primarily manual, lacking standardization for decision quality assessment. The importance of measuring decision quality is discussed in [14]. While a list of evaluation measures is available on the Ottawa Hospital Research Institute website [17], many published papers do not follow it for assessing decision quality outcomes. Moreover, the available tools focus on patient evaluation, neglecting other stakeholder in Patient- Centered Healthcare (PCH).

The Delphi method [18–20], known for its standardized steps in collaborative decision- making, offers a potential means to enhance decision quality assessment in PCH. An innovative approach could involve adapting the evaluation list [17] for other stakeholders and combining it with the Delphi method. However, this method remains manual. While total automation of multi-criteria decision-making in patient-centered care is neither feasible nor desirable, we propose a methodology to serve as the foundation for an interactive, fair algorithm that engages all stakeholders in decision-making.

Decision Making Needs: Different decision-making scenarios, such as treatment options, post-treatment residence, medical device choices, and healthcare trends awareness, require distinct sets of decision-making criteria. For our discussion, we consider medical devices and senior homes as examples.

The eight principles stated in Sect. 3.2 lead us to consider *Trust, Knowledge, Economics, Emotional Support*, and *Value* as primary criteria for decision-making. Stakeholders must have mutual trust, and patients need trust in medical experts to share personal preferences. Economic factors include both patient and service provider capacity. Emotional support involves reliability from social groups, friends, family, and infrastructure. The specific features supporting mobility, safety, privacy, and physical activities in senior homes measure emotional support. Disagreements on *Value* between patients and doctors require moderation using *Knowledge* and *Trust* to reach a consensus in patient-centered care.

4 Fair Decision Making Principle

Fairness is defined in [21] from two perspectives. We base our work here on this definition, however we customize it with regard to patient-centered selection. So, we put patient's input as "ideal" for the patient, and all other input from stakeholders as "available options recommended by the experts". We first discuss in Sect. 4.1 how the criteria and their weights are to be chosen from expert (domain) knowledge for patient- centric fairness. Next in Sect. 4.2 we explain how patients can introduce their desires in a semantically meaningful way to algorithmic construction. That is, "semantics from patient as additional input" is to be accepted by a fair algorithm. So, functions in the algorithm are intelligent due to expert knowledge and due to the integration of user semantics.

4.1 Expert Knowledge for Fair Decision Making

We propose the following steps to construct criteria at "atomic attribute level" and to determine the *types* of attributes that can lead to "units of measurements". The selection problem is "medical and infrastructure devices" for patient care.

1. Patient constructs the set of needs in the dimensions "Knowledge, Emotional Support", gets expert opinions from trusted experts and with their help selects the set of medical devices (body-ware), and develops data sharing policies, and ethical/cultural needs. Patient lists the comfort requirements and gets the help of experts in dimensions "Economics, Value, Knowledge" in creating a list of infrastructure devices and medical devices.
2. The patient gets a trusted expert to develop the criteria for selection at attribute level. That is, on behalf of the patient a finite set of trusted experts develop the list of criteria that satisfies the need of the patient.

3. The experts collaborate in choosing the *type* for each criteria chosen, and assigning a "preference value" (weight consistent with the level of importance) for each attribute. The collaboration will discuss questions like the following:

   – What are the attributes necessary to describe the entity being examined?
   – Which attributes are required to evaluate a Purpose being examined?
   – Which attributes are needed for analysis?
   – Which attributes "from the examination viewpoint" are required for this analysis goal?
   – Which attributes are required to describe the examination environment?
   – Is time frame essential for this analysis issues?
   – Is there any specific geographical area specific for this analysis task?
   – Any other attributes required for analysis goal achievement?

4. Patient shares this list with all trusted stakeholders.
5. Every stakeholder will generate similar lists of their attribute choices and weights for attributes.
6. This information is fed to Algorithm P (described below) that will choose the final set of attributes (criteria) and their respective weights to be submitted to the fair Algorithm F described in Sect. 5.

Algorithm P

Let $E = \{E_1, E_2, \cdots, E_k\}$ denote the set of experts nominated by the stakeholder (patients, physicians, etc;). Because every decision must be patient centric, we let the patient assign a weight σ_j for expert $E_j \in E$ that reflects the patient's level of trust on the expert E_j. Let $\sigma = \sigma_1 + \sigma_2 + \cdots, + \sigma_k$. Let C_{Ei} denote the set of criteria chosen by expert E_i, and $C = _{Ei \in E} C_{Ei}$. For the sake of clarity let $|C| = p$. That is, there are p criteria. For each criteria c_i in C a weight w_i has been assigned by the expert who chose it. So, we have the set $W = \{w_i \mid c_i \in C\}$ of weights for the criteria in set C. In the previous work [6] on medical device selection the use of "geometric mean" is suggested to calculate the weight for a criteria from the weights assigned by stakeholders. In our work we improve on this suggestion by using *weighted geometric mean*. That is, we use the patient-assigned trust weight σ_j to the expert E_j along with the expert-assigned weight w_j to the criteria c of that expert's choice in computing the weighted geometric mean in the formula below:

$$w_c = w_1^{\sigma 1} w_2^{\sigma 2} \ldots w_p^{\sigma p} \frac{1}{\sigma}$$

This weight w_c is a *fair* weight because it satisfies the patient's wish, and the expert's knowledge-based input. We can prune further the set C if the number of criteria is too many by deleting those that have a low weight. This threshold has to be determined by experts. Once done, the set C and the set $W = \{w_c \mid c \in C\}$ are input to Algorithm F. Table 1 is an example of criteria/weight table constructed using Algorithm P for the attributes shown in the table. The number of criteria (attributes) is kept small for clarity of understanding. In principle, the set of attributes and the number of attributes are the choice of the patient and the experts.

Table 1. Criteria Developed Experts - Accepted by Patient

Attributes	Data Type	Measurement/Weight
Clinical Need	Numerical	[0, 1]/5.2
Accuracy	Numerical	Percentage/6.1
Reliability	Ranked Categorical	{Low, Medium, High}/5.8
Risk Factors	Enumerated	{Biological Hazards, Poor Display}/6.3
Safety	Ranked Categorical	{Low, Medium, High}/5.8
Connectivity	Enumerated	{Local Network, Internet Not Possible}/4.3
Release Year	Enumerated Numerical	{1995,..., 2050}/3.4
Expiry Date	Record	⟨Year, Month, Day⟩/5.1
Maintenance	Numerical	[50$, 200$]/4.7
Value	Numerical	Percentage/5.7
Ease of Use	Enumerated	{Easy, Difficult, Very Difficult}/4.3
Compliance	Binary	{Yes, No}/6.3
Connectivity	Enumerated	{Low, Medium, High}/3.1
Privacy	Enumerated	{NK, Limited, Good}/5.9

4.2 Semantics for Patient-Centric Fairness

Patients can specify semantics for *better* choices in selection, introducing a "best match" concept alongside closeness to the patient's requirements, emphasizing a "patient-centric perspective." For attributes like "reliability, safety, and accuracy," where higher' implies better, and for features such as "cost and weight," where lower' is preferable, patient semantics for "lower is better" (LB) and "more is better" (MB) are integrated into scoring functions in the selection algorithm [22]. This approach allows the algorithm to prioritize options aligning with patient preferences on each attribute independently. Running the algorithm with different patient perspectives, including varied weights and attribute semantics, enables patients to consistently identify optimal choices. For instance, patients can determine the most cost-effective option among choices ensuring safety with low risk and maintenance costs [22]. In [22], ontology support is utilized to interpret the "degree of equivalence" between medical terms, enhancing the assessment of similarity between concept terms in the medical field. Semantic distance and similarity measures between sets of concepts contribute to patients' awareness and knowledge of healthcare needs, enabling informed specification of preference levels and ensuring unbiased similarity-based selection and ranking. From the algorithmic perspective a fair ranking will be "tolerant to small values", consistent, flexible and timely. Below is a list of characteristics for an algorithmic perspective of fairness. The algorithm will produce *normalized* results in the interval [0, 1]. Bounded similarity measures are not produced by the widely used distance and cosine functions. The algorithm must be consistent in the sense "reversal of rank" should not happen in any environment in which the same set of inputs are used. The algorithm does not put any restriction on the number of criteria and on the number expert alternatives. For the sake of efficiency and accuracy of result, the algorithm uses only *simple scoring functions* (no exponential, logarithmic or fuzzy arithmetic).

5   A Skeleton of Fair Algorithm

We assume that at every i^{th} attribute level, there exists a sematic scoring function λ_i that compares two given values of the attribute and outputs a score which is a mea- sure of "best match" between the attribute values. In this paper we do not discuss their constructions, because it involves many mathematically sound details. We suggest the readers to construct their own scoring functions that satisfy the semantic criteria explained in Sect. 4.2. We also refer the reader to our recent and ongoing work [21, 22] on the rationale for the choice of scoring functions that we have constructed.

The sets C and W computed by Algorithm P are input to Algorithm F. Let $A = \{A_1, A_2, \cdots, A_m\}$ denote the set of alternatives available. Assume that every A_i has all the attributes of set C (in the same order). Thus, A_is are formalized into "a vector" wherein the attributes are all in the same order, although the types of the attributes of this vector are different. This vector structure helps us to explain precisely the algorithmic steps. The patient inputs a query which consists of three parts, as shown in Table 2.

Table 2. Patient Input - Ideal Choice for the Patient

Attribute Vector:: $Q = \langle Q_1, Q_2, \cdots, Q_n \rangle$ — Semantics:: For each attribute Q_i specifies LB (Less is Better) OR MB: More is better — Mode: For each attribute Q_i specifies EM (exact match) OR BM (best match)

Algorithm F
The algorithm compares each component of the vector Q with the corresponding component of vector A_j. That is, it compares Q_i with A_{ji}, for $i = 1, \cdots, p$. It applies the scoring function λ_i to the pair $(Q_i, A_{ji}$. The value $v_j i = \lambda_i((Q_i, A_{ji})$ is a measure of "closeness", consistent with the semantics and mode specified in the patient query structure. Next, it computes the weighted mean.

$$s_j = \frac{\left(w_1 \times v_{j1} + w_2 \times v_{j2} + \ldots + w_p \times v_{jp}\right)}{\left(w_1 + w_2 + \ldots + w_p\right)}$$

which gives the "similarity measure between the alternate A_j and the patient query Q. Having calculated the measures $s_1, s_2, \cdots, s_m$ for all alternatives, the algorithm ranks the set of alternatives in the non-increasing order. So, the top of the ranked list is the "best match". In case there are other considerations that may prevent this choice being adopted by the patient, the next alternative in the list can be chosen.

6 Conclusion

In our critical survey of existing multi-criteria decision-making approaches in healthcare, we found a lack of specific conclusions on successful methods in the biomedical healthcare domain. The consensus among researchers is the pressing need for more efficient collaborative decision-making methods.

We examined the Patient-Centered Health (PCH) paradigm, highlighting its emerging importance, elucidating its eight fundamental principles, and identifying challenges in employing Multi-Criteria Decision Making (MCR) for shared decision-making. Subsequently, we proposed an approach based on trust, knowledge, value, economics, and social support to collaboratively select significant decision-making attributes, drawing inspiration from our previous work on semantic-centered similarity matching.

To enhance our proposed approach, deeper research is warranted, particularly in formalizing medical treatment selection problems with ontology support. Our three implementations of fair selection and ranking algorithms can be further enriched to handle complex data types of quality attributes in healthcare. Alaa Alsaig's recent thesis introduced a user interface for gathering client requirements, and we are exploring ways to integrate healthcare ontology, making the process more interactive and iterative. This iterative loop allows stakeholders to modify their choices until a consensus is reached, with the fair algorithm serving as a vital component.

References

1. Schey, C., Postma, M.J., Krabbe, P.F.M., Topachevskyi, O., Volovyk, A., Connolly, M.: Assessing the preferences for criteria in multi-criteria decision analysis in treatments for rare diseases. Front. Public Health **8**, 1–10 (2020)
2. Piniazhoko, O., Ne´meth, B.: Practical issues of determining weights for criteria to be used in an MCDA framework - based on a case study. In: ISPOR 22nd Annual International Meeting, Boston, MA, USA, p. 1 (2017)
3. Ozsahin, I., Ozsahin, D.U., Uzun, B., Mustapha, M.T.: Introduction - chapter 1. In: Applications of Multi-Criteria Decision-Making Theories in Healthcare and Biomedical Engineering, Elsevier - Academic Press, pp. 1–2 (2021)
4. Taherdoost, H., Madanchian, M.: Multi-criteria decision making (MCDM) methods and concepts. Encyclopedia **2023**, 77–87 (2023)
5. Dehe, B., Bamford, D.: Development, test and comparison of two multiple criteria decision analysis (MACDA) models: a case of healthcare infrastructure location. Expert Syst. Appl. **2015**, 1–11 (2021)
6. Barkaoui, H., Rejeb, H. B., Barkaoui, A., Tavares, J.M.R.: Multi-criteria decision making for medical device development. Eng. Manag. J. **35**, 1–18 (2022)
7. Stevi´c´, Z., Pamuc˘ar, D., Pus˘ka, A., Chatterjee, P.: Sustainable supplier selection in healthcare industries using a new MCDM method: measurement of alternatives and ranking according to compromise solution (MARCOS). Comput. Ind. Eng. **140**, 106231 (2020)
8. Anvari, A.S.: The application of MCDM methods in covid-19 pandemic: a state of the art review. Appl. Soft Comput.Comput. **126**, 1–40 (2022)
9. Angelis, A., Kanavos, P.: Multiple criteria decision analysis (MCDA) for evaluating new medicines in health technology assessment and beyond: the advance value framework. Soc Sci Medoc. Sci. Med. **188**, 137–156 (2017)

10. Zhao, J., Gao, S., Wang, J., Liu, X., Hao, Y.: Differentiation between two healthcare concepts: person-centered and patient-centered care. Int. J. Nurs. Sci. **3**, 398–402 (2016)
11. Barry, M.J., Edgman-Levitan, S.: Shared decision making - the pinnacle of patient-centered care. N. Engl. J. Med. **366**(9), 78–781 (2012)
12. Constand, M.K., Macdermid, J.C., Bello-Has, V.D., Law, M.: Scoping review of patient-centered care approaches in healthcare. BMC Health Serv. Res. **14**(271), 1–9 (2014)
13. Jayadevappa, R., Chhatre, S.: Patient centered care - a conceptual model and review of the state of the art. Open Health Serv. Policy J. **4**, 15–25 (2011)
14. Kaltoft, M., Cunich, M., Salkeld, G., Dowie, J.: Assessing decision quality in patient- centered care requires a preference-sensitive measure. J. Health Serv. Res. Policy **19**(2), 110–117 (2014)
15. Ogden, K., Barr, J., Greenfield, D.: Determining requirements for patient-centered care: a participatory concept making study. BMC Health Serv. Res. **17**(780), 1–11 (2017)
16. Wan, K., Alagar, V.: Context-aware, knowledge-intensive, and patient-centric mobile health care model. In: 12th International Conference on Natural Computation, Fuzzy Systems and Knowledge Discovery (ICNC-FSKD), Zhangjiajie, China, pp. 2253–2260 (2015)
17. Ottawa Hospital Research Institute, Patient decision aids (2022)
18. Hsu, C.C., Sandford, B.A.: The delphi technique: making sense of consensus. Pract. Assess. Res. Eval. Assess. Res. Eval. **12**(1), 1–10 (2007)
19. Okoli, C., Pawlowski, S.D.: The delphi method as a research tool: an example, design considerations and applications. Inf. Manag. **42**(1), 15–29 (2004)
20. Strozyna, M., Elden, G., Filipiak, W.A.D., Malyszko, J., Wecel, K.: A framework for the quality-based selection and retrieval of open data - a use case from the maritime domain. Electr. Markets **28**, 219–233 (2018)
21. Alsaig, A., Alagar, V., Mohammad, M., Alhalabi, W.: A user-centric semantic-based algorithm for ranking services: design and analysis. Serv. Oriented Comput. Appli. **11**(1), 101–120 (2017)
22. Liu, Y.: A family of algorithms for patient similarity based on electronic health records, Master's thesis, Concordia University, Montreal, Canada (2022). http://spectrum.library.con cordia.ca

A New and Efficient Dormitory Management System

Boyuan Meng[1,2,3], Yushu Zhang[4], Zhujun Li[5], Wendong Yu[6], Hongwei Wei[6], and Lin Yuan[1,2,3]([✉])

[1] Key Laboratory of Computing Power Network and Information Security, Ministry of Education, Shandong Computer Science Center, Qilu University of Technology (Shandong Academy of Sciences), Jinan 250353, Shandong, China
yuanl@qlu.edu.cn
[2] Shandong Engineering Research Center of Big Data Applied Technology, Faculty of Computer Science and Technology, Qilu University of Technology (Shandong Academy of Sciences), Jinan 250353, Shandong, China
[3] Shandong Provincial Key Laboratory of Computer Networks, Shandong Fundamental Research Center for Computer Science, Jinan 250353, Shandong, China
[4] School of Art, Nantong University, Nantong 226019, Jiangsu, China
[5] Jinan Springs Patent and Trademark Office, Jinan 250014, Shandong, China
[6] Shandong Tianyi Information Technology Co., Ltd., Jinan 250101, Shandong, China

Abstract. Today's information trend is unstoppable, and the manual management mode of college dormitory management cannot meet the trend of increasing demand for dormitories year by year. It not only wastes many manpower and material resources, but also is easy to appear many unnecessary low-level errors. A system about student dormitory management not only saves many unnecessary repeated working time, but also reduces the generation of various low - level errors in information management, which makes the grasp and operation of information more convenient. The system adopts B/S architecture (browser / server architecture), mainly based on the development and implementation of JAVA language. The system has 3 user roles: administrator, student, and host administrator. The functional module of the administrator includes user management, student management, dormitory management, grade management and authority management, which can realize the management of dormitory information and other information. Students' function module includes student personal details management, online pre-selection dormitory and view of absence records, students can pre-select the dormitory, and view personal information and absence information. The dormitory manager function module includes dormitory management, pre-selection setting and absence management, dormitory details management and absence management, and pre-selection time settings. After testing, the college dormitory management system can effectively realize the above functions and is interface friendly and easy to operate.

Keywords: Dormitory management · JAVA language · B/S architecture

© The Author(s), under exclusive license to Springer Nature Singapore Pte Ltd. 2024
D.-S. Huang et al. (Eds.): ICAI 2023, CCIS 2015, pp. 101–110, 2024.
https://doi.org/10.1007/978-981-97-0827-7_9

1 Introduction

With the advent of the internet age, information technology has become increasingly widely used in various industries, creating infinite value for human society and affecting human work and lifestyle. Dormitory life is an indispensable part of university life, leading to the increasing status of dormitory management in universities. Therefore, in recent years, more and more domestic universities have shifted their attention to dormitory management. At present, logistics management in some universities is mainly managed through manual organization or simple information systems. For universities with a large amount of student information, manual recording is quite troublesome, especially when adding, deleting, modifying, and querying data. The workload of administrators can be very large, which not only wastes time, but also leads to low work efficiency and unnecessary low-level errors. [1].

The management department of college student dormitories can effectively manage dormitory information, student information, grade information, etc., allocate dormitories, greatly improve the management efficiency of administrators, reduce work pressure, avoid low-level errors in work, and also improve students' accommodation experience.

2 Key Technologies of the System

2.1 JSP Technology

JSP is deployed on a network server and can respond to requests sent by the client, dynamically generate HTML based on the request content, and then return it to the requester. JSP technology uses Java language as a scripting language to provide services for users' HTTP requests and can work with other Java programs on the server to handle complex project requirements [2].

2.2 B/S Architecture

The B/S architecture is currently the mainstream networked structural model. The B/S architecture is to input HTTP requests through a browser on the Internet, then the server accepts the request and parses the URL to generate an HTML file. Then, the browser sends the HTML file, and when the browser accepts the corresponding HTML file, it begins to parse and organize the relevant resource files, ultimately displaying the page to the user. The B/S architecture mode unifies the client and centralizes the core parts of system function implementation on the server, simplifying the development, maintenance, and use of the system [3–5].

Our university student dormitory management system allows users to input information and submit some operations from the page, send requests to the server and wait for the server to respond. The server processes and accepts the user's requests, and then the user can view the data returned by the server on the browser [6–8].

2.3 MYSQL Database

MYSQL is a relational database that is used by our university's student dormitory management system to store and access dormitory information, student information, building information, user information, absence information, and more. Every table in the database is interrelated, and without a database, data storage and invocation cannot be carried out, and programs cannot run normally[9]. Due to the characteristics of developing source code, the MYSQL database is generally chosen for the development of small and medium-sized websites. Therefore, this system uses MYSQL as the backend database[10, 11].

3   System Design

3.1 Overall System Design

The student dormitory management system in this university mainly includes three parts: administrator, student, and dormitory staff. The system structure diagram is shown in Fig. 1.

Fig. 1. System Structure Diagram

3.2 System Detailed Design

3.2.1 Administrator Module

After logging in as an administrator, the following functions are available:

① User management: Add, modify, and delete users, and import users from Excel.
② Grade management: Add grades, batch delete grades, delete single data, and perform fuzzy queries.
③ Organizational management: The organizational structure is displayed as a tree structure and adding an organizational requires selecting a superior column, and modifying the organization, and deleting the organization.
④ Student management: Display student information, export all student information in Excel, add and modify student information, delete students, and batch delete students.
⑤ Permission management: Grant permissions to users, select users, and grant th em permissions.

⑥ Building management: Add buildings, including adding floors, dormitories, beds, deleting buildings, editing buildings, and modifying some simple fields.

⑦ Dormitory management: Display dormitory information [12].

⑧ Preselection management: Add and modify the range of preselection time, display the range of preselection time for grades and classes, allocate dormitories, etc.

⑨ Absence management: Add, delete, modify, and check the information of absent students.

View and modify user information, student information, and grade information, update absence information, dormitory information, and other class diagrams, as shown in Fig. 2.

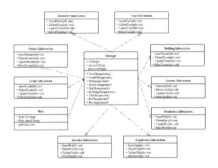

Fig. 2. Administrator Module Class Diagram

3.2.2 Student Module

The functions available after students log in include: [13].

① Personal details management: You can view personal information and modify it.

② Change password: Modify the personal password.

③ Online dormitory selection: Select dormitories based on the time frame set by the administrator.

④ Absence Records: View your own absence records [14, 15].

The management class diagram of the student module is shown in Fig. 3.

3.2.3 Housing Manager Module

The functions of the dormitory administrator after logging in include:

① Building management: Add information such as building, dormitory type, number of floors, and number of rooms.

② Dormitory management: display of data.

③ Absence management: Submit, delete, modify, and query the information of absent students.

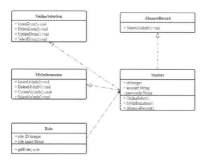

Fig. 3. Student Module Class Diagram

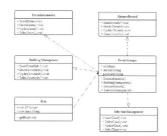

Fig. 4. Dormitory Manager Module Class Diagram

3.3 Database Design

The management system for college student dormitories uses an MTSQL database to store data. Below is the detailed information of each table designed (see Fig. 4).

(1) tb_user table

Used to save user information, including username, password, etc. The structure of this table is shown in Table 1.

Table. 1. tb_user

Field name	Field type	Primary key	Reference key	If null
Id	Int	Y	/	N
User_name	Varchar	/	/	Y
Password	Varchar	/	/	Y
Name	Varchar			Y
Phone	Varchar			Y
Type	Int			Y
remark	Varchar			Y

(2) tb_student table

Used to save student information, including student ID, student ID, etc. The structure of this table is shown in Table 2.

Table. 2. tb_student

Field name	Field type	Primary key	Reference key	If null
Id	Int	Y	/	N
Sno	Varchar	/	/	Y
User_id	Int	/	/	N
Idcard	Varchar			Y
Grade_id	Int			Y
Sex	Int			Y
Class_id	Int			Y

(3) tb_storey table

Used to save floor information, including floor numbers and building IDs, the structure of this table is shown in Table 3.

Table. 3. tb_storey

Field name	Field type	Primary key	Reference key	If null
Id	Int	Y	/	N
Storey_no	Int	/	/	Y
Building_id	Int	/	/	N

(4) tb_selection_dormitory table

Used to save pre-selected dormitory information, including dormitory ID, etc. The structure of this table is shown in Table 4.

(5) tb_slection table

Used to save pre-selected information, including pre-selected names, notes, etc. The structure of this table is shown in Table 5.

(6) tb_role_menu table

Used to save role information, including role ID and menu ID, the structure of this table is shown in Table 6.

Table. 4. tb_selection_dormitory

Field name	Field type	Primary key	Reference key	If null
Id	Int	Y	/	N
Dormitory_id	Int	/	/	N
Clsss_id	Int	/	/	N
Grade_id	Int			Y
Selection_id	Int			Y

Table. 5. tb_selection

Field name	Field type	Primary key	Reference key	If null
Id	Int	Y	/	N
Name	Varchar	/	/	Y
Start_time	Date	/	/	Y
End_time	Date			Y
Grade_id	Int			Y
Class_id	Int			Y
remark	Varchar			Y

Table. 6. tb_role_menu

Field name	Field type	Primary key	Reference key	If null
Id	Int	Y	/	N
R_ID	Int	/	/	N
P_ID	Int	/	/	N

(7) tb_org table

Used to save organizational information, including name, organizational type, etc. The structure of this table is shown in Table 7.

(8) tb_menu table

Used to save character menu information, including titles, icons, etc. The structure of this table is shown in Table 8.

(9) tb_grade table

Used to save grade information, including grade ID and grade, the structure of this table is shown in Table 9.

Table. 7. tb_org

Field name	Field type	Primary key	Reference key	If null
Id	Int	Y	/	N
name	Int	/	/	N
type	Int	/	/	N
Parent_id	Int			Y
remark	Varchar			Y

Table. 8. tb_menu

Field name	Field type	Primary key	Reference key	If null
Id	Int	Y	/	N
Title	Varchar	/	/	Y
Icon	Varchar	/	/	Y
Href	Int			Y
Target	Varchar			Y
Parent_id	Int			Y

Table. 9. tb_grade

Field name	Field type	Primary key	Reference key	If null
Id	Int	Y	/	N
Name	Varchar	/	/	N

(10) tb_dormitory table

Used to save dormitory information, the structure of this table is shown in Table 10.

Table. 10. tb_dormitory

Field name	Field type	Primary key	Reference key	If null
Id	Int	Y	/	N
No	Int	/	/	N
Storey_id	Int	/	/	Y
Building_id	Int			Y

4   System Implementation

When the user enters the system, a login interface will appear. You must select different roles to enter this interface. The accommodation staff edits the pre-selection time in the pre-selection setting interface.

5   Conclusion

The development of this system aims to provide a convenient dormitory management system for universities, mainly using JSP technology, JAVA language, MYSQL database, and B/S architecture technology to design and implement a university student dormitory management system. The main objects of the system are administrators, students, and dormitory administrators. The main functions of the administrator include building management, user management, permission management, dormitory management, etc. The main functions of the student include personal details management, online pre-selection of dormitories, and absence query. The main management of the dormitory administrator includes pre-selection settings, dormitory details management, etc. [16].

Acknowledgement. This work was supported by the Natural Science Foundation of Shandong Province, China (No. ZR2020QF038), the Ability Improvement Project of Science and Technology SMES in Shandong Province (No. 2023TSGC0279), the Youth Innovation Team of Colleges and Universities in Shandong Province (2023KJ329), and the Qilu University of Technology (Shandong Academy of Sciences) Talent Scientific Research Project (No. 2023RCKY128).

References

1. Yang, Y., Chen, S.: Design and implementation of college dormitory management system. In: 2022 Fourth International Conference on Emerging Research in Electronics, Computer Science and Technology (ICERECT), pp. 1–5 (2022)
2. Yang, C.-Y., et al.: Design of high school dormitory management system based on IoT technology. In: 2021 13th International Conference on Wireless Communications and Signal Processing (WCSP), pp. 1–5 (2021)
3. Peng, Z., Liu, T., Mai, L.: Design and implementation of dormitory management system based on SSM framework. In: 2020 International Conference on Information Science, Parallel and Distributed Systems (ISPDS), pp. 321–325 (2020)
4. Yuan, L., et al.: Nonconvex penalty based low-rank representation and sparse regression for eQTL mapping. IEEE/ACM Trans. Comput. Biol. Bioinf. **14**, 1154–1164 (2016)
5. Yuan, L., et al.: ICircDA-NEAE: accelerated attribute network embedding and dynamic convolutional autoencoder for circRNA-disease associations prediction. PLoS Comput. Biol. **19**, e1011344 (2023)
6. Zeng, X., Deng, L.: Research on college counselor training system based on computer virtual reality technology. In: 2022 IEEE Conference on Telecommunications, Optics and Computer Science (TOCS), pp. 1323–1326 (2022)
7. Yuan, L., et al.: Integration of multi-omics data for gene regulatory network inference and application to breast cancer. IEEE/ACM Trans. Comput. Biol. Bioinf. **16**, 782–791 (2018)

8. Yuan, L., et al.: Pan-Cancer Bioinformatics Analysis of Gene UBE2C, Frontiers in genetics 2022;13

9. Yuan, L., et al.: Path-ATT-CNN: a novel deep neural network method for key pathway identification of lung cancer. Front. Genet. **13**, 893358 (2022)

10. Yuan, L., et al.: A novel computational framework to predict disease-related copy number variations by integrating multiple data sources. Front. Genet. **12**, 696956 (2021)

11. Yuan, L., et al.: A machine learning framework that integrates multi-omics data predicts cancer-related LncRNAs. BMC Bioinform. **22**, 1–18 (2021)

12. Li, J., et al.: Design of intelligent dormitory management system based on raspberry Pi. In: 2022 6th International Conference on Wireless Communications and Applications (ICWCAPP), pp. 205–208 (2022)

13. Zhang, Y.: Design of grid information platform based on new media IP information modeling algorithm. In: 2022 Sixth International Conference on I-SMAC (IoT in Social, Mobile, Analytics and Cloud) (I-SMAC), pp. 223–226 (2022)

14. Wang, L.: Construction and application of college student safety management system based on 4R theory. In: 2022 International Conference on Education, Network and Information Technology (ICENIT), pp. 195–199 (2022)

15. Shen, Z., et al.: Prediction of Back-splicing sites for CircRNA formation based on convolutional neural networks. BMC Genomics **23**, 581 (2022)

16. Alhari, M.I., Lubis, M.: Quality of service (QoS) Wifi network study case: Telkom university dormitory hall. In: 2023 IEEE International Conference on Industry 4.0, Artificial Intelligence, and Communications Technology (IAICT), pp. 345–349 (2023)

Edge Collaborative Assisted Caching Content Placement Optimization Strategy Based on DDSG

Taoshen Li[1,2]([✉]), Ling You[2], and Zhihui Ge[2]

[1] School of Information Engineering, Nanning University, 8 Longting Road, Nanning, People's Republic of China
tshli@gxu.edu.cn
[2] School of Computer, Electronics and Information, Guangxi University, 100 Daxue Road, Nanning, People's Republic of China

Abstract. To improve the pre-fetching accuracy of mobile autonomous vehicles in highly dynamic network, we study the cache content update and placement optimization in edge collaborative caching systems, and propose an edge collaborative assisted cache content placement optimization scheme. This scheme leverages the mobility of intelligent driving vehicles as an edge node to create an effective cache content placement strategy, and solve the cooperation problem between roadside units (RSUs) that dynamically adapt to vehicle requests and intelligent driving vehicles. Specifically, by jointly optimizing vehicle scheduling, cache content task unloading ratio, and content placement decisions, the problem of minimizing cache task processing delay is formulated. Considering the non-convexity of the problem, high-dimensional state space and non-convexity of continuous action space, a cache content placement optimization algorithm based on deep deterministic policy gradient (DDPG) is proposed. Using this algorithm, the optimal cache content placement strategy can be obtained in an uncontrollable dynamic environment. Simulation results show that the proposed algorithm can converge quickly and has significant improvements in processing delay compared to other baseline algorithms.

Keywords: Edge Collaborative · Caching Content Placement · Deep Deterministic Strategy Gradient · Autonomous Vehicle · Roadside Unit (RSU)

1 Introduction

In order to realize the safety, comfortable and efficiency of autonomous driving, it is necessary to obtain high-precision surrounding environment information in real time. To reduce transmission delay and network load pressure, mobile edge caching technology can be used to pre-fetch popular content in roadside unit (RSU) in advance. However, the vehicle's environment is complex and rapidly changing, the request data surges and the success rate of content acquisition is low. How to design a cache solution to meet the high requirements of vehicles for communication quality has become a key technology [1].

© The Author(s), under exclusive license to Springer Nature Singapore Pte Ltd. 2024
D.-S. Huang et al. (Eds.): ICAI 2023, CCIS 2015, pp. 111–122, 2024.
https://doi.org/10.1007/978-981-97-0827-7_10

Although caching popular content on edge servers can provide users with data-intensive services and improve network performance, due to the dynamic nature of the vehicle's environment, data content is temporary and changes over time [2]. Especially, the high mobility of vehicles, intermittency of information transmissions, high dynamics of user requests, limited caching capacities, extreme complexity of business, and data freshness scenarios pose an enormous challenge to content caching and distribution in vehicular networks [3].

In order to improve performance and quality of service in intelligent driving vehicle environments, and promote content delivery, some researchers considered the intermittent connections of mobile intelligent driving vehicles and design a content placement scheme for edge caching servers [4–7]. However, these studies mainly focus on improving the performance of edge caching schemes/algorithms and have not fully utilized the relationship between vehicle user preferences and the overall preferences of the entire region, thus unable to obtain complete content delivery services.

Edge collaborative caching is a method to meet the service delay requirements of different edge nodes (such as RSUs) [8]. The authors of [9] designed a cooperative and demand-aware caching strategy to maximize the cache hit ratio. [10] proposed a group caching scheme for 6G mmWave vehicular networks. In [11], the authors designed the collaborative caching scheme between RSUs and intelligent driving vehicles. [12] proposed an edge caching scheme that allows multiple RSUs to collaborate in caching partial content in a distributed manner and transmit encoded data packets using fountain codes. [13] presented a collaborative edge caching scheme based on multi-layer edge caching servers. The authors of [3] constructed the system model to predict the vehicle trajectory and content popularity.

Although cross layer collaboration between different RSUs can provide seamless connectivity to facilitate content delivery, content caching needs to consider the differential communication characteristics and caching capabilities of multiple edge servers. Therefore, to improve the efficiency of active caching for autonomous vehicle users, it is necessary to study fine-grained cache content placement methods and solutions. This paper aims to improve the pre-fetching accuracy of mobile autonomous vehicles in highly dynamic network topology, studies the optimization problem of collaborative perception content placement in vehicular connected networks, and proposes an edge collaborative caching content placement optimization scheme based on deep deterministic strategy gradient (DDPG) in the vehicle assisted edge caching architecture.Finally, the effectiveness and superiority of the proposed scheme are verified through simulation experiments.

2   System Model

2.1   Network Model

Consider the scenario within the coverage range of cellular macro station (MBS) as shown in Fig. 1, which includes a central network MBS, several RSUs, and assisted caching vehicles with unused storage resources. MBS is deployed at the center of the road network to provide continuity of service, while RSUs are randomly located at the center of different transportation areas. Considering the capacity of storage space, it is assumed that all available content is cached on MBS, and each edge collaborative caching node RSU and assisted caching vehicle are equipped with content caching units with limited storage resources.

In Fig. 1, RSU can provide content delivery services for all vehicles within the coverage range. However, due to limited cache capacity space, RSUs can only store a portion of the copy of the content, which will affect the quality of content delivery. Therefore, RSU offloads some cache tasks to assisted caching vehicles, which can significantly reduce the content cache burden of RSU. This layered caching strategy greatly reduces the pressure on the backbone network. This is because even in harsh network communication environments, the vehicles requested content can retrieve popular content from the nearest cache node.

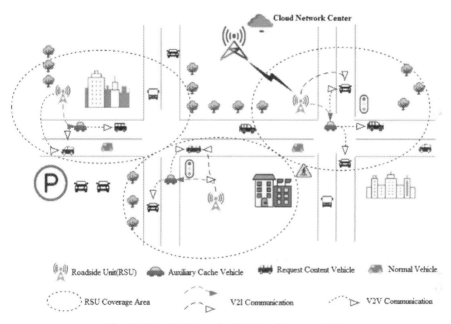

Fig. 1. Edge Collaborative Caching System Model

2.2 Communication Model

Let $\mathbf{N} = \{1,2,\ldots,k,k+1,\ldots,N\}$ be the index set of edge collaborative caching nodes, 0 be the index set of MBS, $\mathbf{K} = \{1,2,\ldots,i,\ldots,N\}$ be the index set of RSUs, and $\mathbf{V} = \{k+1,\ldots,j,\ldots,N\}$ be the index set of assisted caching vehicles that provide content through V2V communication. Assuming that the assisted cache vehicle moves randomly at low speed in the RSU coverage area. Let $\mathbf{M} = \{1,2,\ldots,m,\ldots,M\}$ be the index set of the content request vehicle, and $\mathbf{Q} = \{1, 2,\ldots,q,\ldots,Q\}$ be the size of the content data that needs to be cached. The entire time period of cached content is divided into multiple time slots $t \in \{1,2,3,\ldots,T\}$. The coordinates of the assisted caching vehicle at time slot t are represented as: $l_j(i) = \{x(i),y(i)\}$. In each time slot, after the RSU offloads some cache tasks to the assisted caching vehicle, it will execute the remaining cache tasks locally.

Assuming that RSUs use different communication channels and do not consider interference from other RSUs, the signal-to-noise ratio (SINR) of communication between RSU R_i and assisted caching vehicle V_j can be defined as:

$$r_{i,j}(t) = \frac{p_{i,j}(t)g_{i,j}(t)}{\varepsilon_{i,j}(t)d_{i,j}^{-1}(t)^k \sigma_{i,j}(t)^2} \tag{1}$$

where, $p_{i,j}(t)$ devotes the transmission power between R_i and V_j in the upload link, $g_{i,j}(t)$ devotes the channel gain between R_i and V_j, $\sigma_{i,j}(t)^2$ is additive Gaussian white noise, $\varepsilon_{i,j}(t)$ devotes the path loss at the reference unit distance, k devotes the path loss index, and $d_{i,j}^{-1}(t)$ devotes the euclidean distance between R_i and V_j.

According to Shannon's formula, the transmission rate between R_i and V_j is:

$$V_{i,j}(t) = B\log_2(1 + r_{i,j}(t)) \tag{2}$$

where, B devotes communication bandwidth.

2.3 Caching Model

Let the cache space size of RSU be C_{Ri}, and the cache space size of assisted caching vehicles be C_{Vi}. Because the capacity of cache tasks cannot exceed the maximum space limit, the size of cache content on RSUs is constrained:

$$\sum_{i=1}^{Q} q \leq C_{R_i} \tag{3}$$

The size of the cache content on the assisted caching vehicle is also constrained by the cache capacity:

$$\sum_{i=1}^{Q} q \leq C_{V_i} \tag{4}$$

Let $\beta(t) \in \{0,1\}$ devote the binary matrix where the content is placed at time slot t, where $\beta(t) = 1$ indicates caching the content, $\beta(t) = 0$ indicates that the content has not been cached. In order to make reasonable use of cache capacity and avoid cache redundancy, it is necessary to ensure that content q cannot be cached in multiple assisted caching vehicles and RSUs. Therefore, there is:

$$\sum_{i=1}^{N} \beta(t) = 1 \tag{5}$$

2.4 Computing Model

In system, the caching task R_i of RSU in each time slot uses a partial caching strategy. The delay of R_i's local processing cache task at time slot t can be expressed as:

$$t_{rsu}(t) = \frac{(1 - \alpha_m(t))Q_{i,j}(t)s}{f_{RSU}} \tag{6}$$

where, $\alpha_M(t) \in \{0,1\}$ devote the proportion of tasks offloaded to assisted caching vehicles, $Q_{i,j}(t)$ devotes the size of cache tasks, s devotes the CPU cycle required to cache each unit byte, and f_{RSU} devotes the RSU's ability to handle cache tasks.

In system, the delay of retrieval results provided by servers is usually very small and can be ignored. Therefore, the processing task delay on assisted caching vehicles can be divided into two parts. The transmission delay can be expressed as:

$$t_{transmission}(t) = \frac{\alpha_m(t))Q_{i,j}(t)}{v_{i,j}(t)} \qquad (7)$$

where, $Q_{i,j}$ is the size of cache content that is cached on the assisted caching vehicle.

The other part is the delay generated by processing cache tasks on assisted cache vehicles. It can be expressed as:

$$t_{vehicle}(t) = \frac{\alpha_m(t)Q_{i,j}(t)s}{f_{vehicle}} \qquad (8)$$

where, $f_{vehicle}$ devotes the ability of assisted caching vehicles to handle caching tasks.

2.5 Optimization Description of the Problem

Our optimization objective of this paper is to minimize the maximum cache delay of edge collaborative cache nodes by jointly optimizing cache task scheduling, task offloading ratio, and content cache decisions. The optimization problem is:

$$\min \sum_{t=1}^{T} \sum_{n=1}^{N} \beta_n(t) \max\{t_{rsu}(i), t_{transmission}(i) + t_{vehicle,j}(i)\} \qquad (9)$$

s.t. (3), (4), (5)

$$\beta(t) \in \{0, 1\}, \forall t \in \{1, 2, ..., T\}, \forall n \in \{1, 2, ..., N\} \qquad (9a)$$

$$l_j \in \{(x(i), y(i)) | x(i) \in [0, X], y(i) \in [0, Y]\}, \forall t, \forall j \in \{k + 1, ..., N\} \qquad (9b)$$

$$\sum_{t=1}^{T} \sum_{n=1}^{N} \beta_n(t)Q_{i,j}(t) = Q \qquad (9c)$$

where, the constraint (5) is to ensure that each content cannot be cached in multiple edge collaborative cache nodes within the RSU coverage range; the constraint (9a) means that each content can be cached or uncached, represented by 1 or 0; the constraint (9b) ensures that the area where the assisted caching vehicle travels is within the RSU coverage range to meet the task offloading requirements; the constraint (9c) specifies all cache tasks to be processed and completed throughout the entire time period.

3 Our Optimization Algorithm for Caching Content Placement

Based on the vehicle aided edge cache architecture, we propose an optimization scheme for the content placement of the edge collaborative assisted cache based on the depth deterministic policy gradient (DDPG). The technical idea of the algorithm is to use DDPG to train an effective content placement optimization scheme in the edge collaborative assisted caching system, and optimize the system performance by determining the proportion of cached content on RSUs and assisted caching vehicles.

3.1 The Application of DDPG Algorithm

DDPG is a classic offline reinforcement learning algorithm. The goal of algorithm is to maximize the expected total discount reward by learning the optimal strategy, that is, the weighted sum of reward values obtained by all behaviors from the current state to a certain future state, which can be expressed as:

$$R_t = \sum_{i-t}^{T} \gamma^{i-t} r(s_i, a_i) \tag{10}$$

where, t is the time step size, γ is a reward discount factor, and $r(s_i,a_i)$ is the reward for taking action a_i in state s_i.

Considering that the system proposed in this paper must handle continuous action execution and cached content offloading transmission, we adopt a deterministic strategy and use the action value function Q to represent the long-term reward of the system. The Q function is represented by the Bellman equation as:

$$Q^u(s_i, a_i) = E[r(s_i, a_i) + \gamma Q^u(s_{i+1}, a_{i+1})] \tag{11}$$

where, μ devotes a certain strategy.

DDPG uses two different neural networks to approximate the strategy function $\mu(s |\theta^\mu)$ of the Actor network and the q value function $Q(s,a|\theta^Q)$ of the Critical network, respectively. We use an expected return function to measure the performance of a strategy μ, and define it as:

$$J = E_{R_i, S_i \sim E, A_i \sim \pi}[R_t] \tag{12}$$

where, E devotes the environments.

3.2 Content Placement Optimization

The current proportion of content offloaded to the assisted caching vehicle is not only related to the previous RSU, assisted caching vehicle and its environment, but also the vehicle scheduling and content proportion offloading decisions taken by the previous system. In order to solve the optimization problem (9), we modeled the non convex computing offloading problem as a Markov decision process (MDP). Therefore, the algorithm's design mainly involves state space **S**, action space **A**, state transition probability, and reward function **R**.

1) State space

In the autonomous driving vehicle assisted caching system, the state space of the requested content vehicle is jointly determined by the RSU, assisted caching vehicle, and its environment. The system state space $S(t)$ in time slot t can be described as

$$S(t) = \left(T(t), q(t), l_{V_{k+1}}(t), ..., l_{V_N}(t), Q_{remain}(t), Q_1(t), ..., Q_K(t) \right) \tag{13}$$

where, $T(t)$ devotes the remaining stayable time of the assisted cache vehicle at the t-th time slot; $q(t)$ devotes the position coordinate information of the RSU;l_{V_j}(t) devotes the position coordinate information of the assisted cache vehicle V_j within the RSU coverage range; $Q_{remain}(t)$ devotes the size of the remaining cache tasks that the system needs to complete throughout at the entire time period; $Q_K(t)$ devotes the cache task size randomly generated by RSU in the t-th time slot.

2) Action space

According to the current state of the system and the observed environment, the agent searches for assisted cache vehicles that meet time constraints within the RSU's coverage range and determines how many cache tasks need to be unloaded onto the assisted cache vehicles. Therefore, the action space of the system includes:

① $V_j(t)$: Assisted cache vehicles selected by agents for cache task offloading;
② $v\_vehicle(t)$: The driving speed of the assisted cache vehicle in the RSU range;
③ $\alpha_{Vj}(t)$: the offloading ratio of cache tasks from RSUs to assisted cache vehicles.

The federated system state a_t in the our system can be expressed as:

$$a_t = (V_j(t), v\_vehicle(t), ..., \alpha_{Vj}(t)) \tag{14}$$

3) Reward function

The behavior of intelligent agents is based on rewards, and appropriate rewards play a crucial role in the DDPG framework. The reward function is defined as:

$$r(t) = R(s_t, a_t) = -T_{delay}(t) \tag{15}$$

The cache task processing delay at time slot t is:

$$T_{delay}(t) = \sum_{i=1}^{N} \beta_n(t) \max\left\{ T_{R_i}(t), T_{transmission}(t) + T_{Vj}(t) \right\} \tag{16}$$

3.3 Implementation of Algorithm

Our algorithm is divided into a training phase and a testing phase. The training process of the algorithm is as follows:

Step1: After the previous training step, the Actor network $\mu(s|\theta^\mu)$ outputs action $a_t = \mu(s_t|\theta^\mu)$;

Step2: Adding behavioral noise n_t to construct the action space and obtain action $a_t = \mu(s_t|\theta^\mu) + n_t$, where n_t follows a Gaussian distribution $n_t \sim N(\mu_e, \sigma_{e,t}^2)$, μ_e is the mean, and $\delta_{e,t}$ is the standard deviation;

Step3: After executing action a_t, the agent observes a new state s_{t+1} and obtains an immediate reward r_t;

Step4: Store experience data tuples (s_t, a_t, r_t, s_{t+1}) in the experience buffer pool;

Step5: Randomly select N data groups (s_i, a_i, r_i, s_{i+1}) from the experience buffer pool to form a small batch sample, and input them into the participant network and Critical network;

Step6: Using small batch samples for training, the participant's target network $\mu' = (s|\theta^{\mu'})$ outputs actions $a_{i+1} = \mu'(s_{i+1}|\theta^{\mu'})$ to the critical target network $Q'(s, a|\theta^{Q'})$;

Step7: Based on the training of small batch samples and action a_{t+1}, the Critical network calculates the target value $y_i = r_i + \gamma(1\text{-}done)Q_{i+1}(s_{i+1},a_{i+1}|\theta^{Q'})$;

Step8: The Actor network $\mu(s|\theta^\mu)$ inputs actions into the Critical network to calculate the gradient of the action: $\nabla_a Q(s, a|\theta^Q)|_{s=s_i,a=\mu(s_i)}$, where the parameters can.

be derived by its own optimizer;

Step9: According to the calculated gradient, update the Actor network using approximate values: $\nabla_{\theta^\mu}\mu|s_i \approx \frac{1}{N}\nabla_a Q(s, a|\theta^Q)|_{s=s_i,a=\mu(s_i)}\nabla_{\theta^\mu}\mu(s|\theta^\mu)|_{s=s_i}$;

Step10: DDPG agent uses a small constant τ to update the criticism target network and participant target network: $\theta^{\mu'} = \gamma\tau + (1\text{-}\tau)\theta^{\mu'}, \theta^{Q'} = \tau\theta^Q + (1\text{-}\tau)\theta^{Q'}$;

In our scheme, vehicle scheduling and content placement proportion decisions are made by the DDPG based cache content placement optimization algorithm. The testing part of the algorithm is described as follows:

Algorithm 1: DDPG based cache content placement optimization algorithm

Input: Training cycle E; Test length N;Trained Actor network $\mu\,(\tilde{s} \mid \theta^\mu)$; Current status: $T(t),q(t),l_{V_{k+1}}(t),...,l_{V_N}(t),Q_{remain}(t),Q_1(t),...,Q_K(t)$.

Output: Reward R_i.

1: **for** each episodes e=1,2,...,E **do** //Perform Iteration
2:    Reset the simulation parameters of the edge collaborative assisted caching system and observe the current state s_1;
3:    **for** i=1,2,..., I **do**
4:       Select action: $a_t = \mu\,(\tilde{s} \mid \theta^\mu)$;
5:       Perform action a_t and obtain rewards R_i according to formula (15);
6:    **end for**
7: **end for**
8: return R_i.

4   Simulation Experiments and Performance Analysis

In this section, we verify the effectiveness and feasibility of the DDPG algorithm proposed in this paper. The experimental parameters are set as follows: the coverage area of RSU is 1 km, the number of assisted caching vehicles $K = 4$, the maximum driving speed $V_{max} = 50$m/s; channel power gain $\alpha_0 = 50$dB, channel transmission bandwidth $B = 1$MHz, the receiver's noise power is $\alpha_r = -100$dBm, the maximum transmission power is 0.1W, the cache task processing capabilities of RSU and assisted caching vehicles are $f_{RSU} = 0.6$GHz and $f_{vehicle} = 0.8$GHz, respectively.

In experiment, three baseline methods were used as comparative schemes: (1) Cache all content schemes in the local RSU (LocalOnly); (2) Cache all content on the assisted cache vehicle (VehicleOnly); (3) DQN based cache task offloading (DQN)[14]. During simulation experiments, the average values obtained from multiple runs of the comparative algorithm are used as performance comparison data.

In Fig. 2 and 3, we analyzed the convergence performance of the DDPG algorithm under different learning rates and effects of different discount factorsγon the convergence. When $\alpha_{actor} = 0.001$ and $\alpha_{critic} = 0.001$, the DDPG-based algorithm converges faster and quickly reaches a convergence state. When $\gamma = 0.6$, the convergence performance of algorithm is the best. Therefore, in the subsequent experiments, we selected $\alpha_{actor} = 0.001, \alpha_{critic} = 0.001$ and $\gamma = 0.6$.

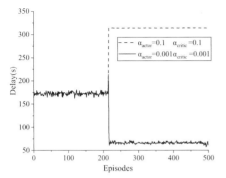

Fig. 2. Convergence of DDPG Algorithm at different learning rates

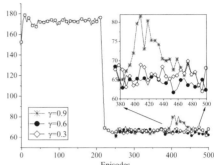

Fig. 3. Effects of different discount factors γ on the convergence of algorithm

Figure 4 compared the delay results obtained after convergence of different algorithms. It can be seen that after the algorithm converges, the DDPG algorithm has the smallest processing delay and better performance than other schemes. The reason is that DDPG algorithm can handle continuous action spaces and explore the spaces neglected in DQN algorithm, thus accurately finding the optimal unloading strategy and significantly reducing delay.

Figure 5 compared and analyzed the average processing delay of the four comparison algorithms for different assisted cache vehicles in the RSU coverage area. Experimental results show that except for the DQN algorithm, the average processing delay of the other three algorithms fluctuates very little. The proposed DDPG algorithm has the minimum delay, as it can find the optimal value and obtain the optimal control strategy in continuous actions.

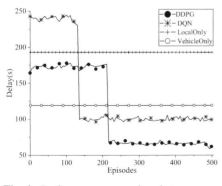

Fig. 4. Performance comparison between different algorithms

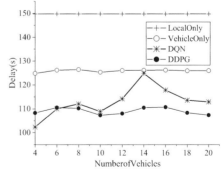

Fig. 5. Effect of different vehicle numbers on delay

Figure 6 observed the influence of different caching task processing capabilities of vehicles on algorithm convergence. When $f_{vehicle} = 0.8$Ghz, the delay of the proposed algorithm is significantly lower than that of $f_{vehicle} = 0.2$Ghz and $f_{vehicle} = 0.5$Ghz. This is because the stronger the ability of assisted cache vehicles to handle cache tasks, the faster the vehicles can process cache tasks, and RSUs will be more inclined to offload cache tasks to vehicles to reduce the burden of local slow execution tasks on RSUs.

In Fig. 7, we compared the average processing delay of four algorithms under.

different cache task sizes. Experimental results show that for the same task size, the proposed DDPG algorithm has the lowest delay. This is because the proposed algorithm can explore a continuous action space and take precise actions to accurately find the optimal unloading strategy, significantly reducing delay. In addition, the RSU local offloading algorithm and the algorithm for offloading all to assisted cache vehicles cannot fully utilize the computing resources of the entire system, so the processing latency is relatively large in normal times.

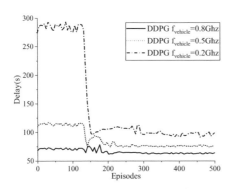

Fig. 6. Influence of different caching task processing capabilities of vehicles on algorithm convergence

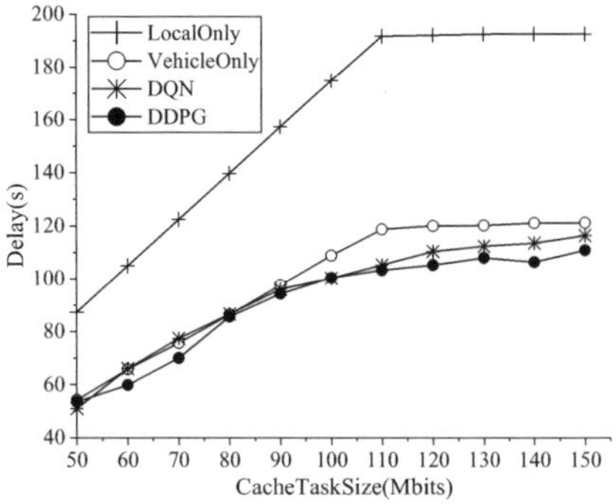

Fig. 7. Effect of different cache task sizes on delay

5 Conclusions

This paper proposes an edge collaboration assisted cache content placement optimization scheme using the trilateral collaboration between cloud center network, roadside unit RSU, and intelligent driving vehicles. This scheme collaboratively caches popular content on edge caching servers (vehicle caching devices) to reduce the personalized needs of content requesting vehicle users for the entire backbone network. Even in harsh network communication environments, vehicle users can obtain popular content from RSUs or the nearest assisted cache vehicles. By jointly optimizing vehicle scheduling, caching content task offloading ratio, and content placement decisions, the optimal caching mode is selected to determine the location and proportion of content placement, greatly reducing network congestion and transmission delay. Considering the non convexity of the original optimization problem, high-dimensional state space and continuous action space, a DDPG-based cache content placement optimization algorithm is proposed. The algorithm overcomes the disadvantage of frequent switching of RSUs by vehicles, ensures the integrity of content delivery, reduces the waste of RSU storage space, and improves the performance of vehicle content response delay, throughput, and scalability. The simulation results show that the proposed algorithm can quickly converge to the optimal, and has significant improvements in processing delay compared with other baseline algorithms.

References

1. Han, X., et al.: Reliability-aware joint optimization for cooperative vehicular communication and computing. IEEE Trans. Intell. Transp. Syst. **22**, 5437–5446 (2021). https://doi.org/10.1109/TITS.2020.3038558

2. Nan, Z.J., Jia, Y.J., Ren, Z., Chen, Z.C., Liang, L.: Delay-aware content delivery with deep reinforcement learning in internet of vehicles. IEEE Trans. Intell. Transp. Syst. **23**, 8918–8929 (2022). https://doi.org/10.1109/TITS.2021.3087833

3. Feng, B.Q., Feng, C.Y., Feng, D.Q., Wu, Y.P., Xia, X.G.: Proactive content caching scheme in urban vehicular networks. IEEE Trans. Commun. **71**, 4165–4180 (2023). https://doi.org/10.1109/TCOMM.2023.3277530

4. Zhuang, W.H., Ye, Q., Lyu, F., Cheng, N., Ren, J.: SDN/NFV-empowered future IoV with enhanced communication computing and caching. Proc. IEEE **108**, 274–291 (2020). https://doi.org/10.1109/JPROC.2019.2951169

5. Du, H., Leng, S.P., Wu, F., Zhou, L.Y.: A Communication scheme for delay sensitive perception tasks of autonomous vehicles. In: 2020 IEEE 20th International Conference on Communication Technology (ICCT), pp. 687–691, IEEE Press (2020). https://doi.org/10.1109/ICCT50939.2020.9295766

6. Yao, L., Xu, X.Y., Deng, J., Wu, G.W., Li, Z.Y.: A cooperative caching scheme for VCCN with mobility prediction and consistent hashing. IEEE Trans. Intell. Transp. Syst. **23**, 20230–20242 (2022). https://doi.org/10.1109/TITS.2022.3171071

7. Zhao, Z.L., Guardalben, L., Karimzadeh, M., Silva, J., Braun, T., Sargento, S.: Mobility prediction-assisted over-the-top edge prefetching for hierarchical VANETs. IEEE J. Sel. Areas Commun. **36**, 1786–1801 (2018). https://doi.org/10.1109/JSAC.2018.2844681

8. Zhang, K., Leng, S.P., He, Y.J., Maharjan, S., Zhang, Y.: Cooperative content caching in 5g networks with mobile edge computing. IEEE Wirel. Commun. **25**, 80–87 (2018). https://doi.org/10.1109/MWC.2018.1700303

9. Ayenew, T.M., Xenakis, D., Alonso, L., Passas, N., Merakos, L.: Demand-aware cooperative content caching in 5G/6G networks with MEC-enabled edges. IEEE Networking Lett. **4**, 118–122 (2022). https://doi.org/10.1109/LNET.2022.3192173

10. Lin, Z.J., Fang, Y., Chen, P.P., Chen, F., Zhang, G.H.: Modeling and analysis of edge caching for 6g mmWave vehicular networks. IEEE Trans. Intell. Transp. Syst. **24**, 7422–7434 (2023). https://doi.org/10.1109/TITS.2022.3147696

11. Qiao, G.H., Leng, S.P., Maharjan, S., Zhang, Y., Ansari, N.: Deep reinforcement learning for cooperative content caching in vehicular edge computing and networks. IEEE Internet Things J. **7**, 247–257 (2020). https://doi.org/10.1109/JIOT.2019.2945640

12. Li, Y.Y., Du, Q.H.: A cooperative multi-RSU caching scheme in vehicular networks with fountain codes. In: 2021 IEEE 32nd Annual International Symposium on Personal, Indoor and Mobile Radio Communications, pp. 1445–1450. IEEE Press, (2021). https://doi.org/10.1109/PIMRC50174.2021.9569666

13. Chen, J., Wu. H.Q., Yang. P., Lyu, F., Shen, X.M.: Cooperative edge caching with location-based and popular contents for vehicular networks. IEEE Trans. Veh. Technol. **69**, 10291–10305 (2020). https://doi.org/10.1109/TVT.2020.3004720
14. Jiang, F., Zhu, X.L., Sun, C.Y.: Double DQN based computing offloading scheme for fog radio access networks. In: 2021 IEEE/CIC International Conference on Communications in China, pp. 1131–1136. IEEE Press (2021). https://doi.org/10.1109/ICCC52777.2021.9580391

Design and Implementation of CNN-Based DTTB Image Intelligent Fault Diagnosis Platform

Qian Wei📷, Xianbin Xie(✉)📷, and Long Yu📷

Guangxi Radio and Television Broadcasting Technology Center, Nanning 530016, China
w48943668@qq.com

Abstract. Conventional methodologies for monitoring digital terrestrial television broadcasting (DTTB) predominantly concentrate on assessing transmitted signal indices and transport stream (TS) quality. Alarms are activated when the monitored parameters exceed predefined thresholds. Nonetheless, these monitoring techniques fall short in their capacity to comprehensively gauge the end-users' viewing experience. In a concerted effort to rectify this issue, an DTTB image intelligent fault diagnosis platform has been meticulously devised to oversee and diagnose broadcast quality through visual representation. The platform is meticulously engineered upon a Browser/Server (B/S) three-tier architecture and leverages convolutional neural network (CNN) as the core technology to establish a virtual digital artificial intelligence (AI) inspector. It harnesses streaming media services to facilitate swift and concurrent inspections of stations spanning the entire province. Experimental and deployment findings underscore the profound enhancement in monitoring precision and efficiency conferred by this platform, markedly augmenting the support capabilities of the DTTB operation and maintenance system.

Keywords: CNN · DTTB · TS · Intelligent fault diagnosis · Virtual digital AI Inspector

1 Introduction

Presently, the applications of AI within the radio and television sector encompass video restoration, content appraisal [1], virtual digital human [2], and program recommendations [3]. Nevertheless, a conspicuous lacuna exists in the realm of innovative AI application concerning the monitoring and diagnosis of DTTB broadcast quality.

The prevailing methods for monitoring radio and television broadcasting predominantly consist of signal surveillance and stream analysis. Signal monitoring entails the evaluation of parameters such as signal lock status, carrier levels, modulation error ratios (MER), bit error rates (BER), and carrier-to-noise ratios (CNR) [4]. Stream monitoring verifies key data bits according to TR101–290 standard. These methodologies, while indispensable, are disadvantaged by their inability to directly monitor the screen and thus fail to provide an immediate reflection of the user's viewing experience.

In a concerted endeavor to address this inherent challenge, this paper designs and implements the DTTB image intelligent fault diagnosis platform, thereby empowering computers to supplant the manual inspection of television (TV) fault images across various stations. While image fault diagnosis technology is a recognized tool in mechanical fault analysis [5] and medical image recognition, a significant research void remains in the context of TV fault image recognition. This paper ingeniously leverages a CNN-based image fault diagnosis methodology for the identification of TV fault images, effectively bridging the chasm in the application of this technology within the domain of DTTB monitoring.

2   Design of the DTTB Image Fault Diagnosis Platform

2.1   Key Hardware Device Topology of the Platform

As depicted in Fig. 1, the platform's hardware structure includes provincial monitoring centers and DTTB stations. In each station, the DTTB receiver receives and converts on-air signals into TS. These TS are sent to the central monitoring center's streaming media server. The AI server in the central hub, equipped with web services, background interfaces, and AI image recognition engines, monitors and diagnoses faults in live broadcast images from target stations.

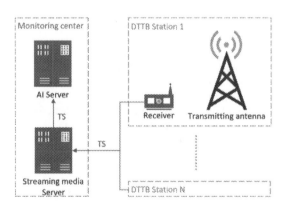

Fig. 1. Key hardware device topology of the platform.

2.2   Software Architecture of the Platform

The platform in this study adopts a B/S three-tier architecture, encompassing the user interface layer, the business logic layer, and the data access layer. As depicted in Fig. 2, the user interface layer is presented as a web page, structured with HTML and JavaScript code. It can be further divided into the display and control layers. The business logic layer encompasses functions such as fault diagnosis, anomaly detection, anomaly tracking, TV image evaluation, computational resource allocation, and alarm generation. This layer is implemented using Python code and serves as the virtual digital AI inspector within the platform. The data access layer manages alarm storage, fault TS storage, and data storage.

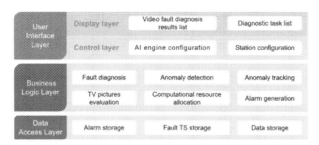

Fig. 2. Software Architecture of the Platform.

2.3 The Virtual Digital AI Inspector

The core function of the platform involves emulating the role of an inspector, responsible for cyclically monitoring and diagnosing live TV images from DTTB stations throughout the province. We have christened this essential component as the "Virtual Digital AI Inspector". As depicted in Fig. 3, the program flow of this virtual digital AI inspector unfolds as follows: It commences by procuring the diagnostic task from the designated task list, encompassing essential information such as the target station's ID and TV program number. Subsequently, based on this task information, the target TS is acquired from the streaming media server. The inspector then undertakes the tasks of decoding, capturing key frame images, and conducting identification and diagnosis of these images. Finally, the outcomes and any ensuing alarms are disseminated to the platform's data interface.

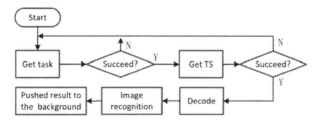

Fig. 3. Program flow chart of virtual digital AI inspector.

2.4 Task Allocation Strategy

Presently, the platform deploys 60 concurrent virtual digital AI inspector processes. Within the province, there exist 87 pivotal stations, each broadcasting 12 TV programs, resulting in a total of 1,044 programs. Given that outfitting each program with an AI inspector is infeasible, the AI inspector is constrained to inspecting programs from each station in a sequential manner. Thus, the formulation of an effective patrol task allocation strategy becomes crucial to ensure the swiftest and most precise response to broadcast anomalies, given the constraints of limited monitoring resources.

Tasks are classified into three categories: normal inspection tasks, anomaly tracking tasks, and manually assigned tasks. As depicted in Fig. 4, normal inspection tasks are systematically appended to the normal task queue. On the other hand, anomaly tracking tasks and manually added tasks are directed to the urgent task queue. AI inspectors accord priority to tasks from the urgent task queue for immediate execution.

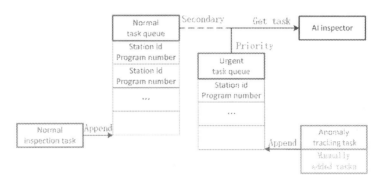

Fig. 4. The strategy for task allocation.

2.5 Fault Image Diagnosis

The term "fault image diagnosis" pertains to the identification and categorization of live television images into four distinct classes: normal images, mosaic images, black screen images, and still frame images. An array of aberrant images is exemplified in Fig. 5. The image recognition procedure commences with the preprocessing of input images. Subsequently, the preprocessing outcomes are successively channeled into the black screen feature recognition algorithm, the still frame feature recognition algorithm, and the mosaic recognition algorithm (CNN-based), culminating in the determination of the ultimate diagnosis. Figure 5 provides various examples of fault images: (a) represents a mosaic anomaly image, (b) illustrates a black screen anomaly image, and (c) portrays a still frame anomaly image.

Image Preprocessing. The typical TV program resolution for DTTB stands at 720*576. Experimental observations have indicated that low-quality image elements, such as mosaics, exhibit a pronounced likelihood of manifesting in the lower and middle sections of the program image. Feature extraction from this specific region is also instrumental in identifying black screen faults and still frame anomalies within the image. As depicted in Fig. 6, the preprocessing methodology devised in this platform entails the direct extraction of a 448*448 resolution image from the lower and middle sections

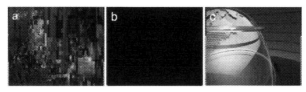

Fig. 5. Diverse fault anomaly images. a: mosaic anomaly image; b: black screen anomaly image; c: still frame anomaly image.

of the program image. Subsequently, this image is scaled down to 224*224 resolution before being sequentially fed into the black screen image, still frame, and mosaic image recognition algorithms. This approach serves a dual purpose: firstly, it diminishes interference from redundant image data, and secondly, it reduces computational load on CNN and other computer vision algorithms. The image preprocessing strategy outlined in this paper yields expedited processing speed, all while preserving the integrity of recognition accuracy.

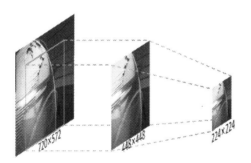

Fig. 6. Image preprocessing.

Mosaic Detection and Diagnosis. Among the DTTB fault images, mosaic anomalies pose a considerable challenge in terms of identification. As illustrated in Fig. 7, mosaic anomalies exhibit diverse characteristics in terms of color, shape, area, and positioning. Their patterns are inherently random, making it arduous to articulate and encapsulate these traits with conventional computer vision recognition algorithms. Hence, this paper endeavors to address this challenge by employing a deep CNN model within the framework of machine learning algorithms. The training and selection of CNN models emerge as pivotal and technically demanding tasks. The efficacy of the entire system hinges on the judicious choice and training of CNN models.

Comparison of CNN Models. DTTB stations typically broadcast around 12 TV programs, necessitating the monitoring of up to 1,044 programs within the province. Consequently, this paper places a premium on opting for a lightweight network model that demands fewer computational resources. By consulting the ImageNet dataset's recognition outcomes, several promising lightweight network models have been identified,

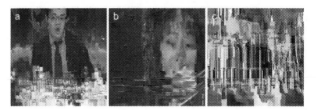

Fig. 7. Samples of mosaic anomaly images.

including MobileNetV2 [6], Squeezenet1.0 [7], Resnet18 [8], and ShuffleNetV2 [9]. Figure 8 illustrates the error rates predicted by these lightweight network models on the ImageNet test set. Upon careful evaluation, it becomes evident that MobileNetV2 outperforms the other lightweight network models in terms of classification performance, both in the Top-1 and Top-5 categories within the ImageNet dataset. Consequently, this paper elects to adopt the MobileNetV2 model.

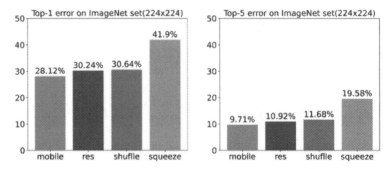

Fig. 8. The error rates of lightweight network models on the ImageNet test set.

Datasets and Training. To amass a comprehensive array of program image samples, this paper has developed a TS degradation software capable of generating diverse mosaic fault images on the player terminal through various degradation methods. These images collectively constitute a dataset. Recognizing the substantial differences that exist across various TV programs, and with the aim of enhancing the model's generalization capacity, this study has meticulously gathered samples from a wide spectrum of TV programs. The image samples have been manually classified into two categories: normal images and mosaic images. Consequently, a dataset named "TV_Image set" has been formulated, comprising 485 mosaic images and 472 normal images.

In the course of this research, the TV_Image set has been partitioned into training and validation sets employing a 5-fold cross-validation approach. The hyperparameters for model training are as follows: *batch size*: 16, *epochs*: 50, *learning rate*: 0.001, *momentum*: 0.9, *weight_decay*: 0, *width_multiplier*: 1.0, *optimizer*: Stochastic Gradient Descent (SGD), and *loss_function*: CrossEntropyLoss. In binary classification applications, the cross-entropy loss function is articulated as follows in formula (1).

$$L = [ylog\hat{y} + (1-y)log(1-\hat{y})] \tag{1}$$

Based on the TV_Image set assembled in this study and following the establishment and training of the aforementioned hyperparameters, the prediction results of the network model are presented in Table 1.

Table 1. The performance of MobileNetV2 after training with the TV_Image dataset.

Net	Precision	Recall	F1	Accuracy
MobileNetV2	99.23%	96.39%	97.79%	98.03%

Black Screen Recognition. A black screen fault pertains to a TV image that remains entirely black for an extended period, rendering the screen devoid of any meaningful visual information. In this state, the RGB pixel value matrix of the image comprises entirely zero values.

In comparison to full-pixel image detection, dilated convolution offers a computational advantage. As depicted in Fig. 9: a) Represents a two-dimensional void matrix with dimensions 3*3 and a dilation rate of 1; b) Depicts a two-dimensional void matrix with dimensions 5*5 and a dilation rate of 2; c) Illustrates a two-dimensional void matrix with dimensions 7*7 and a dilation rate of 3.

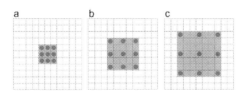

Fig. 9. The schematic diagram of void matrices.

The black screen detection algorithm devised in this study initiates by generating a three-dimensional void matrix with dimensions (h, w, c) to serve as an operator. The dilation rate of the void matrix is denoted as 'd', implying that the matrix elements consist solely of 0s and 1s. Between any two '1' elements, there are 'd-1' zeros. In this algorithm, the parameters for the three-dimensional void matrix are as follows: $h = 224$, $w = 224$, $c = 3$, and $d = 4$.

The algorithm proceeds by performing convolution operations between the three-dimensional void matrix and multiple images obtained from detection. If the resultant values are all less than or equal to the predefined threshold, it is deemed indicative of a black screen fault. Given that the R, G, and B values of each pixel within a black screen are uniformly set to 0, the convolution operation should yield a value of 0 as well. Consequently, the threshold is established at 0.

Still Frame Recognition. The still frame fault is characterized by a prolonged freeze in the TV picture, resulting in a static image for an extended duration. To detect this condition, when consecutive input frames remain in a still frame state, the pixel matrices of the image frames before and after can be subtracted. This subtraction process will yield a 224*224*3 matrix containing all-zero values. This resultant matrix can then be input into the black screen recognition algorithm. If the values obtained from this operation are all less than 0, it serves as an indicator of a still frame fault.

3   Test Experiment

3.1   Experimental Design

Existing platforms typically rely on the standard TR101–290 for evaluating the broadcast quality based on stream transmission parameters, but their accuracy is generally modest. To assess the monitoring and diagnostic prowess of the platform developed in this study in comparison to other similar platforms, an experiment was meticulously devised. For this experiment, a video file encompassing various prevalent DTTB image faults was assembled. VideoLAN Client (VLC) media player was utilized to read this video file, convert it into a TS, and subsequently feed it to each respective platform for fault identification. The experimental design is illustrated in Fig. 10.

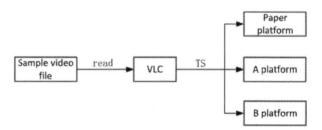

Fig. 10.   The design of a comparative experiment.

3.2   Sample Video Compilation for the Experiment

The test video samples have been skillfully assembled by concatenating segments featuring three prevalent image faults: mosaic faults, black screen faults, and still frame faults. Each segment has a duration of one minute. The mosaic fault test encompasses both a mild mosaic segment and a severe mosaic segment. The black screen fault test incorporates one genuine black screen segment and two simulated black screen segments. The still frame fault test comprises an authentic still frame segment and two simulated still frame segments. The composition of the test video content is detailed in Table 2.

3.3 Experimental Results

The experimental outcomes are delineated in Table 2, exhibiting four distinct prediction and recognition categories: true positive (TP) and true negative (TN) denote accurate identifications, while false positive (FP) signifies the generation of false alarms, and false negative (FN) denotes missed alarms.

Remarkably, despite both platform A and platform B possessing mosaic fault diagnosis capabilities, they failed to detect mosaic faults during the actual testing. Both platforms generated false alarms in response to simulated still frame faults. In the case of platform B, it also triggered false alarms in response to simulated black screen faults.

Conversely, the platform developed in this study achieved both correct identification and diagnosis in this experimental test, marking a notable contrast to the performance of other platforms.

Table 2. Results of comparative testing with similar platforms

Test items	Video segment type	Paper platform	A platform	B platform
Mosaic Faults	Mild Mosaic	TP ($\checkmark$)	FN ($\times$)	FN ($\times$)
	Severe Mosaic	TP ($\checkmark$)	FN ($\times$)	FN ($\times$)
Black Screen Faults	True Black Screen	TP ($\checkmark$)	TP ($\checkmark$)	TP ($\checkmark$)
	False Black Screen 1	TN ($\checkmark$)	TN ($\checkmark$)	FP ($\times$)
	False Black Screen 2	TN ($\checkmark$)	TN ($\checkmark$)	TN ($\checkmark$)
Still Frame Faults	True still frame	TP ($\checkmark$)	TP ($\checkmark$)	TP ($\checkmark$)
	False still frame 1	TN ($\checkmark$)	FP ($\times$)	FP ($\times$)
	False still frame 2	TN ($\checkmark$)	TN ($\checkmark$)	TN ($\checkmark$)

4 Deployment and Performance

The platform is now deployed in 87 pivotal stations, providing coverage to 14 major cities. Over the course of more than eight months, it has effectively identified and diagnosed a total of 1,703 program broadcast faults. These faults encompassed 985 mosaic faults, 557 black screen faults, and 158 still frame faults, along with three false alarms.

Notably, these faults were pinpointed and diagnosed within minutes of their occurrence, facilitating timely resolution. In comparison to manual inspection methods, AI-driven inspections have accelerated by a factor of 36, with the number of inspected programs increasing sixfold. The volume of work accomplished now equals that of 216 human inspectors.

The introduction of this platform has led to a significant enhancement in the efficiency of DTTB image fault diagnosis, ultimately resulting in a reduction in labor-intensive monitoring costs.

5 Conclusion

This paper successfully introduces and implements a CNN-based DTTB image intelligent fault diagnosis platform. The platform proficiently identifies and diagnoses mosaic faults, black screen faults, and still frame faults utilizing the MobileNetV2 network model and the dilated convolution algorithm. Through a series of experiments and deployments, this platform has demonstrated its innovation and practicality.

The platform significantly enhances monitoring accuracy and efficiency while also reducing labor costs and minimizing false alarms. Moreover, it plays a pivotal role in fortifying the support capabilities of the intelligent radio and television operation and maintenance system.

However, an existing limitation of the platform is the prolonged diagnostic cycle, mainly due to the extensive number of stations and programs requiring inspection. To address this, future optimizations will be directed towards algorithm and platform architecture enhancements to further expedite diagnostic efficiency. Potential measures include augmenting the number of AI server Graphic Processing Units (GPUs), enhancing server network bandwidth, and transferring decoding, screenshots, and preprocessing processes to the primary control computers of each station, among other strategies.

References

1. Sun, J.: The application of artificial intelligence technology in radio and television engineering. Video Eng. **47**(7), 62–64 (2023)
2. Wan, X., Jiao, J.L.S., Xia, M.G.T.: The application of artificial intelligence-based virtual digital human in radio and television engineering. Video Eng. **47**(7), 172–174 (2023)
3. Chen, X.: Application of artificial intelligence in radio and television program recommendation system. Video Eng. **47**(7), 163–165 (2023)
4. Li, J.: Research on DTTV monitoring technology. Radio TV Broadcast Eng. **42**(10), 120–122 (2015)
5. Neupane, D., Seok, J.S.: Bearing fault detection and diagnosis using case western reserve university dataset with deep learning approaches: a review. IEEE Access **8**, 93155–93178 (2020)
6. Sandler, M., et al.: Mobilenetv2: inverted residuals and linear bottlenecks. In: 31st Meeting of the IEEE/CVF Conference on Computer Vision and Pattern Recognition, CVPR 2018, pp. 4510–4520. IEEE Computer Society, Salt Lake City, UT, United states (2018)
7. Iandola, N., et al. SqueezeNet: AlexNet-level accuracy with 50x fewer parameters and< 0.5 MB model size. arXiv preprint: arXiv:1602.07360 (2016)
8. He, K., et al.: Deep residual learning for image recognition. In: 29th IEEE Conference on Computer Vision and Pattern Recognition, CVPR 2016, pp. 770–778. IEEE Computer Society, Las Vegas, NV, United states (2016)
9. Ma, N., et al.: Shufflenet V2: practical guidelines for efficient CNN architecture design. In: Ferrari, V., Hebert, M., Sminchisescu, C., Weiss, Y. (eds.) Computer Vision – ECCV 2018. ECCV 2018. Lecture Notes in Computer Science(), vol. 11218, pp. 122–138. Springer, Cham (2018). https://doi.org/10.1007/978-3-030-01264-9_8

Smart Power Safety Hazard Inspection System Based on YOLOv7

Yiheng Liang[1], Xiaoming Li[2], and Zhenrong Deng[1,2(✉)]

[1] School of Computer and Information Security, Guilin University of Electronic Technology, Guilin 541004, China
799349175@qq.com

[2] Guangxi Key Laboratory of Image and Graphic Intelligent Processing, Guilin University of Electronic Technology, Guilin 541004, China

Abstract. Power safety is closely related to people's well-being. Electric power sectors regularly inspect and maintain power lines to guarantee people's safe and stable use of electricity, and the current mainstream inspection method is to use UAVs for power inspection. Traditional inspection and detection methods have many shortcomings, such as high labor costs, slow inspection speed, low detection efficiency, and single detection targets. In this paper, a set of smart power inspection systems based on YOLOv7 was designed to check the common potential security risks of power lines such as bird nests, insulators, garbage, and hardware. After the detection, the detection report could be generated. The system adopted the YOLOv7 target detection algorithm. The accuracy rate of all detection targets was up to 93.9%, and the mAP @.5% reached 95.9%. Therefore, the algorithm could greatly improve the detection efficiency and greatly facilitate the electric power sectors to check power lines, so it has a high use value.

Keywords: Power inspection · Target detection · YOLOv7

1 Introduction

As the industrialization process progresses, the people's demand for electricity is also increasing, and their requirements for the reliability, safety, and stability of power lines are also increasing. Since power safety is closely related to people's well-being, thousands of power towers and ultra-high-voltage power lines have been built in China. Power lines and power towers are the key media for transmitting power [1]. Nevertheless, given that the majority of high-voltage lines are built in the wild and are exposed to the weather for a long time, there are numerous potential security risks, such as bird nests, insulator shedding, garbage on lines, and damper damage [2–4]. Electric power sectors are required to regularly inspect and maintain power towers and power lines to ensure that people can use electricity stably. However, there are tens of thousands of kilometers of power lines in China, and if only relying on manual inspection, there will be a large workload and low efficiency, and they are also susceptible to harsh environments. In the "Twelfth Five-Year" smart grid development strategy, the state proposed to improve the intelligence standard of the power grid. Consequently, it is necessary to replace part of the manual inspection with smart algorithms and equipment in order to respond to the national strategy and meet people's happy lives.

As time goes by, UAV technology is widely used in various aspects of people's lives, and electric power sectors have also begun to use UAVs to inspect power lines[5]. The combination of UAV technology and existing communication technology can realize automatic cruise. Compared with manual and helicopter inspection schemes, UAV inspection is characterized by low maintenance costs and high inspection efficiency [6, 7]. A large number of photos will be taken when the inspection is conducted by UAVs, and then the detection software will identify these photos to determine whether there are targets to be detected.

To this end, a set of smart power inspection systems based on YOLOv7 was designed to detect whether there are power safety hazard targets in the photos taken by UAVs during inspection. Compared with the traditional detection methods, this system has many advantages, such as high detection accuracy, fast detection speed, and diversified detection targets.

2   YOLOv7 Detection Model

YOLOv7 is one of the target detection algorithms with the best comprehensive performance[8]. The YOLOv7 model was proposed by the former staff of the YOLOv4 team. The network of the YOLOv7 is mainly composed of Input, Backbone, and Head.

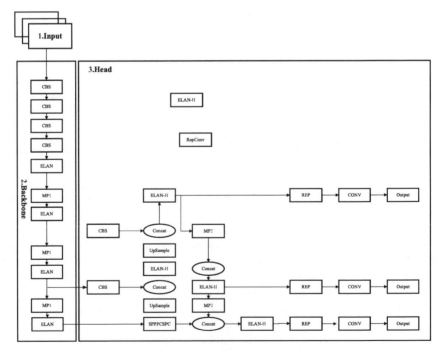

Fig. 1. Structure of YOLOv7 network

The Input reuses the logic of YOLOv5 as a whole. The role of Input is to preprocess the input image and enhance the data. YOLOv7 uses Mosaic data enhancement [9], adaptive anchor frame calculation, and adaptive image scaling. Backbone mainly uses CBS, ELAN, and MP structures. The ELAN structure, designed based on ELAN, uses expand, shuffle, and merge cardinality to realize the ability to continuously enhance network learning ability without destroying the original gradient path. The MP structure is composed of conventional convolution and maxpool dual paths, and it aims to increase the model's ability to extract and fuse features. The Head is composed of the SPPCSPC module, PAFPN structure, and REP module (Fig. 1).

3 Research on Target Detection Algorithm of Hidden Power Safety Hazard

3.1 Experimental Environment and Parameters

The operating system is Windows 10, the processor is Intel(R) Xeon(R) Bronze 3104 CPU @ 1.70GHZ, the memory is 32G, and the graphics card is NVIDIA GeForce RTX 3060. The CUDA version is 12.1, the torch version is 2.0.1, and the Python version is 3.10. The batch-size is set to 16, the epoch is set to 100, and the remaining parameters are default values.

3.2 Dataset Selection and Processing

During the UAV inspection of a power line in Guangxi, 300 photos of bird nest data were captured, encompassing 290 photos of Insulator defect data, 70 photos of garbage data, and 300 photos of hardware data. Then, the data set was expanded by data enhancement methods such as translation, deformation, random cropping, and random rotation. Finally, there were 1178 pictures in the bird nest picture training set, and 112 were tested; 1157 pictures in the missing insulator picture training set, and 117 were tested; 280 pictures in the garbage picture training set, and 25 were tested; 1153 pictures in the hardware picture training set, and 113 were tested. The LabelImg annotation tool was adopted to label the dataset. Some of the labeled data sets are shown in Fig. 2 below. Where Fig. 2(a) is the bird's nest, Fig. 2(b) is the insulator, Fig. 2(c) is the hardware, and Fig. 2(d) is garbage.

3.3 Evaluation Indexes

Precision, Recall, and average accuracy mAP@0.5 were used as evaluation indexes to display the model effect[10].

The calculation formula of Precision is as follows:

$$\text{Pr}ecision = \frac{TP}{TP + FP} \tag{1}$$

The calculation formula of Recall rate is as follows:

$$\text{Re}call = \frac{TP}{TP + FN} \tag{2}$$

(a)Bird nest

(b)Insulator

(c)Various hardware

(d)Garbage

Fig. 2. Part of labeled data

When calculating mAP, the average accuracy of a single category should be calculated first, and the calculation formula is as follows:

$$AP = \int_0^1 p(r)dr \tag{3}$$

$$mAP = \frac{1}{n}\sum_k^n AP_k \tag{4}$$

where TP represents true positive; FP represents false positive; FN represents false negative; P (r) represents the curve drawn with recall rate and accuracy; n represents the total number of classes for target detection; mAP @ 0.5 is the average mAP when the intersection ratio is 0.5.

3.4 Analysis of Experimental Results

The test data sets in different detection targets were taken to verify the model. The accuracy-recall curves of different detection targets are shown in Fig. 3 below.

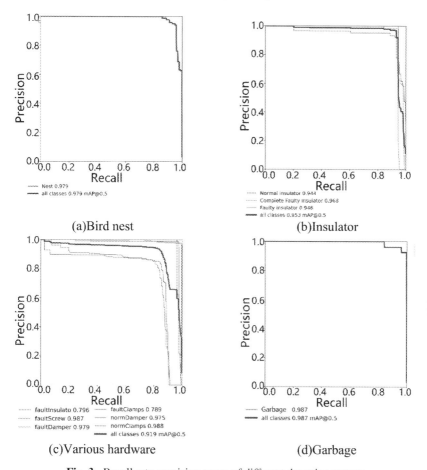

Fig. 3. Recall rate-precision curve of different detection targets

From Fig. 3, a and b, it can be seen that the model has high detection accuracy for large targets such as bird nests and garbage, with the detection accuracy of map@.5 for bird nests reaching 97.9% and map@.5 for garbage detection reaching 98.7%. From Fig. 3, c and d, it can be seen that the detection accuracy of the model for dense small targets such as insulator and hardware is slightly lower. The detection rate of all types of insulators map@.5 was 95.3%, and that of all types of hardware map@.5 was 91.9%.

To compare the effect of the model, the YOLOv7 and YOLOv5 models were used to compare the test sets of different detection target data sets, respectively. The comparison results are shown in Table 1.

The comparison shows that the average accuracy of YOLOv7 is 3.3% higher than that of YOLOv5, and the average recall rate is 1.5% higher than that of YOLOv5. The average mAP@.5% of all detection targets is up to 95.9%, which is 1.1% higher than that of YOLOv5, proving that the detection method used in this paper has better detection ability.

Table 1. Comparative experiment of YOLOv5 and YOLOv7

	YOLOv5			YOLOv7		
Detection target	Accuracy%	Recall rate%	mAP@.5%	Accuracy%	Recall rate%	mAP@.5%
Bird nest	90.2	96	97.8	94	95.9	97.9
Insulator	92.6	92.2	92.0	96.7	93	95.3
Garbage	87.3	99	99	92.5	99	98.7
Hardware	91.4	87.9	90.7	92.4	93	91.9
Average value	90.3	93.7	94.8	93.9	95.2	95.9

4   System Design and Implementation

4.1   Overall Design of System

The design of the smart power inspection system based on YOLOv7 aims to design a set of safe, efficient, convenient, and concise systems that can allow users to find bird nests, hardware, missing insulators, garbage, and other power safety hazards more quickly.

The system interface was developed and designed by Python language combined with PyQt5. The smart power inspection system based on YOLOv7 mainly includes the inspection target detection module, the search and detection result display module, and the detection report generation module. It mainly consists of an interaction layer, a logic processing layer, and a data layer. The interaction layer is open to users and can operate relevant data, freely select detection targets, and display relevant information to users. The logic processing layer receives the request from the interaction layer and completes the corresponding business logic processing operation. The operation performed by users must be processed by the business logic layer before it can be displayed to the users, including the function implementation in the target detection module, file searching and detection result statistics, and detection report generation. The data layer includes the processing of the data to be detected, the data generation after detection, and the storage of the detection report. The system architecture is shown below (Fig. 4).

4.2   Function Module Design and Effect Display

The function module mainly includes the target detection module, the search and detection result display module, and the detection report generation module. The specific design of the system module is as follows (Fig. 5):

(1) **Target detection module:** It mainly detects various common power safety hazards such as bird nests, hardware, garbage, and insulators in power inspection. During the detection, statistics and detection will be performed according to different lines and different power towers. The system will call sub-threads to query and receive the running status of the program and update the front-end progress bar in real

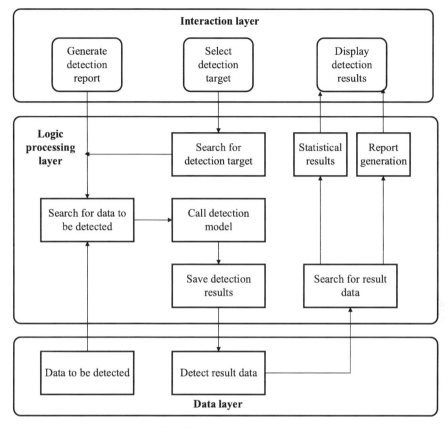

Fig. 4. System architecture

time. After the detection starts, the system will call the YOLOv7 target detection algorithm trained in the back end to automatically mark the targets contained in the photo, including bird nests, hardware, garbage, and insulator defect, and then save the photos containing the targets in the specified folder.

(2) **Search and detection result display module:** It counts the number of photos detected and the number of power towers containing targets, and display the photos containing the detection targets on the front-end interface so that users can view the photos freely.

(3) **Detection report generation module:** It mainly realizes report generation, counts the detection results and the photos with targets, and summarizes the statistical results to generate the report documents.

4.3 System Operation Effect

Running the program, after entering the system, target detection categories and the target to be detected, such as bird nest detection, and hardware detection can be chosen freely. After the selection, the folder of photos to be detected and the output path of the detection result can be chosen. The running effect is as follows (Fig. 6):

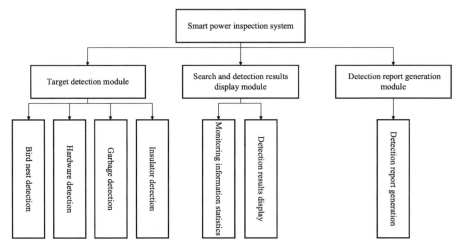

Fig. 5. Structure of system module function

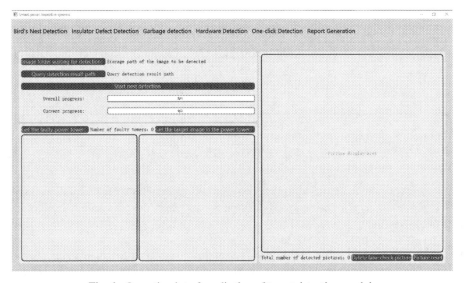

Fig. 6. Operation interface display of target detection module

After completing the detection, the detection results will be output to the front end, and the number of photos detected and the number of fault power towers, including target photos, will be automatically counted. The specific effect is as follows (Fig. 7):

After completing the detection or when there has already been local data containing the detection results, report generation can be selected, and the report generation module can freely choose to generate what kind of detection target report. The specific effect is as follows (Fig. 8).

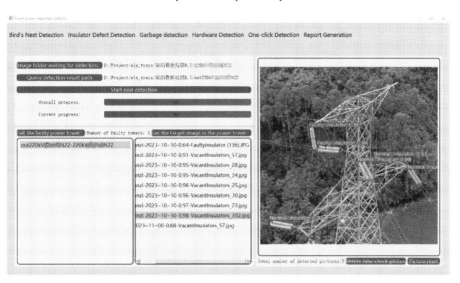

Fig. 7. Display of detection results interface

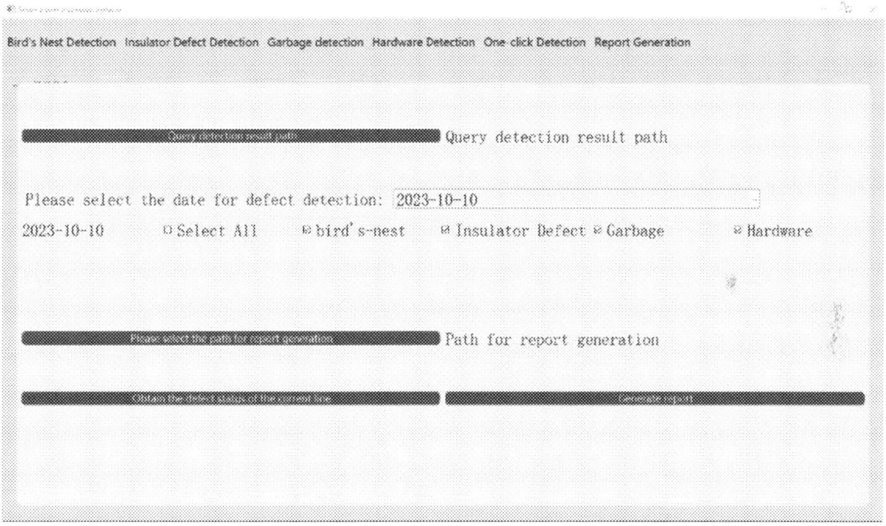

Bird's Nest Detection Insulator Defect Detection Garbage detection Hardware Detection One-click Detection Report Generation

Query detection result path

Please select the date for defect detection: 2023-10-10

2023-10-10   □ Select All   ⊠ bird's-nest   ⊠ Insulator Defect ⊠ Garbage   ⊠ Hardware

Path for report generation

Generate report

Fig. 8. Display of detection report generation interface

Taking the detection of bird nests as an example, the following diagram displays the style of the generated report, including line information, date information, fault details, and list (Fig. 9).

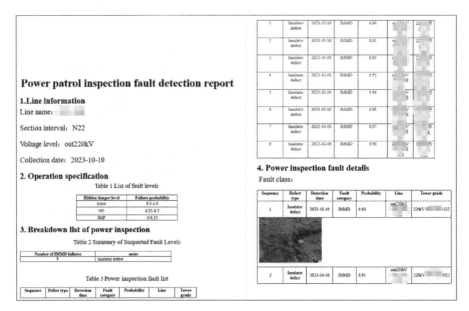

Fig. 9. Detection report display

5 Conclusion

In this paper, a set of smart power inspection systems based on YOLOv7 was designed to rapidly detect various power safety hazards, including but not limited to multiple bird nests, insulator defect, garbage, and hardware. The average accuracy of all detection targets is 93.9%, and the mAP @.5% is up to 95.9%. Upon the completion of the detection, a detection report can be generated directly. The design of the smart inspection system can significantly improve the detection efficiency of power line targets and guarantee the safe and stable power consumption of the people.

Acknowledgments. This work was partially supported by the National Natural Science Foundation of China (No.6202780103), by the Guangxi Science and Technology Project (No.AB22035052), Guangxi Key Laboratory of Image and Graphic Intelligent Processing Project (Nos. GIIP2211, GIIP2308).

References

1. Li, N., et al.: Detection of defects in transmission line based on the unmanned aerial vehicle image recognition technology. Electr. Des. Eng. **27**(10), 102–106 (2019)
2. Sheng, L., et al.: Research and implementation of birds and floating objects target detection technology in transmission lines. In: 2022 IEEE 10th Joint International Information Technology and Artificial Intelligence Conference (ITAIC), vol. 10, pp. 2528–2532). IEEE (2022)

3. Liu,Y., Dong, S., Chen, Z., Lv, D., Yu, B.: Research on bolt visual recognition of high voltage transmission line shockproof hammer resetting robot. In: 2022 IEEE 6th Information Technology and Mechatronics Engineering Conference (ITOEC), Chongqing, China, pp. 1329-1333 (2022).https://doi.org/10.1109/ITOEC53115.2022.9734637

4. Cao, G., et al.: Research on small-scale defect identification and detection of smart grid transmission lines based on image recognition. In: 2021 IEEE 4th International Conference on Automation, Electronics and Electrical Engineering (AUTEEE), Shenyang, China, pp. 423–4272021. https://doi.org/10.1109/AUTEEE52864.2021.9668671

5. Jinju, Q., Zhengwei, H., Lin, Y.: Application of image processing technology in UAV power line inspection. Electr. Technol. Softw. Eng. **15**, 72–73 (2017)

6. Yao, P.F., Geng, B., Yang, M., Cai, Y.M., Wang, T.: Research on technology of autonomous inspection system for UAV based on improved Yolov4. In: 2020 5th International Conference on Mechanical, Control and Computer Engineering (ICMCCE), Harbin, China, pp. 664–668 (2020). https://doi.org/10.1109/ICMCCE51767.2020.00146

7. Liu,K., Zhong, L.: Object detection of UAV power line inspection images based on federated learning. In: 2022 IEEE 5th International Electrical and Energy Conference (CIEEC), Nangjing, China, pp. 2372-2377 (2022). https://doi.org/10.1109/CIEEC54735.2022.9846340

8. Wang, C.Y., Bochkovskiy, A., Liao, H.Y.M.: YOLOv7: Trainable bag-of-freebies sets new state-of-the-art for real-time object detectors. arXiv 2022. arXiv preprint: arXiv:2207.02696 (2022)

9. Bochkovskiy, A., Wang, C.Y., Liao, H.Y.M.: YOLOv4: optimal speed and accuracy of object detection (2020). https://doi.org/10.48550/arXiv.2004.10934

10. Padilla, R., Netto, S.L., Da Silva, E.A.: A survey on performance metrics for object-detection algorithms. In: 2020 International Conference on Systems, Signals and Image Processing (IWSSIP). IEEE (2020)

A Defect Detection Method of Drainage Pipe Based on Improved YOLOv5s

Yusheng Sun, Weibo Zhong[✉], Yuhua Li, Xiao Cui, Zhe Zhao, and Weihai Chen

Zhengzhou University of Light Industry, 136 Science Avenue, Zhengzhou 450000, China
2325882181@qq.com

Abstract. In response to the existing challenges associated with manual interpretation, low efficiency, high leakage, and misdetection rates in detecting defects in urban underground drainage pipes, this study presents a defect detection method of drainage pipe based on improved YOLOv5s. The proposed method improves the detection of large target defects and reduces the leakage detection rate by increasing a deep target detection layer. Additionally, the introduction of deformable convolutional networks (DCN) allows for more accurate feature extraction from targets with complex shapes. Furthermore, the loss function is improved by employing MPDIoU as the bounding box loss function, which not only accelerates the convergence speed of bounding boxes but also enhances target recognition accuracy. Experimental results demonstrate that the improved model surpasses the performance of the original YOLOv5s, exhibiting an improvement of 3.8% in accuracy, 1.9% in recall, and 2.1% in average precision. Additionally, the proposed method achieves an impressive inspection speed of up to 54.64 FPS (frames per second), enabling real-time and efficient drain defect detection. This method is highly practical as it provides technical support for the future deployment of CCTV pipeline robots.

Keywords: Drainage Pipe · Defect Detection · YOLOv5s · DCN · MPDIoU

1 Introduction

Drainage pipes within urban environments constitute an integral facet of the city's infrastructure. Their primary function lies in efficiently managing the treatment of sewage and rainwater, thereby ensuring unimpeded urban thoroughfares, facilitating convenient living conditions, and fostering sustainable urban development and social stability. Nonetheless, the presence of pipeline defects, such as cracks, misalignments, disconnections, and obstructions, has become apparent due to factors such as pipeline aging, urban waterlogging, and the execution of road and bridge construction projects. These defects bear the potential to instigate environmental contamination, waterlogging, and traffic predicaments, as well as pose a substantial threat to the structural integrity of buildings. Furthermore, they may engender nuisances such as odors and pest infestations, thereby significantly impinging upon the progress of cities and the quality of life for its inhabitants. Hence, it becomes imperative to promptly and effectively undertake pipeline defect detection measures to safeguard the integrity of urban construction.

© The Author(s), under exclusive license to Springer Nature Singapore Pte Ltd. 2024
D.-S. Huang et al. (Eds.): ICAI 2023, CCIS 2015, pp. 144–155, 2024.
https://doi.org/10.1007/978-981-97-0827-7_13

Presently, the common methods employed for unmanned detection of drainage pipe defects include sonar detection, periscope detection, and closed-circuit television detection (CCTV) [1]. Among them, CCTV detection stands as one of the most extensively utilized approaches in engineering sites. Common CCTV pipe robots are shown in Fig. 1. However, this method excessively relies on manual interpretation during the defect recognition process, which is not only a complicated process and a large workload, but also has a high misjudgment rate [2]. Consequently, in recent years, the realization of automatic identification of pipeline defects based on machine vision and artificial intelligence has become a research hotspot in this field. Dong [3] used support vector mechanism to build a multi-class classifier model to extract the parameters such as grayscale difference, equivalent area, and circularity from pipeline weld images to build a feature database. Through training, they achieved a remarkable accuracy of 90% in identifying weld defects using the MSVM (Multicategory Support Vector Machines) classifier. Hawari A [4] employed morphological processing, Gabor filtering, and elliptical fitting algorithms for detecting cracks, deformations, and deposits, respectively. Their average accuracy rate was recorded at 75%. Huang [5] combined two algorithms to extract gaps and crack locations in gas pipeline interfaces. By merging morphological processing with the open top-hat algorithm and the MSER (Maximally Stable Extremal Regions) algorithm, they achieved a segmentation accuracy of 61.5% for gap detection and an 86.7% accuracy rate for crack segmentation.

Fig. 1. Two different sizes of CCTV pipeline robots.

As deep learning continues to advance, a range of deep learning-based target detection algorithms have found application in pipeline defect detection research. These algorithms can be broadly classified into two types: one-stage and two-stage methods. Notable examples of one-stage target detection algorithms include the YOLO series [6–9] and the SSD network [10]. On the other hand, the two-stage algorithms include Faster R-CNN [11] and Mask R-CNN [12]. Wang [13] employed the Faster R-CNN network to detect and recognize six types of defects in underground drainage pipes with an average accuracy of 88.99%. Li [14] designed a new two-stage object detection algorithm for defect detection in underground drainage pipes. They utilized a multi-layer global feature fusion technique and achieved a model mAP value of 50.8%. Lu [15] employed StyleGAN2 for preprocessing the original images and made improvements to the feature fusion layer of YOLOX. They also modified the loss function to CIOU, resulting in an impressive mAP value of 68.76% for recognizing five types of defects.

The research achievements of these esteemed experts and scholars demonstrate a certain level of progress in pipeline defect detection through the integration of machine vision and artificial intelligence. However, it is worth noting that there are still some limitations, such as limited variety of defect recognition, lower accuracy rates, slower detection speed and other shortcomings.

To achieve a more efficient pipeline defect detection, the present study proposes a method that builds upon the improved YOLOv5s framework. The method incorporates several key enhancements. Firstly, a deep target detection layer is introduced to enhance the system's ability to detect large-area defects. Additionally, the introduction of DCN enables the extraction of more comprehensive feature information for irregular defects. Lastly, the optimization of the bounding box loss function to MPDIoU results in faster convergence speed and more accurate regression results. These advancements collectively contribute to a more efficient and accurate pipeline defect detection methodology.

2   Methods

The YOLOv5 algorithm framework comprises three components: the Backbone, Neck, and Head networks. The Backbone network is composed of the CBS, C3, and SPPF modules. The CBS module is responsible for extracting local features and performing downsampling operations. The C3 module utilizes residual structures to extract features while enhancing computational speed. The SPPF module achieves the fusion of local and global features. On the other hand, the Neck network incorporates the Pyramid Attention Network (PAN) to facilitate multi-scale feature fusion and enhancement. Lastly, the Head network performs prediction and filtering on the feature maps generated by the Neck network, ultimately enabling the detection of defect locations and categories.

At present, YOLOv5 has been updated to version 7.0, and it is divided into YOLOv5n, YOLOv5s, YOLOv5m, YOLOv5l, and YOLOv5x based on the model's width and depth. To meet the requirements of the pipeline defects detection task and achieve faster detection results without compromising accuracy, this study selects YOLOv5s as the foundational model due to its smaller parameter count. The proposed improvements include the addition of a deep target detection layer, replacing the last C3 convolutional layer in the Backbone with deformable conv, and modifying the bounding box loss function to MPDIoU. These enhancements contribute to the development of a pipeline defect detection algorithm that combines high accuracy and efficiency. The overall network structure is illustrated in Fig. 2.

2.1   Addition of a Deep Target Detection Layer

The original YOLOv5s model consists of three detection heads in the Head layer, each responsible for detecting different sizes of objects: large, medium, and small. However, in pipeline defect images captured by CCTV pipe robots, we frequently encounter numerous large-sized defects. To improve the detection accuracy of the model for large targets, we introduce a deeper target detection layer to refine the YOLOv5s model. This improvement involves incorporating new CBS and C3 layers into the Backbone network,

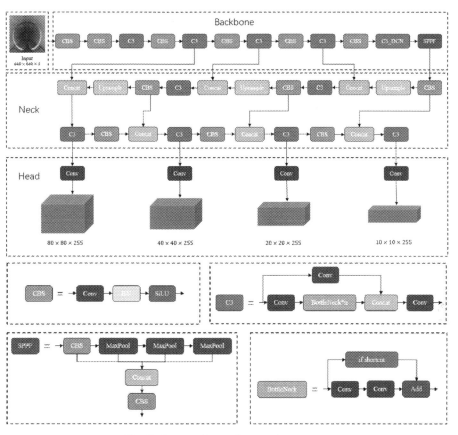

Fig. 2. Improved overall network structure.

increasing the final downsampling factor from 32x to 64x. As a result, the model becomes more adept at extracting feature information for larger-sized defects, thus improving its overall performance. Furthermore, an additional upsampling and convolution operation is introduced in the Neck network, resulting in four different sizes of feature maps: 80px × 80px, 40px × 40px, 20px × 20px, and 10px × 10px. Subsequently, these feature maps are inputted into the Head network for prediction and filtering. With the inclusion of the new detection layer, it is necessary to reconfigure the detection anchor boxes. In this study, the K-means clustering method is employed to obtain the priori anchor boxes for the dataset, and the specific configuration is outlined in Table 1.

Through the extraction of deeper features, we can achieve more precise capture of the features and intricacies of oversized targets, thereby significantly enhancing the overall accuracy of our detection system. This improvement enables the new model to better address larger defects such as disconnections and misalignments in pipeline defect detection tasks.

Table 1. Anchor box configuration corresponding to different size feature maps.

Feature map size	Receptive field size	Priori anchor box size
80px × 80px	small	(44,31) (132,56) (59,139)
40px × 40px	middle	(117,106) (71,293) (244,127)
20px × 20px	large	(132,257) (592,104) (283,245)
10px × 10px	extra large	(630,149) (625,209) (468,343)

2.2 Deformable Convolutional Network Module

The traditional convolution operation in CNN involves dividing the feature map into segments of the same size as the convolution kernel, with fixed positions for each segment on the feature map. However, due to the diverse shapes of pipe defects, this convolution method yields suboptimal results. To address this complexity in target types, this study introduces Deformable Convolution Networks (DCN) [16]. DCN incorporates a learnable offset to the sampling positions in standard convolution, enabling the convolution kernel to expand its range during the training process. This adjustment allows the kernel to better conform to the shape of the target, as depicted in Fig. 3. In Fig. 3, the green dots in (a) represent the standard convolution kernel, while the blue dots in (b), (c), and (d) represent the updated kernel positions after incorporating the offset. It is evident that the inclusion of the offset enables the kernel to adapt to various scenarios, including target movement, size scaling, rotation, and more.

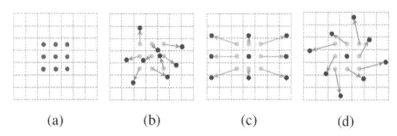

(a)   (b)   (c)   (d)

Fig. 3. Comparison of deformable convolution and standard convolution.

The traditional convolution structure can be defined by Formula 1, where p_0 represents each point in the output feature map, corresponding to the center point of the convolution kernel, and p_n represents each offset of p_0 within the range of the convolution kernel.

$$y(p_0) = \sum_{p_n \in R} w(p_n) \cdot x(p_0 + p_n) \tag{1}$$

In the case of deformable convolution, each point is introduced with an offset, which is generated by another convolution from the input feature map. This can be represented by Formula 2.

$$y(p_0) = \sum_{p_n \in R} w(p_n) \cdot x(p_0 + p_n + \Delta p_n) \tag{2}$$

As the introduced offsets in deformable convolution are typically non-integer values and do not correspond to actual pixel points on the feature map, it becomes necessary to employ bilinear interpolation to obtain the final pixel values after the offset. The diagram illustrating deformable convolution is depicted in Fig. 4.

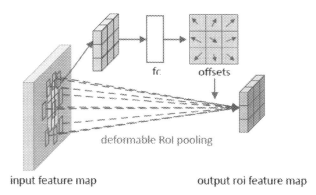

input feature map   output roi feature map

Fig. 4. Illustration of 3×3 deformable convolution.

The introduction of offsets and control points in deformable convolution allows for its adaptability to non-rigidly deformed targets, concurrently enhancing the receptive field, improving target localization accuracy, and reducing computational requirements. By precisely adjusting the position of the sampling point, deformable convolution can effectively capture the intricate details and boundaries of the target, thereby significantly improving the model's accuracy in detecting defective targets.

2.3 Improvement of Loss Function

The loss function of the YOLOv5s model comprises classification loss, bounding box loss and confidence loss, with the total loss being the sum of these three components. Within this framework, the localization loss function employs the CIoU (Complete Intersection over Union) metric, which considers parameters such as the distance between the predicted box and the real box, overlap rate, scale, penalty term, etc. However, this type of loss function struggles to optimize effectively when the predicted box and the ground truth bounding box have the same aspect ratio but vastly different width and height values.

To address the above issue, this study introduces a novel similarity comparison metric called Minimum Point Distance-based IoU (MPDIoU) as the measurement method for the new model's bounding box loss function. This metric considers the distance between the top-left and bottom-right points of the predicted box and the ground truth bounding box, in addition to the original IoU calculation, while simplifying the calculation process. The formula for MPDIoU is as follows:

$$MPDIoU = \frac{A \cap B}{A \cup B} - \frac{d_1^2}{w^2 + h^2} - \frac{d_2^2}{w^2 + h^2} \tag{3}$$

$$d_1^2 = \left(x_1^B - x_1^A\right)^2 + \left(y_1^B - y_1^A\right)^2 \tag{4}$$

$$d_2^2 = \left(x_2^B - x_2^A\right)^2 + \left(y_2^B - y_2^A\right)^2 \tag{5}$$

Let A and B represent the ground truth bounding box and predicted bounding box, respectively. The coordinates (x_1^A, y_1^A) and (x_2^A, y_2^A) represent the top-left and bottom-right coordinates of the ground truth bounding box, while (x_1^B, y_1^B) and (x_2^B, y_2^B) represent the top-left and bottom-right coordinates of the predicted bounding box. The MPDIoU metric incorporates all relevant factors considered in the existing loss function, including overlapping or non-overlapping regions, center point distance, and deviations in width and height.

Compared to traditional loss functions, MPDIoU offers a more efficient and concise computation process, effectively improving the accuracy and efficiency of bounding box regression in defect detection tasks.

3 Experimental Results and Analysis

3.1 Dataset Construction and Environment Configuration

The dataset used in this study is sourced from the Video Pipe ICPR2022 Video Pipeline Challenge. It consists of videos captured by CCTV pipeline robots in multiple urban underground drainage pipeline inspection projects, totaling 575 videos. From these videos, we extracted and selected 1659 images depicting various types of common pipeline defects, including misalignment (CK), crack (PL), leakage (SL), disconnection (TJ), shedding (TL), and obstruction (ZW).

To ensure the robustness and generalization capabilities of the model, we employed techniques such as rotation, flipping, and brightness adjustment to augment the dataset, resulting in a final count of 3000 defect images. The annotated samples were divided into training, validation, and test sets in an 8:1:1 ratio.

The specifications of the software and hardware devices employed in this experiment are presented in Table 2. Additionally, Table 3 provides an overview of the hyperparameters utilized during the training process.

Table 2. Software and Hardware Device Specifications.

Device	Environmental parameters
operating system	Ubuntu 20.04
CPU	Xeon(R) Platinum 8255C
GPU	NVIDIA RTX 3080 (10 GB)
memory	40 GB
programming Language	Python 3.8
deep learning framework	PyTorch 1.11.0、CUDA 11.3

Table 3. Training Hyperparameters.

Hyperparameters	Value
image size	640×640
epoch	100
batch size	16
initial learning rate	0.01
momentum	0.937

3.2   Evaluation Metrics

In order to assess and evaluate the performance of the improved YOLOv5s model, this study utilizes several performance evaluation metrics, including Precision (P), Recall (R), mean Average Precision (mAP) and Frames Per Second (FPS). The formulas for calculating these metrics are as follows:

$$P = \frac{T_p}{T_p + F_p} \tag{6}$$

$$R = \frac{T_p}{T_P + F_N} \tag{7}$$

$$mAP = \frac{\int_0^1 p(r) dr}{N_{classes}} \tag{8}$$

$$FPS = \frac{T_T}{N_{figure}} \tag{9}$$

Among them, T_p denotes the number of positive samples recognized correctly by the model, F_p denotes the number of positive samples recognized incorrectly by the model, F_N denotes the number of negative samples recognized incorrectly by the model, $p(r)$ is the PR curve, $N_{classes}$ denotes the number of classes of all defects, T_T denotes the total time of detection, and N_{figure} denotes the number of detection images.

3.3   Analysis of Experimental Results

Compared to the Original YOLOv5s. By inputting the pipeline defect dataset into the YOLOv5s algorithm model before and after the improvement and incorporating the pre-trained weights from YOLOv5s on the COCO dataset, we conducted training for 100 epochs. We obtained the mean Average Precision, Precision, and Recall curves, as depicted in Fig. 5. The figure shows that the improved mAP curve stabilizes after approximately 70 iterations, exhibiting an obvious enhancement compared to the pre-improvement stage. Ultimately, the improved YOLOv5s model achieved a mAP value of 88.0%, surpassing the original YOLOv5s model by 2.1%. Moreover, the precision and recall rates improved by 3.8% and 1.9% respectively.

Due to the addition of the deep target detection layer and DCN module in the model of this paper's algorithm, the size of the model weights is increased from 14.1 MB to 24.7 MB. Nevertheless, the detection speed of our algorithm is 54.6 FPS, which is comparable to the 54.9 FPS achieved by YOLOv5s. Both algorithms enable real-time detection capabilities.

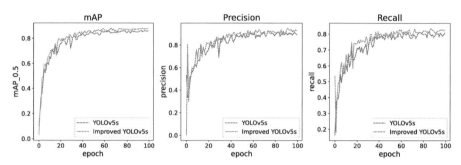

Fig. 5. Comparison of curves before and after algorithm model improvement.

Comparison of Other Algorithms. To further validate the superiority of the algorithm proposed in this study, it is compared with the common target detection algorithms Faster-RCNN, Mask-RCNN, YOLOv3, YOLOv4, and YOLOv7 on the same dataset and calculate the average accuracy of six types of defects separately. The detailed detection results are shown in Table 4. Notably, the improved algorithm showcased obvious enhancements in detecting three specific types of defects: misalignment, crack, and obstruction, with respective increases of 5.9%, 3.5%, and 3.1% compared to the top-performing alternatives. Moreover, the accuracy of detecting the remaining three defect types was comparable to the other leading algorithms. In terms of overall performance, our algorithm outperforms all others.

Table 4. Comparison results of different detection algorithms.

Model	AP/%						P (%)	R (%)	mAP (%)	FPS (f/s)
	CK	PL	SL	TJ	TL	ZW				
Faster-RCNN	81.5	57.8	82.2	66.7	76.9	90.3	82.3	69.7	75.9	54.2
Mask-RCNN	84.0	60.6	76.1	70.2	75.6	91.1	83.2	68.3	76.2	53.2
YOLOv3	85.8	53.8	81.5	68.8	77.3	92.1	84.5	70.2	76.6	55.8
YOLOv4	90.1	68.2	81.8	96.2	79.9	90.4	88.6	78.3	84.4	51.3
YOLOv7	86.9	71.6	82.9	96.6	80.7	89.1	90.0	77.6	84.6	51.6
Improved YOLOv5s	96.0	75.1	82.6	97.8	81.2	95.2	93.0	82.8	88.0	54.6

Ablation Experiments. In order to verify the effects of different improved methods on the model performance and experimental results, based on the original YOLOv5s model, this study conducted six sets of ablation experiments on the three improved methods, and the results are shown in Table 5. The mAP value of the original YOLOv5s model is 85.9%. By incorporating a deep object detection layer, the mAP improved by 1.0%. Introducing deformable convolution led to a further mAP improvement of 0.9%. By changing the bounding box loss function to MPDIoU, the mAP increased by 1.5%. Additionally, integrating the deep object detection layer with the deformable convolution module resulted in mAP improvement of 1.6%. Finally, when all three improvement methods were simultaneously applied to the YOLOv5s model, the mAP increased by 2.2%, demonstrating the best overall performance.

Table 5. Effect of different improvement methods on model performance.

Deep object detection layer	MPDIoU	DCN	P (%)	R (%)	mAP_0.5 (%)	FPS (f/s)
×	×	×	89.2	80.9	85.9	54.9
√	×	×	94.3	81.6	86.9	53.1
×	√	×	91.2	82.1	86.8	52.6
×	×	√	93.1	82.2	87.4	54.6
√	√	×	91.6	83.1	87.5	51.5
√	√	√	93.0	82.8	88.0	54.6

Visualization Result Analysis. To present a more intuitive illustration of the algorithm's performance before and after improvement in pipeline defect detection, this study conducted tests on the original YOLOv5s model and the improved YOLOv5s model using the test dataset. The detection results are depicted in Fig. 6. The first row showcases the original pipeline defect images, followed by the detection results using the YOLOv5s model in the second row, and finally the detection results using the improved YOLOv5s model as described in this paper, displayed in the third row. The figures are annotated with the detected defect types and corresponding confidence scores.

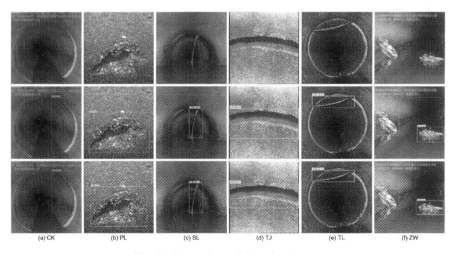

(a) CK   (b) PL   (c) SL   (d) TJ   (e) TL   (f) ZW

Fig. 6. Comparison of detection effects.

The detection effect graphs clearly demonstrate that both the original YOLOv5s model and the enhanced model can effectively identify the correct pipeline defects. However, it is evident that the improved model yields higher confidence scores compared to the original model, and the improved model exhibits enhanced precision in localizing the detected defects.

4 Conclusion

To address the challenges in urban drainage pipeline defect detection, enhancing both speed and accuracy, this paper proposes an improved YOLOv5s algorithm model. This model enhances the precision of defect target detection and reduces the missed detection rate through means of incorporating a deep target detection layer, introducing deformable convolutions, and refining the loss function. As a result of these improvements, the enhanced model demonstrates a 3.8% increase in accuracy, a 1.9% increase in recall rate, and a 2.1% increase in mean average precision. And the detection speed is similar to that of the original model, meeting the standards for real-time detection. This algorithmic model effectively solves the problems of low detection accuracy, high missed detections, and false positives in drainage pipeline defect detection, making it highly practical. Moving forward, our future research will involve collecting a more diverse range of pipeline defect images to further expand the defect dataset. We will also focus on building a more lightweight network model, reducing parameter size and model complexity. Furthermore, we plan to deploy the model on CCTV pipeline robots to facilitate improved pipeline defect detection in collaboration with industry professionals.

Acknowledgment. This research is jointly supported by the National Natural Science Foundation of China (62072414).

References

1. Haurum, J.B., Moeslund, T.B.: Sewer-ML: a multi-label sewer defect classification dataset and benchmark. In: IEEE Computer Society. Virtual, Online, United States (2021)
2. Moradi, S., Zayed, T., Golkhoo, F.: Review on computer aided sewer pipeline defect detection and condition assessment. Infrastructures **4**(1) (2019)
3. Shaohua, D., Xuan, S., Shuyi, X., et al.: Automatic defect identification technology of digital image of pipeline weld. Nat. Gas Ind. B **6**(4) (2018)
4. Hawari, A., Alamin, M., Alkadour, F., et al.: Automated defect detection tool for closed circuit television (CCTV) inspected sewer pipelines. Autom. Constr. **89** (2018)
5. Huang, Y.L.: Research on pipeline crack defect detection method based on video images. Xi'an University of Technology (2018)
6. Redmon, J., Farhadi, A.: You only look once: unified, real-time object detection. In: 2016 IEEE Conference on Computer Vision and Pattern Recognition (CVPR), pp. 779–788 (2016)
7. Redmon, J., Farhadi, A.: YOLO9000: better, faster, stronger. In: 2017 IEEE Conference on Computer Vision and Pattern Recognition (CVPR), pp. 6517–6525 (2017)
8. Redmon, J., Farhadi, A.: YOLOv3: an incremental improvement (2018)

9. Bochkovskiy, A., Wang, C., Liao, H.M.: YOLOv4: optimal speed and accuracy of object detection. arXiv (2020)

10. Liu, W., Anguelov, D., Erhan, D., Szegedy, C., Reed, S., Fu, C.-Y., Berg, A.C.: SSD: single shot multibox detector. In: Leibe, B., Matas, J., Sebe, N., Welling, M. (eds.) ECCV 2016. LNCS, vol. 9905, pp. 21–37. Springer, Cham (2016). https://doi.org/10.1007/978-3-319-464 48-0_2

11. Liu, Y., Zhu, S., Qiu, W., et al.: A lightweight faster R-CNN for ship detection in SAR images. IEEE Geosci. Remote Sens. Lett. **19**, 1–5 (2022)

12. Yu, W., Ren, Y., Hu, C., et al.: Using the improved mask R-CNN and softer-NMS for target segmentation of remote sensing image. In: Proceedings of 2021 4th International Conference on Pattern Recognition and Artificial Intelligence (PRAI), pp. 1–6 (2021)

13. Wang, A.M., Lei, B.H., Chen, J.C.P.C.: Towards an automated condition assessment framework of underground sewer pipes based on closed-circuit television (CCTV) images. Tunnel. Underground Space Technol. **110** (2021)

14. Li, D., Xie, Q., Yu, Z., et al.: Sewer pipe defect detection via deep learning with local and global feature fusion. Autom. Constr. **129**(2), 103823 (2021)

15. Lu, Q.R., Ding, X., Liang, Y.W.: Underground drainage pipe defect recognition algorithm based on improved YOLOX. Electron. Meas. Technol. **45**(21), 161–168 (2022)

16. Dai, J., Qi, H., Xiong, Y., et al.: Deformable convolutional networks. In: 2017 IEEE International Conference on Computer Vision (ICCV), pp. 764–773 (2017)

Visual Servo Control System for AUV Stabilization

A. V. Zuev[1,2(✉)], L. S. Shchetkov[1], E. Sh. Mursalimov[1], V. F. Filaretov[2], and Changan Yuan[3]

[1] Academician M.D. Ageev Institute of Marine Technology Problems Far Eastern Branch of RAS, Vladivostok, Russian Federation
alvzuev@yandex.ru
[2] Far Eastern Federal University, Vladivostok, Russian Federation
[3] Guangxi Academy of Science, Nanning 530007, China

Abstract. The article describes a method for stabilizing an autonomous underwater vehicle (AUV) using a visual servo control system. A monocular video camera is placed on the bottom of AUV, which takes pictures of the seabed under the device. A special visual marker, represented by an ArUco(Augmented Reality University of Cordoba) marker, is pre-installed on the seabed. The proposed method makes it possible to stabilize the control object in the hover mode without calculating the 3D position of the AUV or the visual marker. The control action is calculated based on the position of the feature points of the visual mark on the image from the camera. Also, the control does not need to take into account the dynamic and kinematic features of the AUV. This makes it possible to apply the proposed method on a wide range of types of underwater vehicles.

The analysis of the proposed control system using the AUV mathematical model in the CoppeliaSim modeling environment demonstrated its efficiency and effectiveness.

Keywords: Visual servoing · visual marker · autonomous underwater vehicle · hovering AUV · stabilization AUV

1 Introduction

The AUVs are used to solve various problems underwater. Often it is necessary to carry out work near any objects. To do this, it is necessary to stabilize the AUV near such an object for a long period of time. For stabilization, you need to use feedback, which contains direct or indirect information about the position or displacement of the AUV. In this case, various means of navigation are used, the most frequent of which are inertial navigation and external beacons. However, inertial navigation cannot provide a long-term accurate position calculation due to the gradual accumulation of errors, and when using beacons, the accuracy of position determination deteriorates with increasing distance between them and the AUV, which is inevitable when working at great depths.

In such cases, it is advisable to use visual feedback. It allows you to detect the displacement of the AUV by tracking objects on the seabed or the seabed itself. In this

D.-S. Huang et al. (Eds.): ICAI 2023, CCIS 2015, pp. 156–166, 2024.
https://doi.org/10.1007/978-981-97-0827-7_14

case, previously unchosen points [1, 2] or points corresponding to prominent visual features of the tracked object [3–9] are determined on the image.

In paper [1], prominent points are determined on the image from the camera of the AUV, which shoots the seabed, using a special algorithm. The number of points may be from several hundred to several thousand. The control action is determined by the offset of these points between the current and reference frames. However, due to the high computational load, the image is sent to a remote powerful computer for processing. Forwarding an image has a negative effect on signal generation delay.

According to the visual data from the camera, the 3D position of the landmark relative to the AUV can be calculated [2, 3, 5, 6, 9, 12]. Thus, the control is carried out according to the calculated 3D position. However, such methods are highly dependent on the quality of the 3D position determination, and also add computational burden to the position determination.

In works [4, 7, 8] the control action is calculated directly from visual data. In works [4, 7], homography between the current and reference images is used, however, to calculate the control signal, data from the inertial navigation system is also needed, which complicates the control algorithm. The work [8] does not describe the recognition of feature points of an object on an image, which are further used in the control system presented by the author.

Unfortunately, there has been little research on this problem lately. No simple algorithm for visual stabilization for a wide range of underwater vehicles was found in the considered works. In each of the above works, either additional sensors are used, or there is an additional computational load, or the proposed method can be used for a narrow range of kinematic circuits of AUVs. This article proposes a method for AUV stabilization without calculating the 3D position of the vehicle or a mark, which uses only the image from the camera to calculate the control signal, which can be applied to a wide range of types of AUV kinematic schemes. This is the main novelty of this work. The computational load is reduced, since the 3d position is not calculated. The use of a minimum number of sensors makes it possible to simplify and reduce the cost of the underwater vehicle. Thus, the novelty of the proposed method is in the simplicity and universality of practical application.

2 Proposed Visual Servo Stabilization System

Visual servo allows you to calculate the desired camera movement, according to the position of the feature points of the object in the image. In this work, one of the IBVS (image based visual servo) methods described in [10] is used as the basis for visual servo control. IBVS provide 6 degrees of freedom control. To calculate the 6 degrees of freedom control, 3 or more feature points on the image are needed. 4 feature points are used in this work. A set of feature points is used to calculate the movement, not each point individually. This allows you to implement 6 degrees of freedom control. This article provides brief information about IBVS for practical use. For more information about visual servo control, see [10].

IBVS uses error $e = S^* - S$ to calculate control signal, where: $S = [x_1, y_1, x_2, y_2, x_3, y_3, x_4, y_4]^T$ is a vector of current pixel coordinates of feature points $(x_i, y_i$ are pixel coordinates of the i-th feature point on the image), $S^* = \left[x_1^*, y_1^*, x_2^*, y_2^*, x_3^*, y_3^*, x_4^*, y_4^*\right]^T$ is a vector of desired pixel coordinates of feature points $(x_i^*, y_i^*$ are desired pixel coordinates of the i-th feature point in the image).

The control vector v is computed using the error e: $v = k \cdot L_e^+ \cdot e$, where $v = [V, \omega]^T$ is vector of the linear $V = \left[V_x, V_y, V_z\right]^T$ (along the x, y, z axes, respectively) and angular $\omega = \left[\omega_x, \omega_y, \omega_z\right]^T$ (around the x, y, z axes, respectively) velocities; k is gain; L_e^+ is pseudoinverse of the interaction matrix defined by the Eqs. (1).

$$L_e^+ = \left(L_e^T \cdot L_e\right)^{-1} L_e^T; \, L_e = \begin{bmatrix} L_{p1} \\ L_{p2} \\ L_{p3} \\ L_{p4} \end{bmatrix}; L_{pi} = \begin{bmatrix} \frac{-1}{Z_i^*} & 0 & \frac{x_i^*}{Z_i^*} & x_i^* y_i^* & -\left(1 + x_i^{*2}\right) & y_i^* \\ 0 & \frac{-1}{Z_i^*} & \frac{y_i^*}{Z_i^*} & 1 + y_i^{*2} & -x_i^* y_i^* & -x_i^* \end{bmatrix}$$

$$(1)$$

where x_i^*, y_i^* are desired pixel coordinates of the i-th feature point in the image,

Z_i^* is the desired distance from the camera to the i-th feature point.

Then each component of the vector v is used for a separate.

proportional–integral–derivative controller (and Fig. 1): $u_i = PID_i(v_i), i \in [1, 6]$, [1, 6], where PID is a well-known proportional–integral–derivative (PID) controller with an error input, i is the index. For each element of the vector u a separate PID controller is used. Later, the u vector will be converted for use in the thrusters control system.

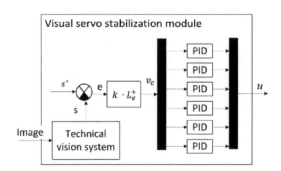

Fig. 1. Block diagram of the visual stabilization module

The ArUco marker is used as a visual marker, and its corners are considered feature points. ArUco marker is recognized using the ArUco software module included in the OpenCV library. After the marker is recognized, the coordinates of its corners on the image are written to the vector S. If the system was just launched and the marker was recognized for the first time, then the coordinates of its corners are written to the vector S^*.

The block diagram of the control system is shown in Fig. 2.

Figure 3 shows the configuration of frames for the vehicle: {O} is global frame, {B} is frame associated with the vehicle, {C} is camera frame.

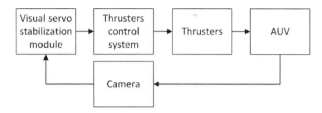

Fig. 2. Structural diagram of the AUV visual servo stabilization system

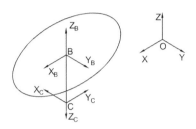

Fig. 3. Configuration of frames

The AUV has 6 degrees of freedom. Stabilization is carried out along the axes X_B, Y_B, Z_B, as well as along the yaw (the roll and pitch of the AUV stabilizes due to the location of the center of displacement above the center of gravity). The vector u is calculated in frame $\{C\}$. Since the frame $\{C\}$ and the frame $\{B\}$ have different orientations, the vector u must be transformed according to the frame $\{B\}$: $u_B = \left[F_x, F_y, F_z, M_x, M_y, M_z\right]^T = [-u_2, -u_1, -u_3, 0, 0, -u_6]^T$, where F_x, F_y, F_z are traction force along the X_B, Y_B, Z_B, respectively, M_x, M_y, M_z are moments around the X_B, Y_B, Z_B axes, respectively, u_i is i-th element in the vector u.

The vector u_B is the input signal for the thrusters control system.

3   AUV Turn in the Direction of Water Flow

The system also has the ability to turn the AUV along the water flow, since the bow resistance is less than the lateral resistance. To detect the flow, vectors u calculated in n previous iterations are used (for example, the calculation time for 50 iterations is approximately 3 s. That is, this is the history of the vector u for the last 3 s). Vectors u calculated in n previous iterations are added to the array U. According to the lengths and angles of the linear components of these vectors, their average values are calculated, as well as the deviation from the average values of each vector u:

$$M_l = \frac{\sum l_i}{n}, \; M_a = \frac{\sum a_i}{n}, N_l = \sqrt{\frac{1}{n}\sum(l_i - M_l)^2}, N_a = \sqrt{\frac{1}{n}\sum(a_i - M_a)^2}, \quad (2)$$

where M_l, M_a are average length and angle respectively, N_l, N_a are standard deviations of lengths and angles respectively, $l_i = \sqrt{U_{i,1}^2 + U_{i,2}^2}$ and $a_i = \mathrm{atan}\frac{U_{i,2}}{U_{i,1}} = atan2(U_{i,2}, U_{i,1})$ are the length and angle, respectively, in polar coordinates of the i-th vector u in the horizontal plane, $U_{i,j}$ is the j-th element of i-th vector u from U.

If N_l and N_a do not exceed the threshold value, it is considered that the vector fluctuations do not go beyond the specified range over time, that is, it retains some constancy. Thus, the system responds to a constant perturbation, which is what the flow is considered to be. Then the feature points on the image in the vector S^* are rotated relative to their common geometric center (similar to the rotation of the visual marker around itself) by an angle M_a, which is close to the true angle between the flow direction and the longitudinal axis of the AUV.

The new position of desired feature points in the image is calculated in Eqs. (3)–(4):

$$c = \begin{bmatrix} c_x \\ c_y \end{bmatrix} = \begin{bmatrix} \frac{\sum S^*_{x,i}}{4} \\ \frac{\sum S^*_{y,i}}{4} \end{bmatrix}; S^*_l = S^* - \begin{bmatrix} c_x \\ c_y \\ c_x \\ c_y \\ \vdots \end{bmatrix} = \begin{bmatrix} x^*_1 - c_x \\ y^*_1 - c_y \\ x^*_2 - c_x \\ y^*_2 - c_y \\ \vdots \end{bmatrix} \in R^8; S^*_{lp} = \begin{bmatrix} \sqrt{\left(S^*_{l1}\right)^2 + \left(S^*_{l2}\right)^2} \\ \text{atan}\left(\frac{S^*_{l2}}{S^*_{l1}}\right) \\ \sqrt{\left(S^*_{l3}\right)^2 + \left(S^*_{l4}\right)^2} \\ \text{atan}\left(\frac{S^*_{l4}}{S^*_{l3}}\right) \\ \vdots \end{bmatrix} \in R^8; \quad (3)$$

$$\overline{S^*_{lp}} = \begin{bmatrix} S^*_{l1} \\ S^*_{l2} - M_a \\ S^*_{l3} \\ S^*_{l4} - M_a \\ \vdots \end{bmatrix} \in R^8; \overline{S^*_l} = \begin{bmatrix} \overline{S^*_{lp1}} \cdot \cos \overline{S^*_{lp2}} \\ \overline{S^*_{lp1}} \cdot \sin \overline{S^*_{lp2}} \\ \overline{S^*_{lp3}} \cdot \cos \overline{S^*_{lp4}} \\ \overline{S^*_{lp3}} \cdot \cos \overline{S^*_{lp4}} \\ \vdots \end{bmatrix} \in R^8; S^*_r = \begin{bmatrix} \overline{S^*_{l1}} + c_x \\ \overline{S^*_{l2}} + c_y \\ \overline{S^*_{l3}} + c_x \\ \overline{S^*_{l4}} + c_x \\ \vdots \end{bmatrix} \in R^8;$$

$$(4)$$

where: μ_j is j-th element if vector μ; c is the center of the desired marker position in the image pixel coordinate system; S^*_l is the coordinate vector of the feature points x^*_i, y^*_i relative to c; S^*_{lp} is the coordinate vector of the feature points x^*_i, y^*_i relative to c in the polar coordinate system; $\overline{S^*_{lp}}$ is the coordinate vector of the feature points x^*_i, y^*_i relative to c rotated by an angle M_a represented in the polar coordinate system; $\overline{S^*_l}$ is the coordinate vector of the feature points x^*_i, y^*_i relative to c rotated by an angle M_a represented in the Cartesian coordinate system; S^*_r is the coordinate vector of the feature points x^*_i, y^*_i relative to c rotated by an angle M_a represented in the image pixel coordinate system.

Rotated feature points are considered desired points: $S^* = S^*_r$, the matrix L_e^+ (1) is recalculated according to the updated coordinates of the desired feature points.

Since the desired feature points have new coordinates, they do not coincide with the current feature points coordinates. This will affect the vector e. Therefore, the algorithm of visual servo control will rotate the AUV to reduce the error e.

4   Search for a Lost Visual Marker

The visual servo stabilization system works when the visual marker is located in the camera's field of view. However, the thrusters have limited power, and short-term disturbing effects of such force are not excluded, which will shift the AUV so much that the visual marker will be out of the camera's field of view. An algorithm for searching for a lost visual marker has been developed for this case.

The visual servo stabilization system is implemented as a software infinite loop. In each iteration, the position of the marker on the camera image is calculated, stored and updated. Thus, in each iteration, the position of the visual marker in the previous iteration is known.

If the marker is not found in the current iteration (i.e. it is partially or completely outside the camera's field of view), then the algorithm for its search is launched. To calculate the direction in which the lost marker is located, the last calculated position of the visual marker on the camera image is used.

4.1   First Iteration Where the Marker is not Found

Since the visual marker was not found in the current iteration, in the previous iteration it was close to some border of the image. When calculating the marker offset direction, the assumption is used that the position of the marker on the image in the current iteration is on a straight line passing through the center of the image and the position of the marker on the image in the previous iteration. The direction of the search for the marker is described by the angle of inclination of this straight line in the image:

$$\alpha = \operatorname{atan} \frac{m_y^{i-1} - \frac{r_y}{2}}{m_x^{i-1} - \frac{r_x}{2}} = atan2\left(m_y^{i-1} - \frac{r_y}{2}, m_x^{i-1} - \frac{r_x}{2}\right), \tag{5}$$

where m_x^{i-1}, m_y^{i-1} are position coordinates of the marker on the image along the x and y axes, respectively, in the previous iteration in image coordinates; r_x, r_y are image resolution in pixels along the x and y axes, respectively.

In the previous equation, there is a correction for the offset of the coordinate axes of the image, since initially the origin of the coordinates of the image is in its upper-left corner, and the control signal must be calculated relative to the center of the image.

To search for the marker, movement is set in a previously calculated direction with a simultaneous increase in height above the ground. This allows you to most quickly increase the camera's field of view in the direction of the lost marker. To move in a previously defined direction, a 3-dimensional vector $d = [d_x, d_y, d_z]^T$ is used, where d_x is the distance that can be overcome horizontally to search for the marker, d_y is the allowable perpendicular displacement (usually $d_y = 0$), d_z is the allowable upward displacement. The vector d is the relative allowable displacement of the AUV when searching for the marker. The desired position of the AUV is calculated relative to the AUV position in the previous iteration using the vector d. The AUV moving to the desired position using an inertial navigation system should detect the lost marker.

The vector d is set when setting up the system in such a way that when the AUV moves to a new position, there is a high probability of detecting a lost marker. In this case, the vector d must be limited so that, in case of an unsuccessful search, the AUV does not move too far from its original position.

Below is the calculation of the desired position using the vector d and the previously calculated angle α:

$$d = \left[d_x, d_y, d_z\right]^T, \ R_1 = \begin{bmatrix} \cos a & -\sin a \\ \sin a & \cos a \end{bmatrix}, \ R_2 = \begin{bmatrix} \cos \varphi & -\sin \varphi \\ \sin \varphi & \cos \varphi \end{bmatrix}, \quad (6)$$

$$N = \begin{bmatrix} N_x \\ N_y \end{bmatrix} = R_2 \cdot R_1 \cdot \begin{bmatrix} d_x \\ d_y \end{bmatrix}, \ D = \begin{bmatrix} N_x \\ N_y \\ d_z \end{bmatrix}, \ P^* = P_{i-1} + D, \quad (7)$$

where:

N is a horizontal two-dimensional vector corresponding to the rotation of the vector $\left[d_x, d_y\right]^T$ by an angle α in the connected system of the AUV and the angle φ(yaw);R_1, R_2 are rotation matrices for angles α and φ respectively; D is the displacement vector of the AUV in the global frame; P_{i-1} is the position of the AUV in the last iteration where the marker was detected; P^* is the desired position of the device, when moving to which the lost marker should be detected.

The movement to the position P^* is regulated using the PID controller for each degree of freedom to which the position error is applied. The orientation of the AUV is maintained unchanged. As soon as the marker is detected, the movement to the position P^* stops and the control of the AUV passes to the visual servo stabilization system.

If the device came to the P^* position without detecting the marker, then its movement stops. Most likely, further movement will not lead to success. Therefore, in this case, the operator should take control.

5 AUV Model

A mathematical model in the CoppeliaSim environment was used to study the system. This article uses the model no of the underwater vehicle proposed in [11].

The vector τ of forces and moments acting on the AUV is defined as:

$$\tau = [(-T_1 + T_2 - T_3 + T_4) \cdot 0.707; \ (T_1 + T_2 + T_3 + T_4) \cdot 0.707; \ T_5 + T_6 + T_7 + T_8;$$
$$(-T_5 + T_6 + T_7 - T_8) \cdot \frac{K}{2}; \ (T_5 + T_6 - T_7 - T_8) \cdot \frac{M}{2}; (-T_1 - T_2 + T_3 + T_4) \cdot \frac{L}{2}]^T; \quad (8)$$

where T_i is the trust of i-th thruster; $K = 0.4, \ M = 0.35, \ L = 0.37$ are geometric parameters for the configuration of thrusters.

All thrusters in the model are described by aperiodic links of the first order with a time constant equal to 0.1: $0.1\dot{T}_i + T_i = T_i^*$, where T_i^* is the desired thrust of i-th thruster. The maximum thrust of each thruster is 30N.

Vector T^* is calculated in the thrusters control system as:

$$T^* = \left[\frac{u_{B2}}{2.828} - \frac{u_{B1}}{2.828} - \frac{u_{B6}}{2L}; \frac{u_{B2}}{2.828} + \frac{u_{B1}}{2.828} - \frac{u_{B6}}{2L}; \frac{u_{B2}}{2.828} - \frac{u_{B1}}{2.828} + \frac{u_{B6}}{2L}; \frac{u_{B2}}{2.828} + \frac{u_{B1}}{2.828} + \frac{u_{B6}}{2L}; \right.$$
$$\left. \frac{u_{B3}}{4} + \frac{u_{B4}}{4K} + \frac{u_{B5}}{2M}; \frac{u_{B3}}{4} - \frac{u_{B4}}{2L} + \frac{u_{B5}}{2M}; \frac{u_{B3}}{4} - \frac{u_{B4}}{2L} - \frac{u_{B5}}{2M}; \frac{u_{B3}}{4} + \frac{u_{B4}}{2L} - \frac{u_{B5}}{2M} \right]^T; \quad (9)$$

The model has these parameters: matrix of added mass $A = -diag\{-15, -185,$ $-185, -5, -19.6, -19.6\}$, AUV mass $m = 170$, inertia matrix $I_b = diag\{10.2,$ $23.4, 23.4\}$, linear D_1 and quadratic D_2 matrices of hydrodynamic coefficients: $D_1 =$ $D_2 = diag\{18, 105, 105, 20, 80, 80\}$.

6 Simulation Results

In the CoppeliaSim environment, the ArUco marker is initially located under the AUV in the center of the camera's field of view. The frequency of processing the scene in the CoppeliaSim environment is 20 Hz, the frequency of generating the control signal is also 20 Hz. The camera resolution is set to 512×512. Limitations of deviations of standard deviations for orientation along the water flow: $limit(N_l) = 230.0$, $limit(N_a) = 0.34906$; $n = 40$. Parameters of the PID controllers of the visual servo stabilization system (Degree of freedom of the PID controller: Proportional gain, Integral gain, Differential gain):$x : \{P : 4.0, I : 1.0, D : 4.0\}, y : \{P : 4.0, I : 1.0, D : 4.0\}, z : \{P : 300.0, I : 20.0, D : 200.0\}, yaw : \{P : 5.0, I : 0.0, D : 5.0\}$. Parameters of PID controllers for marker search: $x : \{P : 5.0, I : 0.0, D : 1.0\}, y : \{P : 17.0, I : 0.0, D : 1.0\}, z : \{P : 10.0, I : 50.0, D : 12.0\}, yaw : \{P : 300, I : 0.0, D : 5.0\}$.

The operation of the visual servo stabilization system was tested with a longitudinal (see Fig. 4a) and transverse (see Fig. 4b) flow at a speed of 0.11 m/s each. In this case, the possibility of AUV turn in the direction of water flow was disabled. The initial position of the device in the frame $\{O\}$: $x_O = 0$, $y_O = 0$, $z_O = 0.5$. When starting the simulation, the AUV is stationary relative to the surrounding water, that is, it moves at the speed of the current.

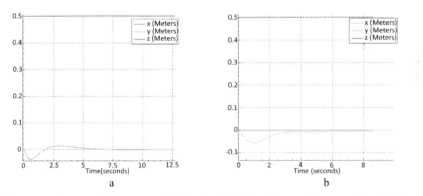

Fig. 4. a) Coordinates of the AUV with longitudinal flow; b) coordinates of the AUV with transverse flow

The operation of the control system when the AUV turn in the direction of water flow is shown in Fig. 5. The transverse direction of the flow is used, the flow velocity is 0.11 m/s. When starting the simulation, the device is stationary relative to the surrounding water, that is, it moves at the speed of the current. The control system detected the current and began to turn at the time of 6.5 s.

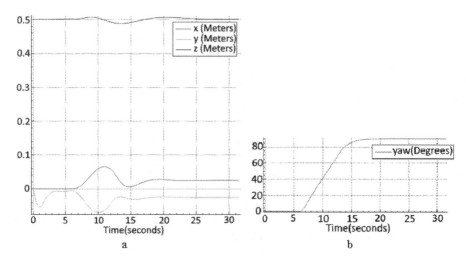

Fig. 5. a) The coordinates of the AUV when oriented in the direction of the current; b) the course of the device when oriented in the direction of the current

The result of the system's operation when searching for the lost marker is shown in Fig. 6. To shift the AUV, an external force $F_d = [-60, -60, 0]^T$ is introduced in the frame {O}, which acts on the device during the first 2 s of the simulation. The vector d to search for the marker is equal to $[2.0, 0.0, 1.0]^T$. The visual marker was lost at a time of 1.3 s, and detected at a time of 10.5 s.

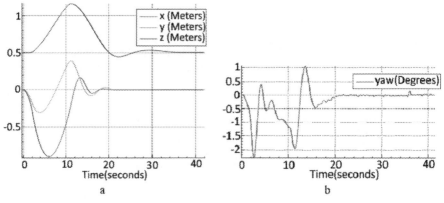

Fig. 6. a) Coordinates of the AUV when searching for the lost marker; b) yaw of the AUV when searching for the lost marker

Figure 7 shows the operation of the system with variable flow. The flow speed varies according to the law: $x : -0.11 \cos(0.5t)$, $y : -0.11 \cos(0.5t + 0.1)$.

These graphs (Fig. 4, Fig. 5, Fig. 6 and Fig. 7) prove the operability of the control system. Figure 4 shows that the AUV returns to its original position despite the drift. Figure 5 shows that the AUV turned in the direction of the current and stabilized. Figure 6

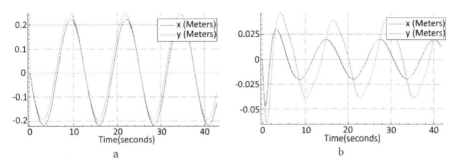

Fig. 7. a) The stabilization system is not running; b) The stabilization system is running

shows that the device successfully detected the marker after the drift, returned to its original position and stabilized. Figure 7 shows that the oscillation amplitude decreases greatly with variable flow.

7 Conclusion

A visual servo control system has been developed to stabilize the AUV near objects by a visual marker. Also, the system has added functionality for determining the flow and reversal along it to reduce resistance and search for the lost visual mark. The study of the proposed system has shown the effectiveness of its work. It is worth noting that using a higher resolution image or a less productive computer (the simulation was carried out on a computer with an Intel Core i5 2450 m central processor) will lead to a decrease in the frequency of control signal generation, which will negatively affect the quality of control. It is also worth noting that the image of a real underwater photography will be different than in a computer simulation. This aspect should be taken into account on a case-by-case basis, but it is beyond the scope of this article. Further work will be aimed at the implementation of this system on a real AUV for the research of Institute of Marine Technology Problems Far Eastern Branch of RAS.

Acknowledgements. The work was supported by the Ministry of Science and Higher Education of the Russian Federation, project no. FZNS-2023-0011.

References

1. Kostenko, V.V., Pavin, A.M.: "Avtomaticheskoe pozicionirovanie neobitaemogo podvodnogo apparata nad ob"ektami morskogo dna s ispol'zovaniem fotoizobrazhenij" [Automatic positioning of an uninhabited underwater vehicle over seabed objects using photographic images]. Underwater Res. Robot. 1(17), 39–47 (2014). (in Russian)
2. Bechlioulis, C.P., Karras, G.C., Nagappa, S., Palomeras, N., Kyriakopoulos, K.J., Carreras, M.: A robust visual servo control scheme with prescribed performance for an autonomous underwater vehicle. 2013 IEEE/RSJ International Conference on Intelligent Robots and Systems (IROS), pp. 3879–3884 (2013)

3. Bian, C., Gao, J., Yang, B., Yan, W., Liang, X., Wu, D.: Visual docking control of an AUV with a moving station using a cross-tracking approach with unscented Kalman filtering. In: Global Oceans 2020: Singapore – U.S. Gulf Coast (2020)

4. Nguyen, L.-H., Hua, M.-D., Allibert, G., Hamel, T.: A homography-based dynamic control approach applied to station keeping of autonomous underwater vehicles without linear velocity measurements. IEEE Trans. Control Syst. Technol. **29**(5), 2065–2078 (2021)

5. Lwin, K.N., Myint, M., Mukada, N., Yamada, D., Matsuno, T., Minami, M.: Robustness of 3D pose estimation against turbidity using dual-eye cameras and active/lighting 3D marker for visual-servoing based AUV. In: 2018 OCEANS - MTS/IEEE Kobe Techno-Oceans (OTO) (2018)

6. Lwin, K.N., Yonemori, K., Myint, M., Yanou, A., Minami, M.: Autonomous docking experiment in the sea for visual-servo type undewater vehicle using three-dimensional marker and dual-eyes cameras. In: 2016 55th Annual Conference of the Society of Instrument and Control Engineers of Japan (SICE), pp. 1359–1365 (2016)

7. Krupinski, S., Allibert, G., Hua, M.-D., Hamel, T.: An inertial-aided homography-based visual servo control approach for (almost) fully actuated autonomous underwater vehicles. IEEE Trans. Robot. **33**(5), 1–20 (2017)

8. Yahya, M.F., Arshad, M.R.: Image-based visual servoing for docking of an autonomous underwater vehicle. In: 2017 IEEE 7th International Conference on Underwater System Technology: Theory and Applications (USYS), pp. 54–59 (2017)

9. Yahya, M.F., Arshad, M.R.: Position-based visual servoing for underwater docking of an autonomous underwater vehicle. In: 2016 IEEE International Conference on Underwater System Technology: Theory and Applications (USYS), pp. 121–126 (2016)

10. Chaumette, F., Hutchinson, S.: Visual servo control, part I: basic approaches. IEEE Robot. Autom. Mag. **13**(4), 82–90 (2006)

11. Fossen, T.I.: Handbook of Marine Craft Hydrodynamics and Motion Control, pp. 1–567. Wiley, Hoboken (2011)

12. Ren, R., Zhang, L., Liu, L., Yuan, Y.: Two AUVs guidance method for self-reconfiguration mission based on monocular vision. IEEE Sens. J. **21**(5), 10082–10090 (2021)

Research on Fuzzy Weighted Controller for Battery Discharge of Dual-Channel Dual-Active Bridge

KaiXin Shu[1], Yu Fang[1(✉)], Sheng Wang[2], Liang Lu[1], YuXuan Fang[1], and Xuehua Wang[3]

[1] College of Information Engineering, Yang Zhou University, Yangzhou, China
yfang@yzu.edu.cn
[2] College of Electrical, Energy and Power Engineering, Yang Zhou University, Yangzhou, China
[3] The State Key Laboratory of Advanced Electromagnetic Engineering and Technology, School of Electrical and Electronic Engineering, Huazhong University of Science and Technology, Wuhan, China

Abstract. Aiming at the dual closed-loop control of dual-active bridge (DAB) charging and discharging circuits in energy storage devices, which is difficult to allocate discharging current reasonably based on battery performance, a fuzzy weighted controller with battery pack voltage and its variation as input is proposed. The domain of the fuzzy controller is designed and the fuzzy solution is achieved through the weighted average method. The obtained output is combined with the weighting law to construct a fuzzy weighted control law, Thus, adaptive allocation of battery discharging current for each channel is achieved. By building a simulation model, the proposed fuzzy weighted control law is applied to a dual channel dual active bridge circuit for simulation verification. The experimental results show that the fuzzy weighted controller based on dual channel and dual active bridge battery discharge can achieve dynamic voltage equalization of battery packs with different performance under discharging conditions, thus verifying the correctness of the theoretical and design methods studied in this paper.

Keywords: Dual Active Bridge · Double Closed-Loop Control · Fuzzy Control · Weighted Law

1 Introduction

In order to realize the comprehensive utilization of batteries with different performance in energy storage equipment, batteries with similar performance are currently packaged together, but multiple such battery packs are not suitable for direct parallel use due to different performance, and battery pack balancers need to be installed between different battery packs to realize voltage-balanced control of two battery packs [1, 4]. However, this method is complicated control, not only increases the burden of the whole system operation and the cost of the system due to one more device, but also reduces the reliability of the whole energy storage system.

In this case, independent power electronic circuits can be used to charge and discharge battery packs with different performance, and the discharging current of each channel battery pack must be allocated according to the battery performance to ensure the battery life. The value of the battery pack discharging current can usually be weighted distribution method, that is, the discharging current of the high-voltage battery pack is greater than the discharging current of the low-voltage battery pack, but this method fails to fully consider the characteristics of the battery, such as the change of the internal resistance may lead to misjudgment of the battery performance. In order to solve this problem, the State of Charge (SOC) of the battery pack in different channels can be monitored [5, 8]. Common SOC estimation methods include ampere-hour method, open-circuit voltage method, neural network method and so on [9, 15], and then allocate the discharging current of the battery pack according to the value of the SOC. Although the above algorithms can be used to accurately evaluate the SOC, they will seriously occupy the CPU resources, which is not conducive to real-time control and increase the cost of the entire energy storage equipment. In this paper, a fuzzy weighted controller of two-channel battery charging and discharging circuit is proposed for discharging battery packs with different performance of battery packs. By detecting the voltage of battery packs and its rate of change in discharging conditions, the discharging current of battery packs in each channel can be distributed adaptively, so as to extend the service life of batteries and reduce the operating cost of batteries.

2   Dual Channel Dual Active Bridge Circuit and Its Control Strategy

The main circuit architecture of energy storage composed of bidirectional DC-DC battery charging and discharging circuit, bidirectional AC-DC inverter circuit, energy storage battery, etc., is shown in Fig. 1. When the battery is in discharging state, the energy in the bidirectional battery charging and discharging circuit flows from the battery side to the bus. When the battery is charged, the energy in the bidirectional battery charging and discharging circuit flows from the bus to the battery side.

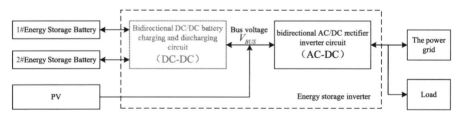

Fig. 1. Structure block diagram of single-phase energy storage inverter

The topology of the bidirectional DC-DC battery charging and discharging path adopts a dual-channel dual-active bridge charge and discharge path, as shown in Fig. 2. Two independent dual-active bridge circuits are used in parallel at high-voltage side, namely 1# dual-active bridge and 2# dual-active bridge.

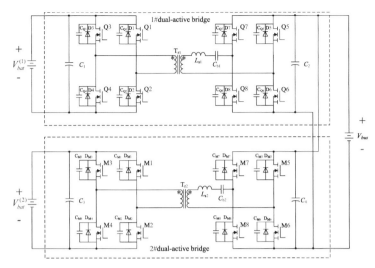

Fig. 2. Independent dual-channel dual-active bridge charging and discharging circuit

As shown in Fig. 3, When the DAB circuit is in the state of battery discharge, the double closed-loop control is adopted, that is, the voltage outer loop and the current inner loop, and the modulation method is phase-shift control. The error between the given value of the bus voltage and the sampled value of the bus voltage is taken as the input of the voltage controller, and the output of the voltage control loop is taken as the current set value, that is, the set of the current controller, and the current controller outputs the pulse modulation signal. In this paper, a fuzzy weighted control method is proposed for the two-channel charge and discharge circuit during battery discharge process. The fuzzy weighted control law is shown in Fig. 3, so as to realize the adaptive distribution of battery discharging current according to the performance of the battery, and finally realize the dynamic voltage balancing of the battery pack.

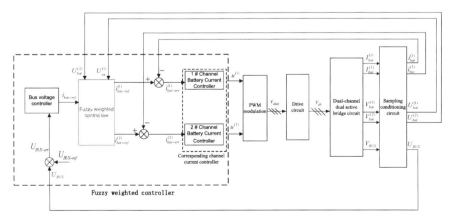

Fig. 3. Double closed-loop control structure during battery discharge

3   Principle of Fuzzy Weighted Control Law for Battery Discharge

On the basis of the traditional weighted distribution control law, the paper introduces a fuzzy controller based on the voltage value of the battery pack and its rate of change to build a fuzzy weighted control law, as shown in Fig. 4.

The input of fuzzy control law is the error $U_e^{(1,2)}$ between 1# channel battery pack sampling voltage $U_{bat}^{(1)}$ and 2# channel battery pack sampling voltage $U_{bat}^{(2)}$ and the error $\Delta U_e^{(1,2)}$ between the change rate of 1# channel battery pack sampling voltage $\Delta U_{bat}^{(1)}$ and the change rate of 2# channel battery pack sampling voltage $\Delta U_{bat}^{(2)}$. In the Fig. 4, $U_{bat0}^{(1)}$ and $U_{bat0}^{(2)}$ are the initial sampling values of the battery pack voltage in channel 1# and channel 2#, $\rho_0^{(1)}$ and $\rho_0^{(2)}$ are the initial weighted coefficients of the given values of the discharging current in channel 1# and channel 2#, $\rho_{k-1}^{(1)}$ and $\rho_{k-1}^{(2)}$ are the last weighted coefficients of the given discharging current in channel 1# and channel 2#, $\rho_k^{(1)}$ and $\rho_k^{(2)}$ are the current weighted coefficients of the given discharging current in channel 1# and channel 2#, $i_{bat-ref}$ is the given value of discharging current, $i_{bat-ref}^{(1)}$ is the set value of discharging current in channel 1#, and $i_{bat-ref}^{(2)}$ is the set value of discharging current in channel 2#. The output of the introduced fuzzy control is the increment $\Delta\rho$ of the weighted coefficient required by the weighted control law.

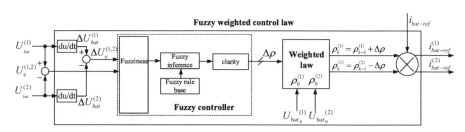

Fig. 4. Fuzzy weighted control law

3.1   Design Method of Fuzzy Controller

The main principle of fuzzy controller is: the error $U_e^{(1,2)}$ between the sampling voltage of the 1# channel battery pack $U_{bat}^{(1)}$ and the sampling voltage in the 2# channel battery pack $U_{bat}^{(2)}$ and the error $\Delta U_e^{(1,2)}$ between the change rate of the sampling voltage of the 1# channel battery pack $\Delta U_{bat}^{(1)}$ and the change rate of sampling voltage of the 2# channel battery pack $\Delta U_{bat}^{(2)}$ are taken as the input of the fuzzy controller. The fuzzy controller firstly quantifies and fuzzies $U_e^{(1,2)}$ and $\Delta U_e^{(1,2)}$, then fuzzy inference is carried out to obtain the fuzzy control quantity, and finally clarity is carried out to obtain the weighted coefficient increment $\Delta\rho$ required by the weighted law.

This paper is equipped with 10 kW dual independent channel DAB DC-DC converter, whose operating range is as follows: battery pack voltage operating range is 40 V–60 V,

BUS voltage range is 390 V–410 V. On the basis of the above operating range, the domain of fuzzy control input is designed, the basic domain of input quantity $U_e^{(1,2)}$ is $[-5,5]$, and the quantization factor is 0.4, then the fuzzy domain is $[-2, 2]$, and the values of the fuzzy domain are $\{-2, -1, 0, 1,2\}$, the corresponding five fuzzy language variables are $\{NB, NS, ZE, PS, PB\}$; the basic domain of the other input $\Delta U_e^{(1,2)}$ is $[-0.2,0.2]$, and if the quantization factor is 10, the fuzzy domain is $[-2, 2]$, and the values of the fuzzy domain are $\{-2, -1, 0, 1,2\}$, corresponding to the five fuzzy language variables $\{NB, NS, ZE, PS, PB\}$. Both $U_e^{(1,2)}$ and $\Delta U_e^{(1,2)}$ adopt triangular dependency functions, as shown in Fig. 5.

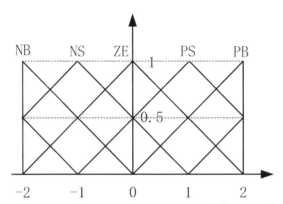

Fig. 5. Triangular membership function of $U_e^{(1,2)}$ and $\Delta U_e^{(1,2)}$

Under the working conditions of power supply batteries with different performance and voltage amplitude, the optimal weighted coefficient can be obtained by combining calculation, simulation and experiment, so as to find out the trend of the weighted coefficient changing with the battery performance and power supply voltage. In order to obtain a reasonable weighted coefficient of the respective discharging current reference signal of the two channels, it is necessary to combine the manual experience and specific debugging results. Actually, fuzzy controller design process is mainly divided into two stages:

1. When the battery pack voltage in each channel is inconsistent, it's main important to eliminate the battery pack voltage error $U_e^{(1,2)}$ between channels, so that the battery pack voltage in the two channels gradually become equal.
2. When the battery voltages in each channel tend to be equal, the discharging current of the battery packs in each channel is distributed adaptively according to the performance of the battery packs in each channel $\Delta U_e^{(1,2)}$.

Based on the above criteria, the following fuzzy control rules for $\Delta\rho$ are designed, as shown in Table 1.

The basic domain of $\Delta\rho$ is $[0.5, 0.5]$, the scaling factor is 0.25, the fuzzy domain is $[-2, 2]$, and the values of the fuzzy domain are $\{-2, -1, 0, 1, 2\}$, the corresponding five fuzzy language variables are $\{NB, NS, ZE, PS, PB\}$.The triangular membership

Table 1. $\Delta\rho$ fuzzy language rules

$\Delta\rho$		$\Delta U_e^{(1,2)}$				
		NB	NS	ZE	PS	PB
$U_e^{(1,2)}$	NB	NB	NB	NB	NS	ZE
	NS	NB	NS	NS	ZE	PS
	ZE	NB	NS	ZE	PS	PB
	PS	NS	ZE	PS	PS	PB
	PB	ZE	PS	PB	PB	PB

function of $\Delta\rho$ is the same as that of $U_e^{(1,2)}$ and $\Delta U_e^{(1,2)}$ in Fig. 5. Language variables output by $\Delta\rho$ can be obtained according to the fuzzy inference control table, and the ambiguity is solved by weighted average method, as shown in Eq. 1.

$$x_0 = \frac{\sum\limits_{i=1}^{n} x_i * u(i)}{\sum\limits_{i=1}^{n} u(i)} \tag{1}$$

In Eq. (1), xi (i = 1, 2, ..., n) is each element in the discourse domain, and u(i) is the membership degree of the fuzzy set after fuzzification, here is the weighted coefficient of the corresponding element.

The fuzzy quantity after clarity is refined by scale factor and the actual value of $\Delta\rho$ is obtained.

3.2 Design of Weighted Law

By substituting the weighted coefficient increment $\Delta\rho$ obtained above into the weighted law, the following iterative formula is obtained, as shown in Eq. 2

$$\begin{cases} \rho_k^{(1)} = \rho_{k-1}^{(1)} + \Delta\rho \\ \rho_k^{(2)} = \rho_{k-1}^{(2)} - \Delta\rho \end{cases} (k = 1, 2, 3 \cdots) \tag{2}$$

In Eq. (2), $\rho_{k-1}^{(1)}$ and $\rho_{k-1}^{(2)}$ are the weighted coefficients of the last discharging current reference signal in channel 1# and channel 2#, $\rho_k^{(1)}$ and $\rho_k^{(2)}$ are the weighted coefficients of the current discharging current reference signal in channel 1# and channel 2#.

The initial weighted coefficient is weighted by the initial sampling values of the battery pack voltage in channel 1# and channel 2#, as shown in Eq. 3

$$\begin{cases} \rho_0^{(1)} = \dfrac{U_{bat_0}^{(1)}}{U_{bat_0}^{(1)} + U_{bat_0}^{(2)}} \\ \rho_0^{(2)} = \dfrac{U_{bat_0}^{(2)}}{U_{bat_0}^{(1)} + U_{bat_0}^{(2)}} \end{cases} \tag{3}$$

In Eq. (3), $\rho_0^{(1)}$ and $\rho_0^{(2)}$ are the weighted coefficients of the last discharging current reference signals in channel 1# and channel 2#; $U_{bat_0}^{(1)}$ and $U_{bat_0}^{(2)}$ are the initial battery voltage sampling signals in channel 1# and channel 2#.

By multiplying the obtained current weighted coefficients $\rho_k^{(1)}$ and $\rho_k^{(2)}$ with the given discharging-current $i_{bat-ref}$ output by the voltage controller, the given discharging-current signals $i_{bat-ref}^{(1)}$ and $i_{bat-ref}^{(2)}$ of the battery pack in channel 1# and channel 2# can be obtained, as shown in Eq. 4

$$\begin{cases} i_{bat-ref}^{(1)} = \rho_k^{(1)} \cdot i_{bat-ref} \\ i_{bat-ref}^{(2)} = \rho_k^{(2)} \cdot i_{bat-ref} \end{cases} (k = 1, 2, 3 \cdots) \qquad (4)$$

The obtained reference signal of battery pack discharging current in each channel is sent to the corresponding current controller, and the actual discharging current of each channel is controlled by the corresponding current controller. To realize the adaptive allocation of the discharging current of each channel battery pack according to the battery performance, and ensure the stability of the BUS voltage V_{BUS}.

4   Analysis of Experimental Results

The simulation model of 10 kW dual-channel DAB charge–discharge circuit is built. The main parameters are as follows: the switching frequency is 40 kHz, the resonant inductance is 2.3 μH, the bus-side filter capacitance is 88.6 mF, the battery-side filter capacitance is 1 mF, the DC isolation capacitor is 3.8 μF, and turns ratio of the transformer is 1:8. Battery voltage range is 40 V–60 V; Bus voltage range is 390 V–410 V; The rated power is 10 kW, and the simulation model of the built 10 kW dual-channel DAB charge–discharge circuit is shown in Fig. 6.

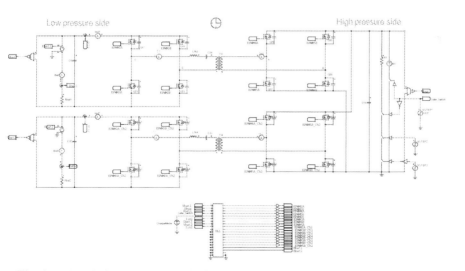

Fig. 6.   10 kW dual-channel dual-active bridge charge–discharge circuit simulation model

In the simulation model, the controlled voltage source is combined with the nonlinear function programmed in the dynamic link library to simulate the characteristic curves Ebat1 and Ebat2 of battery packs with different performance, as shown in Fig. 7.

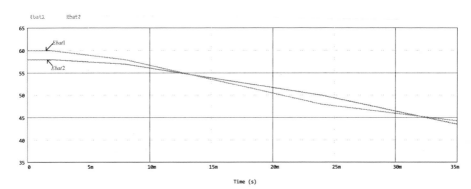

Fig. 7. Characteristic curves simulating different battery performance

In the simulation model, the parameters of the fuzzy controller are designed as follows: the input quantities of the fuzzy controller are $U_e^{(1,2)}$ and $\Delta U_e^{(1,2)}$, and their basic discourse domains are $\{-1, 1\}$ and $\{-0.2, 0.2\}$, the corresponding fuzzy discourse domains are $\{-2, -1, 0, 1, 2\}$, and the quantization factors are $K_{U_e^{(1,2)}} = 2$ and $K_{\Delta U_e^{(1,2)}} = 10$. The basic domain of the output $\Delta\rho$ of the fuzzy controller is $\{-0.5, 0.5\}$, the domain of ambiguity is $\{-2, -1, 0, 1, 2\}$, and the scale factor is $K_{\Delta\rho} = 0.25$.

According to the above parameters, the simulation model based on dynamic link library is built in PSIM simulation software, and the code is written by Visual Studio in dynamic link library. Under the condition that the bus voltage is set to 400 V, when the two battery packs with different battery performance, the dynamic voltage equalization of the battery during discharge is simulated by comparing the discharge effect of the fuzzy weighted controller and the traditional fixed-value weighted controller, as shown in Fig. 8. In Fig. 8, Ebat1 and Ebat2 are the performance curves of the two battery packs. $V_{bat}^{(1)}$ and $V_{bat}^{(2)}$ is the terminal voltage of the battery pack, $I_{bat}^{(1)}$ and $I_{bat}^{(2)}$ are the battery discharging current, and V_{BUS} is the bus voltage. Figure 8(a) shows the simulation waveform of battery discharging with different battery performance when the traditional weighted controller is used, and Fig. 8(b) shows the simulation waveform of battery discharging with different battery performance when the fuzzy weighted controller is used. By comparing the simulation waveforms, the fuzzy weighted controller proposed in this paper shows that battery packs with different performance can achieve obvious voltage balancing effect (finally realized $V_{bat}^{(1)} = V_{bat}^{(2)}$); However, with the traditional weighted controller, the variation trend of battery pack terminal voltage is still consistent with the variation trend of battery performance, and it does not have the effect of voltage equalization. The given current in the traditional weighted controller distributes the given current of each channel according to the terminal voltage of the battery pack, which does not play a role in regulating the battery performance. The given current of the fuzzy weighted controller distributes the given current of each channel through the changing trend of the battery pack terminal voltage, which can inhibit the changing trend of the voltage, to achieve the effect of dynamic voltage balancing on the battery pack voltage, so as to extend the battery life. It can be seen that the simulation results verify the effectiveness of the proposed fuzzy weighted controller in the paper, and the effect of the fuzzy weighted controller is better than that of the traditional weighted controller under the discharge conditions of two different performance battery packs.

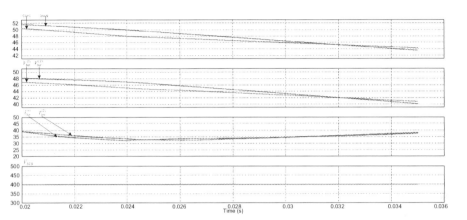

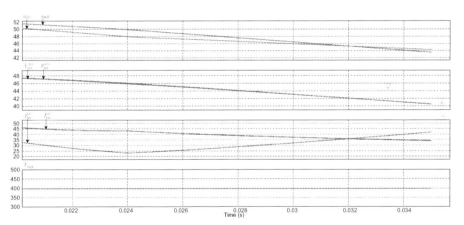

Fig. 8. (a) Discharge simulation waveforms of batteries with different performance using traditional weighted controllers. (b) Discharge simulation waveforms of batteries with different performance using fuzzy weighted controller

5   Conclusion

This paper studies the principle and design method of the fuzzy weighted controller for battery discharge based on double-channel dual-active bridge. According to the performance of the battery reflected by the actual voltage of the battery in each channel, the weighted coefficient required for calculating the given discharging-current value in each channel can be obtained by using the fuzzy weighted controller in this paper. By multiplying the weighted coefficient with the battery discharging current reference signal output by the bus voltage controller, the battery discharging current reference signal of each channel is obtained, and the actual discharging current of each channel is controlled by the current controller. The fuzzy weighted controller adopted in this paper can finally realize the dynamic voltage balancing when two different-performance battery packs discharge, so it can prolong the service life of the battery, reduce the cost of the battery, and improve the cost performance of the energy storage system. The intelligent weighted control law of multi-channel battery discharge will be studied to realize the adaptive allocation of multi-channel battery discharging current.

Acknowledgements. This paper was supported in part by the National Natural Science Foundation of China under Grant 61873346, in part by the Industry Prospect and Common Key Technologies of Yangzhou Science and Technology Plan Project under Grant YZ2022005, and in part by the Intelligent Energy Internet Research Institute Joint Fund of State Grid Yangzhou Power Supply Company and Yangzhou University under Grant SGTYHT/17-JS-202.

References

1. Yuanmao, Y., Junpeng, W., Xiaolin, W.: A multi-winding transformer-based active cell equalizer with self-driven switches for series-connected lithium-ion batteries and super-capacitors. J. Energy Storage **70** (2023)
2. Yuling, L., Jinhao, M., Feng, Y., et al.: A switchable indicator for active balance of the lithium-ion battery pack using a bypass equalizer. J. Energy Storage **68** (2023)
3. Alfredo, A., Adyr, A.E., Juvenal, R., et al.: A high-efficiency capacitor-based battery equalizer for electric vehicles. Sensors (Basel, Switzerland) **23**(11) (2023)
4. Hongrui, L., Hairui, L., Xudong, Y., et al.: A multi-objective parallel layered equalizer for large-scale lithium ion battery system. IET Electr. Power Appl. **16**(12) (2022)
5. Jinglin, H., Xichun, F., Hui, Z., et al.: Adaptive VSG control strategy considering energy storage SOC constraints. Front. Energy Res. **11** (2023)
6. Li, L., Shangfeng, X., Shaojie, Z., et al.: Energy storage steady-state PCS power allocation algorithm based on SOC dynamic sequencing. J. Phys. Conf. Ser. **2584**(1) (2023)
7. Zhang, Z., Shao, J., Li, J., et al.: SOC estimation methods for lithium-ion batteries without current monitoring. Batteries **9**(9) (2023)
8. Zheng, Y., et al.: Capacity and state-of-charge (SOC) estimation for lithium-ion cells based on charging time differences curves. Int. J. Energy Res. **46**(13) (2022)
9. Xiong, X., Wang, S., Fernandez, C., et al.: A novel practical state of charge estimation method: an adaptive improved ampere-hour method based on composite correction factor. Int. J. Energy Res. **44**(14) (2020)
10. Xin, Z., Jiawei, H., Zekun, W., et al.: Study of SOC estimation by the ampere-hour integral method with capacity correction based on LSTM. Batteries **8**(10) (2022)

11. Zhang, M., Fan, X.: Design of battery management system based on improved ampere-hour integration method. Int. J. Electr. Hybrid Veh. **14**(1–2) (2022)
12. Al Hadi, A.M.R., Ekaputri, C., Reza, M.: Estimating the state of charge on lead acid battery using the open circuit voltage method. J. Phys. Conf. Ser. **1367** (2019)
13. Qurthobi, A., Pambudi, A.B.K., Darmawan, D., et al.: Correlation between battery voltage under loaded condition and estimated state of charge at valve-regulated lead acid battery on discharge condition using open circuit voltage method[J]. Int. J. Power Electron. Drive Syst. **9**(1) (2018)
14. Jianbo, D., Jinhao, Z., Yin, L., et al.: Improved BP neural network algorithm to SOC estimation of mine power supply. J. Phys. Conf. Ser. **2562**(1) (2023)
15. Ren, G., Gaojian, R., Yin, L., et al.: SOC estimation of mine power supply based on improved BP neural network algorithm. IOP Conf. Ser. Earth Environ. Sci. **526**(1) (2020)

Multi-scale Texture Network for Industrial Surface Defect Detection

Liqiong Wang[1], Yan Huang[2], and Fanrong Kong[2(✉)]

[1] Hubei Tobacco Industry Co., Ltd., Hubei 430040, China
[2] Shanghai Development Center of Computer Software Technology, Shanghai 201112, China
`nike117@163.com`

Abstract. Automated surface defect detection is crucial for ensuring product quality in industrial settings. This paper presents a multi-scale texture network that addresses this challenge by effectively analyzing textures at various scales. The proposed network incorporates a "Multi-Scale Texture Feature Processing" module to generate multi-scale texture tokens for comprehensive surface analysis. Additionally, a "Multi-Head Feature Encoding" mechanism captures local details and global patterns, leading to improved accuracy. Besides, we introduce a multi-scale perceptual loss function that guides training by optimizing images at different scales while preserving perceptual similarity. Experimental results demonstrate the effectiveness of our approach, offering high accuracy in automated surface defect detection.

Keywords: Industrial Surface Defect Detection · Multi-scale · Texture Feature · Perceptual Loss

1 Introduction

In recent years, the demand for automated surface defect detection in industrial production has been increasing [1, 2]. This is driven by the limitations of traditional manual detection methods, which are time-consuming and prone to errors, resulting in overlooked defects. Automated surface defect detection methods are superior to manual methods as they leverage advanced technologies to enhance detection efficiency and accuracy [3–5]. By automating this process, manufacturers can save time and effort in quality control, resulting in increased productivity and cost savings.

For achieving accurate and automated surface defect detection, many works have been attempted. For example, Wang et al. proposed a real-time steel surface defect detection approach with high accuracy on NEU-DET [6]. Lian et al. developed an enhanced version of YOLOv4, named YOLOv4-Defect, for detecting tiny surface defects in industrial applications. This approach improved efficiency, accuracy, and reduced model size and computation consumption [7]. Similarly, Xie et al. developed a feature-enhanced YOLO algorithm for industrial surface defect detection, surpassing state-of-the-art methods in accuracy [8]. Addressing limited labeled data, Wang et al. introduced a few-shot steel surface defect detection framework, achieving outstanding performance comparable to models with abundant data [1]. Furthermore, Yeung et al. proposed FANet, an

D.-S. Huang et al. (Eds.): ICAI 2023, CCIS 2015, pp. 178–188, 2024.
https://doi.org/10.1007/978-981-97-0827-7_16

efficient fused-attention model that achieved state-of-the-art results on NEU-DET and GC10-DET datasets [9]. Liu et al. proposed an adaptive image segmentation network for surface defect detection, outperforming existing approaches on multiple datasets [10].

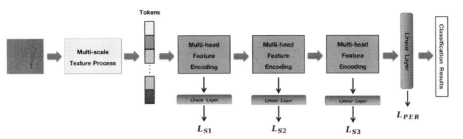

Fig. 1. The main structure of the network.

Besides, for employing advanced techniques to enhance surface defect detection, Jain et al. proposed a data augmentation framework for surface defect detection using GAN-generated synthetic images, demonstrating improved performance compared to existing methods [11]. Li et al. developed a deep learning model for steel surface defect detection, achieving optimal accuracy on the NEU-DET dataset through the use of a multiscale feature extraction module and efficient feature fusion [12]. Feng et al. developed an algorithm using the RepVGG architecture with a spatial attention mechanism, outperforming other algorithms in terms of precision, recall, and F1-score [13]. Wang et al. proposed DRCDCT-Net, a network structure consisting of a Feature Attention Defect module (FAD) and a Cross-Domain Joint Learning Defect module (CJLD), achieving high precision in defect classification tasks with various sample data [14]. Li et al. developed a hybrid CNN-T network for strip steel surface defect classification, which outperformed pure Transformer networks and CNNs in terms of accuracy and efficiency on the NEU-CLS dataset [15].

Although previous works have achieved satisfactory performance, there is a lack of emphasis on the importance of multi-scale texture features in industrial automated defect detection. These features provide valuable insights into surface characteristics and patterns. By analyzing textures at multiple scales, the defect detection system gains a comprehensive understanding of inspected surfaces, capturing both fine details and global variations. This enables effective discrimination between defects and background textures, even in the presence of complex surface textures. Furthermore, multi-scale texture features facilitate adaptability to different defect sizes and patterns, ensuring robust detection in diverse production scenarios. Their integration enhances overall quality control and product integrity.

In this paper, we propose a multi-scale texture network for automated surface defect detection in industrial settings. To achieve this, our network incorporates a "Multi-Scale Texture Feature Processing" module that generates multi-scale texture tokens. These tokens enable the analysis of textures at different scales, providing a comprehensive representation of the inspected surfaces. Furthermore, we introduce a "Multi-Head Feature Encoding" mechanism that allows the network to capture both local details and

global patterns. By effectively integrating this information, the network can differentiate between defects and background textures with improved accuracy. Additionally, we
propose a multi-scale texture perceptual loss function that guides the training process.
This loss function combines multi-scale cross-entropy losses and perceptual losses to
optimize the generated images at various scales while preserving perceptual similarity.

Overall, the main contributions of this paper could be summarized as:

- The proposed Multi-scale Texture Feature Processing module generates multi-scale
  texture tokens, enabling texture analysis at various scales and efficiently enhancing
  the network's defect detection accuracy and effectiveness in industrial settings.
- We have designed a multi-scale texture perceptual loss function that combines
  multi-scale cross-entropy losses and perceptual losses. This loss function optimizes
  defect detection in industrial settings by achieving performance improvement across
  different scales while maintaining perceptual similarity.
- Extensive experiments on the NEU-CLS dataset demonstrate that our proposed
  network could gain competitive performance across different evaluation metrics.

The remainder of the paper is organized as follows:

Section 2 presents a comprehensive introduction to the main method employed in
our proposed network. Section 3 covers the experimental data, evaluation metrics, and
conducts a comprehensive analysis of different experiments. Finally, Sect. 4 provides
the conclusion and limitations of this work.

2 Method

2.1 Overview of the Framework

As depicted in Fig. 1, it presents an overview of the proposed network. Initially, the input
image undergoes "Multi-scale Texture Feature Processing" to extract texture features at
different scales, resulting in the generation of "tokens". These tokens are then passed into
the "Multi-head Feature Encoding" module, where they undergo feature extraction to
capture intricate patterns and structures from the input image. Subsequently, the model is
trained using the "Multi-scale Perceptual Loss" function, which measures the perceptual
similarity between the generated output and the ground truth across multiple scales.
Moreover, this loss function guides the optimization process and encourages the model
to generate visually similar images to the target images. Finally, the framework produces
the ultimate output: classification results based on the generated output from the model.
By combining multi-scale texture feature processing, multi-head feature encoding, and
multi-scale perceptual loss, the framework enhances the model's ability to accurately
classify images.

2.2 Multi-scale Texture Feature Processing

An image, represented as a 2D discrete function $f(x, y)$ with coordinates (x, y), can be
processed using Gabor filters. A Gabor filter is obtained by multiplying a sinusoidal

wave with a Gaussian window, expressed in the spatial domain as:

$$g\left(x, y^{'}, \lambda, \theta, \psi, \sigma, \gamma\right) = \exp\left(-\frac{x^{'2} + \gamma^2 y^{'2}}{2\sigma^2}\right)\cos\left(\frac{2\pi x^{'}}{\lambda} + \psi\right) \tag{1}$$

The Gabor filter parameters include wavelength (λ), orientation (θ), phase offset (ψ), standard deviation of the Gaussian window (σ), and aspect ratio (γ). To align the Gabor filter with different orientations, the original coordinates (x, y) are transformed by:

$$x\prime = x\cos\theta + y\sin\theta \tag{2}$$

$$y\prime = -x\sin\theta + y\cos\theta \tag{3}$$

By applying these transformations, the Gabor filter can capture texture features at different angles within the image. Additionally, to extract texture features from an image, multiple Gabor filters with different scales and orientations are applied. Each filter is convolved with the image, resulting in a Gabor response image $I(x, y\prime, \lambda, \theta)$ given by:

$$I\left(x, y^{'}, \lambda, \theta\right) = f(x, y) \neq g\left(x, y^{'}, \lambda, \theta, \psi, \sigma, \gamma\right) \tag{4}$$

Next, for each $I(x, y\prime, \lambda, \theta)$, several features are extracted:

Mean: The average pixel value can be calculated using the formula:

$$F_{Mean} = \frac{1}{N} \sum_{i=1}^{N} I_i \tag{5}$$

Here, N represents the total number of pixels, and I_i denotes the value of the i-th pixel.

2. Variance: The dispersion of pixel values can be measured as:

$$F_{Variance} = \frac{1}{N} \sum_{i=1}^{N} (I_i - F_{Mean})^2 \tag{6}$$

3. Energy: The overall intensity or amplitude of the Gabor response image can be determined by:

$$F_{Energy} = \sum_{i=1}^{N} I_i^2 \tag{7}$$

4. Gradient: The gradient magnitude and direction can be computed to capture texture edges and details. This information is useful for identifying transitions or boundaries within the image:

$$F_{Gradient} = atan2\left(\frac{\partial I_i}{\partial y}, \frac{\partial I_i}{\partial x}\right) \tag{8}$$

Here, $\frac{\partial I_i}{\partial x}$ and $\frac{\partial I_i}{\partial y}$ are the partial derivatives of the Gabor response image I with respect to the x and y directions. Afterwards, we define the concatenated features f_λ as:

$$f_\lambda = \left[F_{Mean}, F_{Variance}, F_{Energy}, F_{Gradient}\right] \tag{9}$$

Moreover, the multi-scale texture feature is constructed by:

$$F_{MS} = Concat(f_1, f_2, f_3 \ldots f_\lambda) \tag{10}$$

where the value of λ is defined as:

$$\lambda = \lambda_{min} + (i - 1) \times \Delta\lambda \tag{11}$$

here, $\Delta\lambda$ is the stride, and λ_{min} is set as 1, the value of i is calculated by dividing λ_{max} by $\Delta\lambda$ and then rounding down to the nearest integer. By applying multi-scale texture feature encoding, various texture characteristics of the image can be effectively captured and utilized in the subsequent process.

2.3   Multi-head Feature Encoding

The multi-head feature encoding consists of three main steps: query, key, and value calculation, attention score computation, and weighted sum operation. The process can be described using the following formulas: Given input features F_{MS}, , the query (Q), key (K), and value (V) matrices are computed as follows:

$$Q = F_{MS}.W_Q \tag{12}$$

$$W = F_{MS}.W_K \tag{13}$$

$$V = F_{MS}.W_K \tag{14}$$

Here, W_Q, W_K, and W_K are learnable weight matrices that project the input features into the query, key, and value spaces, respectively. Next, the attention scores (A) are calculated based on the query and key matrices using a scaled dot-product attention mechanism:

$$A = softmax\left(\frac{QK^T}{\sqrt{d_k}}\right) \tag{15}$$

where d_k represents the dimensionality of the query and key matrices. Finally, the weighted sum of the value matrix, based on the attention scores, gives the output of the multi-head attention:

$$O = AV \tag{16}$$

The output O represents the extracted features obtained from the multi-head feature mechanism. By adopting the multi-head feature encoding process, it enables the model to learn and leverage diverse information from different parts of the input feature space.

2.4 Multi-scale Perceptual Loss

The multi-scale perceptual loss L_{total} consists of two components: the multi-scale loss $L_{Multi-Scale}$ and the perceptual loss $L_{Multi-Scale}$.

$$L_{total} = L_{Multi-Scale} + L_{PER} \tag{17}$$

where $L_{Multi-Scale}$ is defined as:

$$L_{Multi-Scale} = L_{S1} + L_{S2} + L_{S3} \tag{18}$$

In this context, L_{S1}, L_{S2}, and L_{S3} represent the cross-entropy losses for the last three layers, respectively. The cross-entropy CE loss measures the discrepancy between the predicted outputs and the ground truth labels. The formulas for the cross-entropy losses are as follows:

$$L_{S1} = CE(P_{S1}, L) \tag{19}$$

$$L_{S2} = CE(P_{S2}, L) \tag{20}$$

$$L_{S3} = CE(P_{S3}, L) \tag{21}$$

Here, P_{S1}, P_{S2}, and P_{S3} represent the predicted outputs at the last three layers, while L represents the ground truth label or target. By adopting the multi-scale loss, it encourages the model to capture both local and global details, ensuring accurate generation across different levels of granularity. Moreover, the perceptual loss aims to optimize the visual similarity between the generated output and reference images. It is computed by comparing the feature outputs of the last layer of the network with a pre-trained ResNet's last layer features. By minimizing the discrepancy between these feature representations, the model generates perceptually similar images. Mathematically, the perceptual loss is expressed as:

$$L_{perceptual} = \|F_{Output} - F_{ResNet}\| \tag{22}$$

where $L_{perceptual}$, representing the feature representation of the generated output. Similarly, we denote *ResNet Feature* as F_{ResNet}, representing the feature representation of a pre-trained ResNet network. Overall, the multi-scale perceptual loss function integrates both the multi-scale loss and the perceptual loss to guide the training process. This ensures that the model generates visually appealing and accurate images at multiple scales.

3 Experiments

3.1 Dataset

The NEU-CLS dataset, obtained from Northeastern University, China, is employed to evaluate our model's performance. This dataset focuses on surface defects of hot-rolled strip steel and comprises six distinct defect categories: crazing, inclusion, patches, pitted-surface, rolled-in-scale, and scratches. Each defect category consists of 300 images captured at a raw resolution of 200×200 pixels.

3.2 Implementation Details

The model is implemented using the PyTorch framework on an experimental setup comprising an Intel Core i7-10700K CPU with a clock speed of 3.8 GHz (boosting up to 5.1 GHz), 64 GB RAM, and an NVIDIA GeForce GTX 3080 graphics card. The model is trained on the Ubuntu 16 operating system, with the data divided into an 80%training set and a 20% validation set. The Adam optimizer is used for network optimization, and the learning rate is adjusted using the cosine decay schedule method. In addition, the images undergo resizing to a resolution of 256 × 256 pixels. To further improve the training data, diverse data augmentation techniques are implemented, encompassing random horizontal and vertical flipping, as well as subtle adjustments to the image contrast.

3.3 Evaluation Metrics

To evaluate the model's performance, we first define True Positives (TP) as the number of instances where the model correctly predicts the positive class. Similarly, we define True Negatives (TN) as the number of instances where the model correctly predicts the negative class. False Positives (FP) are the instances where the model incorrectly predicts the positive class. Lastly, False Negatives (FN) are the instances where the model incorrectly predicts the negative class. Based on these components, we can define the evaluation metrics as follows:

Accuracy: This metric measures the overall correctness of the model's predictions by calculating the ratio of correctly predicted instances to the total number of instances in the dataset.

$$Accuracy = \frac{TP + TN}{TP + TN + FP + FN} \tag{23}$$

Recall: It quantifies the model's ability to correctly identify positive instances out of all the actual positive instances in the dataset.

$$Recall = \frac{TP}{TP + FN} \tag{24}$$

Precision: It measures the proportion of correctly identified positive instances out of all instances predicted as positive by the model.

$$Precision = \frac{TP}{TP + FP} \tag{25}$$

F1-score: The F1-score combines precision and recall into a single balanced metric.

$$F1 - score = \frac{2 \cdot (Precision \cdot Recall)}{Precision + Recall} \tag{26}$$

3.4   Comparisons with Different Components

Experimental evaluations were conducted to validate the effectiveness of three key components in defect classification: Multi-scale Texture Processing (MSTP), Multi-head Feature Encoding (MHFE), and Multi-scale Perceptual Loss (MSPL). For the network structure without MSTP and MHFE, a pre-trained ResNet network was used as the feature extractor. Results are summarized in Table 1, revealing that the best performance was achieved when all designed components were incorporated. It is worth noting that among the different components tested, MSTP demonstrated the highest performance, emphasizing the significant contribution of multi-scale texture information in defect classification. The MSTP effectively captures texture information at multiple scales, enabling the model to better discern and classify defects based on their textural characteristics. Together, these experimental findings highlight the effectiveness of integrating MSTP, MHFE, and MSPL components in defect classification.

Table 1. Comparisons with Different Components.

Method	Accuracy (%)	Recall (%)	Precision (%)	Fl-score (%)
+MSTP	98.05	95.77	97.32	98.15
+MHFE	97.15	97.33	96.42	96.51
+MSPL	97.32	98.02	97.12	96.21
Ours	**98.91**	**100.0**	**98.30**	**98.27**

3.5   Effectiveness of MUlti-scale Perceptual Loss

In the conducted experiments, the effectiveness of the MSPL component was evaluated. Results presented in Table 2 demonstrated that utilizing the MSPL component improved the generalization ability of the model, leading to the enhancement in classification accuracy. The MSPL effectively leverages the perceptual information present in defect images to guide the training process, enabling the model to better capture the visual characteristics that are important for accurate defect classification. Furthermore, comparisons were made between the Multi-scale Loss $L_{Multi-Scale}$ and Perceptual Loss $L_{Perceptual}$ components. It was observed that the $L_{Multi-Scale}$ yielded better performance in terms of defect classification accuracy. This finding suggests that the multi-scale approach is more effective in capturing and utilizing the perceptual information necessary for accurate defect classification tasks.

3.6   Compared with Other Methods

In this section, we conducted a series of classification experiments to evaluate the performance of our proposed method, comparing it with several other state-of-the-art classification methods including AlexNet, Vgg, ResNet, ResNeXt, VIT, ConvNeXt, and

Table 2. Effectiveness of MUlti-scale Perceptual Loss.

Method	Accuracy (%)	Recall (%)	Precision (%)	Fl-score (%)
$+L_{CE}$	98.12	95.89	98.02	97.12
$+L_{MAlti-scale}$	98.55	98.92	98.11	97.90
$+L_{Perceptual}$	98.22	97.82	98.12	97.55
$+L_{total}$	**98.91**	**100.0**	**98.30**	**98.27**

InceptionNeXt. The evaluation metrics used in our experiments were accuracy, recall, precision, and F1-score. Our proposed method achieved competitive results in the classification experiments, outperforming all other methods with an accuracy of 98.91%, a recall of 100.0%, a precision of 98.30%, and an F1-score of 98.27% (refer to Table 3). These metrics affirm the effectiveness of our approach in accurately classifying samples and highlight its superiority over the comparison methods.

Table 3. Compared with Other Methods.

Method	Accuracy (%)	Recall (%)	Precision (%)	Fl-score (%)
AlexNet [16]	96.21	93.02	98.25	98.25
Vgg [17]	97.02	97.67	96.49	96.49
ResNet [18]	97.12	95.35	98.26	98.24
ResNeXt [19]	98.15	97.67	98.25	98.25
VTT [20]	97.98	97.56	98.27	98.23
ConvNeXt [21]	98.03	97.67	98.24	98.25
InceptionNeXt [22]	98.12	97.69	98.52	98.28
Ours	**98.91**	**100.0**	**98.30**	**98.27**

4   Conclusions

In this paper, a multi-scale texture network is proposed for automated surface defect detection in industrial settings. The network incorporates a "Multi-Scale Texture Feature Processing" module, which generates multi-scale texture tokens to enable the analysis of textures at different scales. To enhance the network's ability to differentiate between defects and background textures, a "Multi-Head Feature Encoding" mechanism is introduced. This mechanism captures both local details and global patterns, allowing the network to capture and encode the necessary information for accurate classification.

Furthermore, to guide the training process, a multi-scale texture perceptual loss function is proposed. This loss function combines multi-scale cross-entropy losses and perceptual losses. By optimizing the generated images at various scales while preserving perceptual similarity, the network is trained to produce accurate defect localizations while maintaining visual fidelity. Experimental evaluations were conducted to assess the performance of the proposed network. The results demonstrate its effectiveness in automated surface defect detection, outperforming existing methods. In the future work, we would extend our work on different datasets for evaluations.

Acknowledgment. The authors declare no conflict of interest exists in the submission of this manuscript.

References

1. Wang, H., Li, Z., Wang, H.: Few-shot steel surface defect detection. IEEE Trans. Instrum. Meas. (2021)
2. Hou, W., Jing, H.: Rc-yolov5s: for tile surface defect detection. Vis. Comput. (2023)
3. Tsai, D.-M., Fan, S.-K.S., Chou, Y.-H.: Auto-annotated deep segmentation for surface defect detection. IEEE Trans. Instrum. Meas. (2021)
4. Xiao, M., Yang, B., Wang, S., Zhang, Z., He, Y.: Fine coordinate attention for surface defect detection. Eng. Appl. Artif. Intell. (2023)
5. Zhou, H., Yang, R., Hu, R., Shu, C., Tang, X., Li, X.: Etdnet: efficient transformer-based detection network for surface defect detection. IEEE Trans. Instrum. Meas. (2023)
6. Wang, W., et al.: A real-time steel surface defect detection approach with high accuracy. IEEE Trans. Instrum. Meas. (2021)
7. Lian, J., He, J., Niu, Y., Wang, T.: Fast and accurate detection of surface defect based on improved yolov4. Robot. Intell. Autom. (2021)
8. Xie, Y., Hu, W., Xie, S., He, L.: Surface defect detection algorithm based on feature-enhanced yolo. Cogn. Comput. (2022)
9. Yeung, C.-C., Lam, K.-M.: Efficient fused-attention model for steel surface defect detection. IEEE Trans. Instrum. Meas. (2022)
10. Liu, T., He, Z., Lin, Z., Cao, G.-Z., Su, W., Xie, S.: An adaptive image segmentation network for surface defect detection. IEEE Trans. Neural Netw. Learn. Syst. (2022)
11. Jain, S., Seth, G., Paruthi, A., Soni, U., Kumar, G.: Synthetic data augmentation for surface defect detection and classification using deep learning. J. Intell. Manuf. (2020)
12. Li, Z., Wei, X., Hassaballah, M., Li, Y., Jiang, X.: A deep learning model for steel surface defect detection. Complex Intell. Syst. (2023)
13. Feng, X., Gao, X., Luo, L.: X-SDD: a new benchmark for hot rolled steel strip surface defects detection. Symmetry (2021)
14. Wang, J., Zhang, Q., Liu, G., DRCDCT-net: a steel surface defect diagnosis method based on a dual-route cross-domain convolution-transformer network. Meas. Sci. Technol. (2022)
15. Li, S., Wu, C., Xiong, N.: Hybrid architecture based on CNN and transformer for strip steel surface defect classification. Electronics (2022)
16. Krizhevsky, A., Sutskever, I., Hinton, G.E.: Imagenet classification with deep convolutional neural networks. In: Advances in Neural Information Processing Systems, vol. 25 (2012)
17. Simonyan, K., Zisserman, A.: Very deep convolutional networks for large-scale image recognition. arXiv preprint arXiv:1409.1556 (2014)
18. He, K., Zhang, X., Ren, S., Sun, J.: Deep residual learning for image recognition. In: Proceedings of the IEEE Conference on Computer Vision and Pattern Recognition, pp. 770–778 (2016)

19. Xie, S., Girshick, R., Dollar, P., Tu, Z., He, K.: Aggregated residual transformations for deep neural networks. In: Proceedings of the IEEE Conference on Computer Vision and Pattern Recognition, pp. 1492–1500 (2017)
20. Dosovitskiy, A., et al.: An image is worth 16x16 words: transformers for image recognition at scale. arXiv preprint arXiv:2010.11929 (2020)
21. Liu, Z., Mao, H., Wu, C.-Y., Feichtenhofer, C., Darrell, T., Xie, S.: A convnet for the 2020s. In: Proceedings of the IEEE/CVF Conference on Computer Vision and Pattern Recognition, pp. 11976–11986 (2022)
22. Yu, W., Zhou, P., Yan, S., Wang, X.: Inceptionnext: when inception meets convnext. arXiv preprint arXiv:2303.16900 (2023)

Machine Learning

Large-Scale Data Challenges: Instability in Statistical Learning

Bo-Yu Chen and Hao Zhang[(✉)]

Purdue University, West Lafayette, IN 47906, USA
{chen2433,zhanghao}@purdue.edu

Abstract. Numerous approximation methods have been developed to approximate both the kernel matrix and its inverse. We investigate one such influential approximation that has recently gained popularity. However, our results indicate that this approximation fails to address the ill-conditioning of the kernel matrix, potentially leading to significantly large biases and highly unstable prediction results.

Keywords: Ill-conditioning · Kernel Matrix · Mercer's Theorem · Vecchia's Approximation

1 Introduction

Machine learning and statistical analysis are two intertwined fields of study that may both rely on the utilization of kernel or covariance matrices. These matrices form the backbone of numerous algorithms and methods, providing an invaluable means to model the underlying relationships and underpinning the powerful predictive capabilities of these fields [2, 3]. The kernel matrix, in particular, is fundamental to kernel-based methods, encapsulating the relationship between every pair of samples in a dataset.

However, as with all powerful tools, the use of kernel and covariance matrices is not without its limitations and challenges. One significant issue arises when the dimension of the matrix or the sample size becomes exceptionally large - the kernel matrix becomes ill-conditioned [3]. This issue can deteriorate the numerical stability and accuracy of the algorithms, causing substantial impediments to their computational efficiency and predictive power.

Over the years, many solutions have been proposed to circumvent this problem, aiming to improve the conditioning of the kernel matrix or devise alternative methods that are less susceptible to this issue. Two key approaches that have gained significant attention in the scientific community include low-rank approximation and the utility of sparse matrices.

Low-rank approximation serves as a prevalent strategy to address the ill-conditioned kernel matrix problem [4–8]. This method involves approximating the original kernel matrix with a matrix of lower rank, preserving the most critical information while discarding less significant details. Low-rank approximation has the potential to overcome

D.-S. Huang et al. (Eds.): ICAI 2023, CCIS 2015, pp. 191–201, 2024.
https://doi.org/10.1007/978-981-97-0827-7_17

the stability issue by effectively reducing the matrix's dimensionality, which can lead to more stable numerical computations. However, the accuracy of the approximation is inherently dependent on the specific structure of the matrix.

Another major approach to dealing with large kernel matrices is to utilize sparse matrices. Several methods have been proposed to leverage the computational efficiency of sparse matrices. One such technique is covariance tapering, which approximates V using a sparse matrix [9–11]. Studies have demonstrated that an appropriate taper leads to asymptotically optimal interpolation and efficient estimation [9, 12]. Another approach involves using block diagonal matrices for sparse covariance matrices [12–14].

More recently, there has been a burgeoning literature on sparse approximations to the precision matrix [15–18]. This sparse approximation to the precision matrix, since there is no need to invert the kernel matrix, is quite efficient computationally. However, an issue that is not well known is that this approximation does not overcome the ill-condition issue.

The primary objective of this paper is to demonstrate how the sparse approximation approach to precision matrices can lead to unstable outcomes, where minor changes in observations may cause significant variations in results. The structure of the paper is as follows: In Sect. 2, we examine findings indicating that kernel matrices become ill-conditioned with large sample sizes, necessitating a more effective approximation method for large kernel matrices. Section 3 delves into and describes the concept of sparse approximation applied to precision matrices. Finally, in Sect. 4, we conduct simulation studies to show that this sparse approximation approach fails to resolve the issue of ill-conditioning. Consequently, it is prone to numerical instability, potentially leading to considerable bias in parameter estimation.

2   Ill-Condition of Kernel Matrix

A symmetric and positive definite matrix is considered ill-conditioned when its condition number, which is the ratio of its largest eigenvalue to its smallest, becomes extremely large. Although the eigenvalues and condition number of the kernel matrix have not been extensively studied, we argue that when the sampling locations s_i represent independent samples from a probability distribution, the implications of the random matrix theorem suggest that the kernel matrix is likely to be ill-conditioned. We refer to [19] for further details.

According to Mercer's Theorem, when K is continuous, it can be expressed as follows:

$$K(s, x) = \sum_{i=1}^{\infty} \lambda_i \phi_i(s) \phi_i(x) \tag{1}$$

where s and x belong to a compact subset $D \subset R^d$, $\lambda_i \geq 0$ decrease to 0 as $i \to \infty$, and the functions $\phi_i(s)$ form an orthonormal system.

Given n points in D, $s_1, s_2, \ldots, s_n$, we denote by V the kernel matrix where the (i, j) th element is $K(s_i, s_j)$. We define $\lambda_{n,i}$ as the i th largest eigenvalue of V. When s_i are an independent random sample drawn from a certain probability distribution, it has been conclusively proven that the ratio $\lambda_{n,i}/n - \lambda$ approaches λ_i. Indeed, probabilistic error bounds for the difference $\lambda_{n,i}/n - \lambda$ have been provided by [20] and [21]. As a result, we can definitively argue that the ratio $\lambda_{n,1}/\lambda_{n,i}$ tends to infinity as both i and n become increasingly large. Consequently, the kernel matrix V becomes ill-conditioned as n grows. Since the precision matrix—the inverse of the kernel matrix—shares the same condition number as the kernel matrix, it too becomes ill-conditioned when n is large.

3 Sparse Approximation to Precision Matrix

Let s_i, where $i = 1, \ldots, n$, be n points in R^d, and $Y(s_i)$ represents observations corresponding to these points. In spatial statistics, $Y(s_i)$ represents observations at location s_i. In machine learning, s_i may be a vector of input variables, and $Y(s_i)$ denotes the associated response variable. In many problems, the ultimate objective is to predict the response at a new location or input denoted by s. We commonly interpret the observations $Y(s_i)$ as a partial realization of a second-order stochastic process, characterized by a covariance or kernel function $K(s, x)$.

Assuming that all variables have mean 0, the best linear unbiased predictor for $Y(s_i)$ is given by the kriging equation:

$$\widehat{Y}(s) = k(s)'V^{-1}Y \tag{2}$$

Here, V is the kernel matrix, with its (i, j)-th element as $K(s_i, s_j)$. The vector $k(s) = (K(s, s_1), K(s, s_2), \ldots, K(s, s_n))'$, and Y is the vector of n observations.

Computing V^{-1} requires $O(n^3)$ operations, which can be computationally expensive. However, a novel approach to directly approximate V^{-1} was proposed in [22] that leverages a fundamental property of probability distributions.

The joint distribution can be expressed as:

$$p(Y) = p(Y(s_1)) \prod_{i=2}^{n} p(Y(s_i)|Y(s_{i-1}), \ldots, Y(s_1)) \tag{3}$$

where $p(\cdot)$ denotes the probability density or conditional probability density. The conditional density $p(Y(s_i)|Y(s_{i-1}), \ldots, Y(s_1))$ can be approximate by the conditional density of $Y(s_i)$ given $Y(s_j)$ for $j < i$ and s_j in a neighborhood of s_i, denoted by O_i. Using this approximation, $\tilde{p}(Y)$ is obtained as follows:

$$\tilde{p}(Y) \approx p(Y(s_1)) \prod_{i=2}^{n} p(Y(s_i)|Y(s_j), j < i, s_j \in O_i) \tag{4}$$

Since the variables are Gaussian, the right-hand side can be expressed as exp $(-Y'QY/2)$, apart from a multiplying constant. Consequently, this Q serves as an approximation to the precision matrix V^{-1} and is likely to be sparse if the neighborhood sizes are small compared to the sample size n. This approach offers an efficient way to compute V^{-1} without incurring the full computational burden of matrix inversion and is particularly advantageous when dealing with large datasets.

The matrix Q is naturally obtained as a product $\Gamma\prime\Gamma$, where Γ is a lower-triangular matrix. This property allows for a straightforward calculation of the determinant of Q by multiplying its diagonal elements together. As a result, the approximation significantly simplifies the computation of the Gaussian likelihood function, which is essential for parameter estimation.

An alternative but equivalent formulation of this approximation is provided in [15, 18] in terms of regression, which was called the Nearest Neighbor Gaussian Process (NNGP) approximation. Consider regressing $Y(s_i)$ on $Y(s_{i-1})$, $Y(s_{i-2})$, ..., $Y(s_1)$, which yields the following equations:

$$Y(s_2) = b_{21} Y(s_1) + \eta_2 \tag{5}$$

$$Y(s_3) = b_{31} Y(s_1) + b_{32} Y(s_2) + \eta_3 \tag{6}$$

$$\cdots \tag{7}$$

$$Y(s_n) = b_{n1} Y(s_1) + \cdots + b_{n,n-1} Y(s_{n-1}) + \eta_n \tag{8}$$

Let $Y(s_1) = \eta_1$ and $\eta = (\eta_1, \eta_2, \ldots, \eta_n)\prime$. It is clear that η_i are independent normal random variables. Equations (5)–(8) can be equivalently written in matrix form as:

$$Y = BY + \eta \tag{9}$$

or

$$(I - B)Y = \eta \tag{10}$$

Then, we have:

$$V = (I - B)^{-1} D(I - B\prime)^{-1} \tag{11}$$

where D is a diagonal matrix representing the covariance matrix of η.

Therefore,

$$V^{-1} = (I - B\prime)D^{-1}(I - B) = \Gamma\prime\Gamma \tag{12}$$

where $\Gamma = D^{-\frac{1}{2}}(I - B)$ is a lower-triangular matrix.

Equation (12) is precise. When $Y(s_i)$ is regressed on $Y(s_j)$ for $s_j \in O_i$, it subsequently substitutes B by another lower-triangular matrix $\tilde{B}$ whose rows have more 0 elements. Then V^{-1} is approximated by

$$Q = (I - \tilde{B}\prime)D^{-1}(I - \tilde{B})$$

By leveraging this approximation, the likelihood function can be efficiently evaluated without the need for inverting the full kernel matrix V. This not only reduces the computational complexity but also improves the numerical stability of the calculations. The proposed approach offers a valuable tool for handling large datasets and conducting statistical inferences in scenarios where traditional methods may become computationally infeasible.

In practical applications, implementing this approximation requires specifying the particular ordering of the n locations and explicitly defining the neighborhoods. Addressing these two issues has been the subject of study in [16], which not only provided numerical algorithms but also offered theoretical justifications to enhance the Kullback-Leibler divergence, a measure often used in statistical inference. The findings presented in [16] offer valuable insights and methodologies to enhance the practical applicability of the sparse approximation to precision matrix, making it a powerful tool for various statistical and machine learning tasks.

Empirical studies have demonstrated that the sparse approximation to the precision matrix yields a matrix that closely resembles the actual precision matrix. However, it also leads to an ill-conditioned matrix. Surprisingly, to the best of our knowledge, there has been no prior investigation into this particular issue. While the implications of ill-conditioning in numerical methods are well-known [23, 24], the specific effects of the ill-conditioning of the precision matrix on kernel learning and statistical inferences remain unexplored. The next section aims to address these gaps and investigate the consequences of the ill-conditioned precision matrix on kernel learning and statistical inferences.

4   Simulation Studies

The objectives of the simulation study in this section are (i) to numerically demonstrate that the precision matrix given by the Vecchia approximation or the nearest neighbor Gaussian process reviewed in the previous section becomes ill-conditioned when n is significantly large and to elucidate the explicit effects this ill-conditioning has on both parameter estimation and prediction, and (ii) to investigate the implications of the ill-conditioning to numerical stability and parameter estimation.

The underlying process to be simulated is assumed to have mean 0 and a kernel from the Matérn family

$$K(s_1, s_2) = \frac{\sigma^2}{\Gamma(v)2^{v-1}} \left(\frac{\|s_1 - s_2\|}{\alpha} \right)^v \mathcal{K}_v \left(\frac{\|s_1 - s_2\|}{\alpha} \right)$$

where $\|s_1 - s_2\|$ is the Euclidean distance between locations s_1 and s_2, $\mathcal{K}_v$ is a modified Bessel function of the second kind with order v. The parameters $\alpha > 0$, $v > 0$ controls the decay in spatial correlation and the process smoothness, respectively. When $v = 0.5$, the covariance function reduces to the exponential covariance. The process become smoother as v increases and become differentiable in the mean square sense when $v > 1$. Indeed, the process is $[v - 1]$ times differentiable where $[x]$ denotes the integer part of $x > 0$.

In the first simulation, the goal is to explore whether the Vecchia approximation can alleviate ill-conditioning. Nine sets of parameter combinations are selected for the simulation, representing varying degrees of correlation strength and process smoothness (Table 1). The 10,000 locations at which the data are simulated are randomly generated from a uniform distribution on the unit square and remain constant across all parameter combinations. Since the sampling locations are randomly selected, existing results regarding the eigenvalues of the kernel matrix [20, 21] are applicable. As such, the kernel matrix for the Matérn kernel becomes ill-conditioned when n is large.

For each parameter set, the corresponding covariance matrix is obtained, and its eigenvalues are computed by the function `eigen` in R. The condition number is subsequently taken as the ratio of the largest eigenvalue to the smallest eigenvalue.

The `GpGP` package in R is used to construct the approximate precision matrices, implementing an efficient algorithm for the Vecchia approximation. Initially, the locations are reordered using the maximum minimum distance ordering with the `order_maxmin` function, according to the suggestion by [16]. Following this, the `find_ordered_nn` function identifies neighbors for each location. A neighbor size of 30 is chosen, as [15] indicates that a neighbor size of 10 can yield results nearly indistinguishable from full models. Finally, the inverse Cholesky decomposition matrix L of the covariance matrix is obtained with the `vecchia_Linv` function, and the approximate precision matrix is constructed by the product $L\prime L$.

The results presented in Table 1 reveal that the NNGP approximation also suffers from ill-conditioning when the true covariance matrix exhibits similar characteristics. This is consistent with previous research, which has indicated that the condition number of a covariance matrix tends to increase with higher values of the range parameter α and the smoothness parameter v [25–27]. Such increases lead to less accurate computation of condition numbers. In fact, the ill-conditioning was so severe that the system failed to compute the condition number when $v = 3.5$, resulting in negative values.

Table 1. Condition Numbers (CN) of the true kernel matrix and the NNGP approximation

σ^2	α	v	Effective Range	CN (NNGP)	CN (true)
1	0.05	0.50	0.150	2.84×10^5	2.81×10^5
1	0.10	0.50	0.300	1.98×10^6	1.96×10^6
1	0.15	0.50	0.449	5.71×10^6	5.65×10^6
1	0.20	0.50	0.599	1.15×10^7	1.14×10^7
1	0.30	0.50	0.899	2.83×10^7	2.81×10^7
1	0.05	1.50	0.237	2.44×10^{10}	2.43×10^{10}
1	0.10	1.50	0.474	6.38×10^{11}	6.36×10^{11}
1	0.15	1.50	0.712	3.86×10^{12}	3.85×10^{12}
1	0.05	2.50	0.296	1.28×10^{13}	1.25×10^{13}
1	0.10	2.50	0.592	1.25×10^{15}	1.23×10^{15}
1	0.15	2.50	0.888	2.15×10^{16}	1.58×10^{16}

The aim of the second simulation is to evaluate the impact of ill-conditioning on estimation and prediction. While there are numerous ways to study this impact, we adopt a simple and direct method here: we perturb the maximum observed value and examine the consequent effect on estimation and prediction. The perturbations involve the addition of $0.01, 0.05, 0.1, 1, 2$, multiplication by a factor of 0.5 and 2, and substitution with the $90\%, 95\%, 99\%$ quantiles of the observed data. Due to the intensive computation required, we employ a single set of parameters $(\sigma^2, \alpha, \nu) = (1, 0.1, 2.5)$ with an effective range of 0.59 for the study. The simulation proceeds as follows.

Simulate $n = 10000$ locations from a uniform distribution on a unit square and randomly divide the data set into $n_o = 9900$ observed locations and $n_p = 100$ predicted locations. These locations remain fixed throughout the entire simulation process. At each simulation step:

1. Simulate $y_1, ..., y_n$ at the n locations. Divide them into observed data and predicted data according to locations. For simplicity, assume $y_1, ..., y_{n_o}$ are the observed data.
2. With the default settings in the GpGp package, estimate model parameters (σ^2, α) using the function fit_model and the observed data $y_1, ..., y_{n_o}$.
3. Use the function predictions in the package GpGp and the observed data to make prediction at the n_p locations. Calculate RMSE:

$$\text{RMSE} = \sqrt{\frac{1}{n_p} \sum_{i=n_o+1}^{n} (y_i - \hat{y}_i)^2} \tag{13}$$

where $\hat{y}_i$ represents the predicted value.
4. Change one value in the observed data set, the maximum, and repeat Steps 2 and 3 to compute the estimation and the prediction using the perturbated dataset.

The above steps are repeated 100 times. We then summarize the results for different objectives. First, we investigate the sensitivity of the estimates to a minor alteration in a single observation value. Our attention will be centered on the microergodic parameter $\theta = \sigma^2/\alpha^{2\nu}$[2]. This parameter can be consistently estimated, whereas the two individual parameters σ^2 and α cannot, under the fixed-domain asymptotic framework [28]. Moreover, it is established that the maximum likelihood estimator $\hat{\theta}$ possesses an asymptotic distribution of $\sqrt{n}(\hat{\theta} - \theta) \sim N(0, 2\theta^2)$[28, 29]. Consequently, it is more insightful to examine how $\hat{\theta}_n$ is influenced by the ill-conditioning.

For each simulated dataset, we calculate the Maximum Likelihood Estimate (MLE) of θ based on both the simulated data and the perturbed data (i.e., the dataset in which the maximum observation has been modified according to various perturbations). The bias and the root mean square error are reported in Table 2. It's remarkable how a single alteration in observation value can greatly bias the estimate. This phenomenon can be attributed to the ill-conditioning of the kernel matrix. Therefore, the sparse Vecchia approximation is extremely susceptible to outliers in the data.

To uncover the effect of perturbation on prediction, for each simulation, we calculate the RMSE by Eq. (13) with and without perturbation and take the difference. The mean and median of 100 such differences are reported in Table 3. These results suggest that even a minor perturbation to the maximum value can significantly impact linear prediction outcomes. The heightened sensitivity of the prediction to minor changes in a single observation value can be attributed to the ill-conditioning of the kernel matrix.

Table 2. Bias and RMSE of MLE of microergodic parameter with and without perturbation

	Bias ($\times 10^4$)	RMSE ($\times 10^4$)
Unperturbed	0.0084	0.15
Add 0.01	2.31	5.82
Add 0.05	186.93	838.20
Add 0.10	779.45	6225.73
Add 1.00	27847.08	90231.28
Add 2.00	203503.68	784167.28
Multiply 0.50	56796.84	171042.91
Multiply 2.00	420850.22	1905289.82
90% quantile	27443.47	90226.73
95% quantile	22906.06	82514.15
99% quantile	2060.06	6951.45

Table 3. Mean and Median of difference between the prediction RMSE with perturbation and RMSE without perturbation (scaled by 10^{-5})

Perturbation	Median	Mean
Add 0.01	0.02	5.34
Add 0.05	0.34	44.70
Add 0.10	0.36	24.66
Add 1.00	62.87	479.05
Add 2.00	251.70	1449.01
Multiply 0.50	48.90	719.48
Multiply 2.00	266.47	1295.43
90% quantile	66.79	498.72
95% quantile	29.82	475.09
99% quantile	4.59	169.14

Since the perturbation influences the parameter estimation, which in turn affects prediction, the differences highlighted in Table 3 demonstrate the confounding effects of ill-conditioning on both estimation and prediction. To isolate the impact of ill-conditioning on prediction only, we repeat the simulation study without estimating the parameters, and calculate predictions using the true parameter values. The results, detailing the difference between the RMSE with perturbation and the RMSE without perturbation, are reported in Table 4. These differences are significantly smaller than those in Table 3. However, considering that the variance of the process is 1, these differences are still considerably large. These results reaffirm that ill-conditioning can cause a significant change even with a small alteration of a single observation value.

Table 4. Mean and Median of difference between the prediction RMSE with perturbation and RMSE without perturbation(scaled by 10^{-5}) while parameters are not estimated

Perturbation	Median	Mean
Add 0.01	0.01	0.90
Add 0.05	0.04	18.13
Add 0.10	0.09	30.80
Add 1.00	0.12	387.14
Add 2.00	0.38	706.66
Multiply 0.50	0.07	753.26
Multiply 2.00	0.09	1041.54
90% quantile	0.09	286.13
95% quantile	0.07	430.89
99% quantile	0.14	259.62

5 Conclusions

The ill-conditioning of the kernel matrix can induce numerical instability and an increased bias for kernel methods. This issue is not extensively studied and may be concealed in practical applications, such as the sparse Vecchia approximation, which does not overcome this problem. When employing a large kernel matrix, it is advisable to ensure the matrix is well-conditioned.

References

1. Schölkopf, B., Smola, A.J.: Learning with Kernels. MIT Press, Cambridge, MA (2002)
2. Stein, M.L.: Interpolation of Spatial Data: Some Theory for Kriging. Springer (1999). https:// doi.org/10.1007/978-1-4612-1494-6
3. Belkin, M.: Approximation beats concentration? An approximation view on inference with smooth radial kernels. Proc. Mach. Learn. Res. **75**, 1–14 (2018)
4. Williams, C., Seeger, M.: Using the Nyström method to speed up kernel machines. In: Leen, T., Dietterich, T., Tresp, V. (eds.) Advances in Neural Information Processing Systems, vol. 13. MIT Press, Cambridge (2000)
5. Drineas, P., Mahoney, M.W.: On the Nystrom method for approximating a gram matrix for improved kernel-based learning. J. Mach. Learn. Res. **6**, 2153–2175 (2005)
6. Bach, F.: Sharp analysis of low-rank kernel matrix approximations. In: JMLR: Workshop and Conference Proceedings, vol. 30, pp. 1–25, May 2013

7. Zhang, K., Tsang, I.W., Kwok, J.T.: Improved Nystrom low-rank approximation and error analysis. In: Proceedings of the 25th International Conference on Machine Learning, pp. 1232–1239. ACM (2008)
8. Gittens, A., Mahoney, M.W.: Revisiting the Nyström method for improved large-scale machine learning. J. Mach. Learn. Res. **17**(1), 3977–4041 (2016)
9. Furrer, R., Genton, M.G., Nychka, D.: Covariance tapering for interpolation of large spatial datasets. J. Comput. Graph. Stat. **15**(3), 502–523 (2006)
10. Kaufman, C.G., Schervish, M.J., Nychka, D.W.: Covariance tapering for likelihood-based estimation in large spatial data sets. J. Am. Stat. Assoc. **103**(484), 1545–1555 (2008). Taylor & Francis
11. Stein, M.L., Chen, J., Anitescu, M., et al.: Stochastic approximation of score functions for Gaussian processes. Ann. Appl. Stat. **7**(2), 1162–1191 (2013). Institute of Mathematical Statistics
12. Du, J., Zhang, H., Mandrekar, V.S.: Fixed-domain asymptotic properties of tapered maximum likelihood estimators. Ann. Stat. **37**(6A), 3330–3361 (2009)
13. Stein, M.L., Chi, Z., Welty, L.J.: Approximating likelihoods for large spatial data sets. J. Roy. Stat. Soc. Ser. B (Stat. Methodol.) **66**(2), 275–296 (2004). Wiley
14. Eidsvik, J., Shaby, B.A., Reich, B.J., Wheeler, M., Niemi, J.: Estimation and prediction in spatial models with block composite likelihoods. J. Comput. Graph. Stat. **23**(2), 295–315 (2014). Taylor & Francis
15. Datta, A., Banerjee, S., Finley, A.O., Gelfand, A.E.: Hierarchical nearest-neighbor gaussian process models for large geostatistical datasets. J. Am. Stat. Assoc. **111**(514), 800–812 (2016)
16. Guinness, J.: Permutation and grouping methods for sharpening gaussian process approximations. Technometrics **60**(4), 415–429 (2018)
17. Datta, A.: Sparse nearest neighbor Cholesky matrices in spatial statistics (2021). arXiv:2102. 13299 [stat]
18. Datta, A.: Nearest-neighbor sparse Cholesky matrices in spatial statistics. WIREs Comput. Stat. **14**(5), e1574 (2022)
19. Zhang, H.: Spatial process approximations: assessing their necessity (2023). arXiv:2311. 03201 [stat.ML]
20. Braun, M.L.: Accurate error bounds for the eigenvalues of the kernel matrix. J. Mach. Learn. Res. **7**(82), 2303–2328 (2006)
21. Jia, L., Liao, S.: Accurate probabilistic error bound for eigenvalues of kernel matrix. In: Zhou, Z.-H., Washio, T. (eds.) ACML 2009. LNCS (LNAI), vol. 5828, pp. 162–175. Springer, Heidelberg (2009). https://doi.org/10.1007/978-3-642-05224-8_14
22. Vecchia, A.V.: Estimation and model identification for continuous spatial processes. J. Roy. Stat. Soc. B **50**, 297–312 (1988)
23. Golub, G.H., Loan, C.F.V.: Matrix Computations, 4th edn. Johns Hopkins University Press, Baltimore (2012)
24. Trefethen, L.N., Bau III, D.: Numerical Linear Algebra. Society for Industrial and Applied Mathematics (SIAM) (1997)
25. O'Dowd, R.: Conditioning of coefficient matrices of ordinary kriging. Math. Geol. **23**, 721–739 (1991)
26. McCourt, M., Fasshauer, G.E.: Stable likelihood computation for gaussian random fields. In: Recent Applications of Harmonic Analysis to Function Spaces, Differential Equations, and Data Science: Novel Methods in Harmonic Analysis, vol. 2, pp. 917–943 (2017)
27. Basak, S., Petit, S., Bect, J., Vazquez, E.: Numerical issues in maximum likelihood parameter estimation for Gaussian process interpolation. In: Nicosia, G., et al. (eds.) LOD 2021. LNCS, vol. 13164, pp. 116–131. Springer, Cham (2021). https://doi.org/10.1007/978-3-030-95470-3_9

28. Zhang, H.: Inconsistent estimation and asymptotically equal interpolations in model-based geostatistics. J. Am. Stat. Assoc. **99**(465), 250–261 (2004)
29. Wang, D., Loh, W.L.: On fixed-domain asymptotics and covariance tapering in Gaussian random field models. Electron. J. Statist **5**, 238–269 (2011)

Deepfake Detection Performance Evaluation and Enhancement Through Parameter Optimization

Bowen Pei[1] , Jingyi Deng[1] , Chenhao Lin[1(✉)] , Pengwei Hu[1,2] ,
and Chao Shen[1,2]

[1] School of Cyber Science and Engineering, Faculty of Electronic and Information Engineering,
Xi'an Jiaotong University, Xi'an 710049, China
linchenhao@xjtu.edu.cn

[2] The Xinjiang Technical Institute of Physics and Chemistry, Chinese Academy of Sciences,
Beijing, China

Abstract. Deepfake technology has become a subject of concern due to its potential for spreading misinformation and facilitating deceptive activities. To address these issues, various deepfake detection approaches have been developed with similar training paradigms. Then a natural question is which parameters are critical to achieving better detection performance. This study aims to evaluate and optimize the performance of existing deepfake detection systems by analyzing key parameters in the training paradigm. Specifically, we systematically analyze four crucial factors: image cropping, sampling rate, data augmentation, and transfer learning. The impact of different image scopes, such as utilizing the entire image or only the cropped face region, is investigated. We also explore how varying the sampling rate and employing data augmentation techniques can enhance the diversity of the training dataset. Additionally, transfer learning with pre-trained models is leveraged to improve detection accuracy. Through comprehensive experiments and evaluations of several popular and state-of-the-art detection methods, optimal configurations within each factor are identified, providing valuable insights to enhance the efficiency and effectiveness of deepfake detection systems. Given the widespread use and potential negative consequences of deepfake technology, reliable detection systems are crucial in combatting the harmful effects of manipulated media.

Keyword: Deepfake detection · Digital image forensics · Generative adversarial network

This research is supported by the National Key Research and Development Program of China (2020AAA0107702), the National Natural Science Foundation of China (62006181, 62161160337, 62132011, U21B2018, U20A20177, 62206217), the Shaanxi Province Key Industry Innovation Program (2023-ZDLGY-38, 2021ZDLGY01-02).

D.-S. Huang et al. (Eds.): ICAI 2023, CCIS 2015, pp. 202–213, 2024.
https://doi.org/10.1007/978-981-97-0827-7_18

1   Introduction

In the past few years, there has been a significant rise in the development of generative models that utilize machine learning algorithms to create realistic images and videos [6, 9, 10]. Among these models, deepfakes have received a lot of attention because of their potential use in computer vision, including in fields such as special effects production [28], gaming [18], and virtual reality [19]. With the ability to replace faces in pictures or videos with those of other individuals, deepfake technology produces highly convincing but entirely artificial media. However, this technology has also been exploited for malicious purposes such as spreading misinformation or committing financial fraud, causing mounting concerns about the future of information sharing and social trust [11].

In this study, our aim is to evaluate and optimize the performance of existing deepfake detection systems by evaluating four key factors. These factors include image scope, sampling rate, transfer learning, and a combination of data augmentation techniques such as Mixup and Cutmix. Image scope refers to the decision of whether the detection process focuses on the entire image or only the face region. The sampling rate determines the frequency at which frames are analyzed for deepfake detection. Transfer learning involves fine-tuning pre-trained models on new datasets to enhance their performance in detecting deepfakes [22]. Additionally, we employ data augmentation techniques, such as Mixup [26] and Cutmix [25], to increase the diversity of the training dataset. These techniques involve merging pairs of images or regions within images to create new samples with labels that are weighted averages of the original labels. The evaluation of these factors provides valuable insights for improving the efficiency and effectiveness of deepfake detection systems.

In summary, the proliferation of deepfake technology has introduced numerous possibilities in computer vision, but it has also raised substantial concerns regarding the detection of manipulated media. This study aims to optimize the performance of existing deepfake detection systems by evaluating key factors that influence their effectiveness. By systematically assessing different strategies related to image scope, sampling rate, data augmentation, and transfer learning within these systems, we provide valuable insights into enhancing their efficiency and accuracy. These findings contribute to the ongoing efforts in terms of mitigating the harmful consequences of manipulated media by improving the existing deepfake detection systems.

2   Related Work

With the rapid development of deep learning, the application of Generative Adversarial Network [6] is more and more widely. There is a kind of application that can change the face of a person in a video, which is called DeepFake technology. This need for face manipulation is common in the production of special effects in movies, but the cost of making special effects is very high, and special effects engineers who master professional technology are needed. Deepfake can replace the specified face in a video or a picture with the target face. It can achieve good results at a very low cost. Some well-made deepfake videos can't even be distinguished by the naked eye.

2.1 DeepFake Creation

Deepfake creation is the process of manipulating videos to replace a person's face with another person's face or a computer-generated face. Two popular technologies for creating deepfakes are Generative Adversarial Networks (GANs) and autoencoders [2]. GANs consist of two neural networks: a generator and a discriminator. The generator network creates new images of faces, while the discriminator network tries to distinguish between real and fake images. The process of using GANs for deepfake creation involves several technical steps. First, a dataset of images of the target person's face is collected and used to train the generator network. The discriminator network is also trained on a separate dataset of images of the target person's face, along with a dataset of images of other people's faces. Once both networks are trained, an image of another person's face is fed into the generator network to generate a new image of the target person's face. The discriminator network then tries to distinguish between the real and generated images and provides feedback to the generator network on how to improve its output.

2.2 DeepFake Detection

Due to the negative effects of deepfake, it is necessary to give an automatic solution to identify whether the video is deepfake or not. Many researchers have done a lot of work on how to identify deepfake videos automatically. Hand-crafting facial features such as eye color and incomplete mirrors [12], 3D head poses [24], and facial expressions [1, 3] were the subject of early face forensic work. These methods, on the other hand, do not scale well to larger and more complex deepfakes. Researchers are using recent developments in deep learning to automatically remove discriminative features for forgery detection to solve this problem [13, 21]. Previous work, such as Xception [14], achieved state-of-the-art by fine-tuning ImageNet-based models.

3 Methodology

This section outlines our method. Firstly, we present the four selected contrast variables and explain why we chose them. Additionally, we describe our experimental setup and the following evaluation criteria.

Regarding data preparation, the data used can either be the original image or the face image processed by face extraction. We used the popular face detection model, MTCNN (Multi-Task Cascaded Convolutional Neural Network) [27], which can detect faces and their landmarks in images for data extraction during the experiment. Figure 1 illustrates the results after applying MTCNN. We selected three different face extraction ratios and compared their performance to determine which one could yield the best detection results.

Secondly, when developing a detection model, it is necessary to consider either single-frame models or multi-frame models. Single-frame models exhibit greater efficiency in terms of processing speed and memory usage. However, multi-frame models are capable of identifying discontinuities between frames.

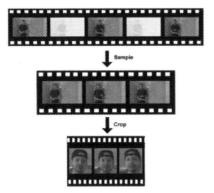

Fig. 1. Pipeline of data preprocessing

Considering the utilization of pre-trained models is an important aspect. The training duration of a neural network can vary from minutes to days, depending on factors such as the dataset size and model parameters. The utilization of transfer learning improves the efficiency of the model training process. Nevertheless, variations in the dataset can impact the performance of the pre-trained model. Consequently, conducting experiments is necessary to evaluate the influence of transfer learning on deepfake detection.

Another crucial aspect to consider is the utilization of data augmentation techniques, specifically Multiple Source Domain Adaptation (MSDA) methods like Mixup and Cutmix. Employing data augmentation methods addresses the issue of class imbalance within the dataset and simultaneously enhances the dataset's size.

We tested the four consideration factors we proposed on several single-frame and multi-frame models and gave our recommended strategies based on the experimental results.

3.1 Models

Through the conducted experiments, we confirmed the validity of the four hypothses. To ensure the generalizability of the experimental findings, we conducted experments on both single-frame CNN (Convolutional Neural Network) and multi-frame CNN models, with an image size of 224×224. The single-frame CNN models utilized are SE-ResNeXt50 and [15]. SE-ResNeXt50 is a variant of ResNext [23] that incorporates squeeze-and-excitation blocks [8] for dynamic channel-wise feature recalibration. SBI adopts EfficientNet-b4 [17] as its classifier and trains the model with self-supervised data. The multi-frame CNN models employed in the experiment include I3D [4], 3D ResNet34 [7] and MC3 [20]. All of these models utilize 3D convolutional filters to extract spatial-temporal discriminative features. In contrast to the traditional single-frame CNN, multi-frame CNN models are capable of capturing inter-frame motion information along the temporal dimension. The CNN models were trained on the DFDC dataset using the Adam optimizer, with an initial learning rate of $1e-5$ and weight decay of $1e-6$.

3.2 Datasets

We test our methods on the Deepfake Detection Challenge (DFDC) dataset [5]. This is a large dataset for the competition. This dataset consists of a large number of videos containing human faces. The length of each video is ten seconds and 300 frames. In this dataset, only 10% videos are real videos, and other videos are fake videos manipulated by various deepfake methods based on the real videos. Therefore, it is an unbalanced dataset.

In the data preprocessing stage, we used MTCNN to extract faces. For the problem of imbalanced data sets, we down-sample fake data randomly in each iteration.

3.3 Metrics

Since we want to design a detection system, we only need to determine whether the input video is real or is generated by deepfake technology. So the output of our system is a number that represents the probability that the input video is real. Therefore, we used the detection accuracy as an indicator to evaluate the effect of the model. The formula is shown in Eq. 1

$$Accuracy = \frac{TP + TN}{TP + TN + FP + FN} \tag{1}$$

where TP, FP, TN, and FN indicate true positive, false positive, true negative, and false negative respectively. The true positive is an outcome where the model correctly predicts the positive class. Similarly, true negative is an outcome where the model correctly predicts the negative class. False positive is an outcome where the model incorrectly predicts the positive class. Additionally, the false negative is an outcome where the model incorrectly predicts the negative class.

We use cross entropy loss in our training. The calculation method of cross entropy loss is shown in Eq. 2

$$Loss = -\frac{1}{n}\sum_{i=1}^{n}\left[y_i log\left(\widehat{y_i}\right) + (1 - y_i)log\left(1 - \widehat{y_i}\right)\right] \tag{2}$$

where n is the number of videos being predicted. $\widehat{y_i}$ is the predicted probability of the video being fake. y_i is 1 if the video is fake, 0 if real. $log()$ is the natural logarithm.

4 Experiment

4.1 Experiment Setup

If not specified, all experiments in this section are conducted on a server with Intel(R) Xeon Silver 4210R 2.4GHz 10-core processors, 188 GB of RAM, and two NVIDIA RTX 3090 GPUs running Ubuntu 20.04 as the operating system.

4.2 Whole Image vs. Clipped Image

We use the idea of hypothesis testing to conduct our experiments. This is a method commonly used in statistics. First, we propose the null hypothesis h_0 and alternative hypothesis h_1 as follows:

h_0: Using the extracted face image is better than using the whole image in the deepfake detection task.

h_1: Using the whole image is better than using the extracted face image in the deepfake detection task.

Experiment Design: The DFDC dataset is officially split into training, validation, and test sets. We use the training set of DFDC dataset for training and adopt the validation and the test sets for performance evaluation of the trained models. For multi-frame CNN models, We conduct experiments on I3D [4], ResNet34 [7], MC3 [20] and SBI [15] models respectively. We resize the image to 224*224 resolution and feed it into the model. We choose Adam optimizer with an initial learning rate of 1e-3 and we choose OneCycleLR [16] as our learning rate scheduler. We set epochs = 20, batch size = 8, and train the model on two GPUs. For single-frame CNN models, we conduct experiments on ResNeXt50 [23] with similar parameter settings. The pipeline of the training process is shown in Fig. 2.

Results: The value of loss and accuracy is shown in Table 1. From the table, we can conclude that the results on the extracted face image are better. So we accept the null hypothesis: using the extracted face image is better than using the whole image. The model using the whole image predicts similar results as a random guess since the loss is greater than 0.693. So the model learned almost nothing from the whole image dataset (Fig. 2).

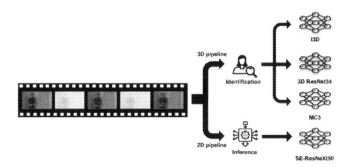

Fig. 2. Pipeline of training models.

Analysis: The results are reasonable because deepfake focuses on facial manipulation, so the main difference between a real image and a deepfake image is in the face area. Therefore the process of extracting face using MTCNN [27] increases the difference and also eliminates the redundant information. So we get a better effect using the extracted face image in the deepfake detection task. Because the whole image is too large, it is

Table 1. Loss of different multi-frame models

Image Extraction	multi-frame Models			
Range	I3D	Resnet34	MC3	SBI
Extracted Image Expanded by 30%	0.377	0.296	0.218	0.762
Extracted Image	0.362	0.272	0.185	0.903
Extracted Image Reduced by 30%	0.383	0.311	0.244	1.010
Whole Image	0.706	0.634	0.692	/

difficult for the model to extract information from the whole image that is conducive to identifying real and fake.

4.3 Discrete Frames vs. Consecutive Frames

First, we propose the null hypothesis h_0 and alternative hypothesis h_1 as follows:

h_0: Using discrete frames is better than using continuous frames in the deepfake detection task.

h_1: Using continuous frames is better than using discrete frames in the deepfake detection task (Table 2).

Table 2. Loss and Accuracy of 2D model

Image Extraction Range	Loss	Accuracy
Extracted Image	0.260	0.89
Whole Image	0.694	0.50

Experiment Design: We use the strategy of extracting one frame every five frames to extract 60 frames for each video. Then adopting MTCNN to extract the face in the image. We use the training set of the DFDC dataset for training, the validation, and the test set to evaluate the performance of the trained models. In the model evaluation phase, 60 frames are extracted continuously and discretely for each video. The trained model is used to infer the label and the binary Cross entropy loss value is calculated.

Results: The value of loss and accuracy is shown in Table 3. From the table, we can conclude that the model performance evaluated on the validation set and the test set are consistent. So we accept the null hypothesis: using continuous frames is better than using discrete frames (Table 4).

Table 3. Loss of different multi-frame models

| Temporal Sampling | multi-frame Models | | | |
Rate	I3D	Resnet34	MC3	SBI
Consecutive	0.362	0.272	0.185	0.903
Discrete	0.524	0.326	0.424	0.924

Table 4. Loss and Accuracy of 2D model

Max Frames	Loss	Accuracy
1	0.32	0.86
2	0.24	0.89
3	0.64	0.63

Analysis: Continuous frame means sampling at 30 frames per second, and discrete frame sampling at 6 frames per second. This experimental result shows that the difference between deepfake and real videos is more obvious at 30 Hz sampling rate than at 6 Hz sampling rate, and the sampling rate is a factor that needs to be considered when designing a deepfake detection system.

4.4 Pretrained Model vs. Model Trained from Scratch

First, we propose the null hypothesis h_0 and alternative hypothesis h_1 as follows:

h_0: Using the pretrained model is better than using the training model from scratch in the deepfake detection task.

h_1: Using the pretrained model is no better than using the training model from scratch in the deepfake detection task.

Experiment Design: Firstly, we used clipped continuous images to train the model from scratch. And then used the trained model from the first step to train our new model.

Results: The value of loss and accuracy is shown in Table 5. From the table, we conclude that for different models, the impact of whether to use the pre-trained model is different. So we accept the alternative hypothesis: using the pretrained model is better than the training model from scratch in the deepfake detection task (Table 6).

Analysis: The result is reasonable, because the pre-training models of different models may be trained on different datasets, and the gap between the dataset used for pre-training and the deepfake data set may be different.

Table 5. Loss of different multi-frame models

Transfer Learning	multi-frame Models			
	I3D	Resnet34	MC3	SBI
Train from Scratch	0.362	0.272	0.185	0.834
Pretrained Model	0.352	/	/	0.903

Table 6. Loss and Accuracy of 2D models

Transfer Learning	Loss	Accuracy
Train from Scratch	0.693	0.50
Pretrained Model	0.260	0.89

4.5 Data Augmentation vs. Not use

First, we propose the null hypothesis h_0 and alternative hypothesis h_1 as follows:

h_0: Using the data augmentation method is better than not using it (Table 8).

h_1: Using the data augmentation method does not increase performance.

Experiment Design: We use CutMix and Mixup methods to augment the original dataset. In CutMix augmentation, we cut and paste random patches between the training images, where the ground truth labels are mixed in proportion to the area of patches in the images. In Mixup augmentation, we implement linear interpolations of random pairs of training images and proportionally mix their ground truth labels. Then we trained our model on the augmented dataset.

Results: The value of loss and accuracy is shown in Table 7. From the table, we can conclude that the results evaluated on the validation set and the test set are consistent. So we accept the null hypothesis: using the Data Augmentation method is better than not using it.

Table 7. Loss of different multi-frame models

Data Augmentation	multi-frame Models			
	I3D	Resnet34	MC3	SBI
Not use	0.362	0.272	0.185	0.903
Data Augmentation	0.325	0.373	/	0.793

Analysis: The augmentation of Cuxmix increases artifact localization ability by making the model focus on less discriminative parts of the object being classified, and Mixup guides the network to explore more fine-grained discriminative features.

Table 8. Loss and Accuracy of 2D models

Data Augmentation	Loss	Accuracy
Not use	0.383	0.88
Data Augmentation	0.260	0.89

5 Discussion

Our study explores several factors that impact the accuracy and efficiency of deepfake detection systems, including image scope, sampling rate, transfer learning, and data augmentation techniques. By comparing different strategies, we aim to identify the most effective approaches for detecting deepfakes and improving the overall performance of deepfake detection systems.

One key finding of our study is that the scope of the image analyzed significantly affects the accuracy of deepfake detection systems. Our results indicate that analyzing only the face region of an image provides a higher accuracy rate than analyzing the entire image. This is likely due to the fact that deepfake techniques primarily manipulate facial features rather than the entire image.

Another finding of our study is that data augmentation techniques, including Mixup and Cutmix, can significantly improve the performance of deepfake detection systems. By combining pairs of images or regions within images, these techniques create new samples with weighted averages of the original labels, which improves the generalization and robustness of the deep learning models. Our experiments show that Cutmix, which involves combining regions within images, is particularly effective in improving the accuracy of deepfake detection.

Finally, we find that transfer learning, which involves fine-tuning a pre-trained model on a new dataset, can improve the performance of deepfake detection systems. This approach enables the model to learn from a larger and more diverse dataset, improving its ability to detect deepfakes. Our results show that transfer learning with a pre-trained face recognition model can significantly improve the accuracy of deepfake detection systems.

6 Conclusion

In conclusion, our study demonstrates that deepfake detection is a challenging task that requires careful consideration of several key factors. By evaluating the impact of image scope, sampling rate, transfer learning, and data augmentation techniques on the performance of deepfake detection systems, we provide valuable insights into effective strategies for detecting deepfakes.

Our findings suggest that analyzing only the face region of an image, employing data augmentation techniques such as Cutmix, and using transfer learning with a pre-trained face recognition model can significantly improve the accuracy and efficiency of deepfake detection systems. These insights can inform the development of more effective deepfake detection methods and contribute to mitigating the potential harm caused by deepfakes in society.

References

1. Agarwal, S., Farid, H., Gu, Y., He, M., Nagano, K., Li, H.: Protecting world leaders against deep fakes. In: CVPR Workshops, pp. 38–45 (2019)
2. Baldi, P.: Autoencoders, unsupervised learning, and deep architectures. In: Guyon, I., Dror, G., Lemaire, V., Taylor, G.W., Silver, D.L. (eds.) Unsupervised and Transfer Learning - Workshop held at ICML 2011, Bellevue, Washington, USA, July 2, 2011. JMLR Proceedings, vol. 27, pp. 37–50. JMLR.org (2012). http://proceedings.mlr.press/v27/baldi12a.html
3. Baltrusaitis, T., Zadeh, A., Lim, Y.C., Morency, L.: Openface 2.0: Facial behavior analysis toolkit. In: 13th IEEE International Conference on Automatic Face & Gesture Recognition, FG 2018, Xi'an, China, May 15–19, 2018, pp. 59–66. IEEE Computer Society (2018). https://doi.org/10.1109/FG.2018.00019
4. Carreira, J., Zisserman, A.: Quo vadis, action recognition? a new model and the kinetics dataset. In: Proceedings of the IEEE Conference on Computer Vision and Pattern Recognition, pp. 6299–6308 (2017)
5. Dolhansky, B., Bitton, J., Pflaum, B., Lu, J., Howes, R., Wang, M., Ferrer, C.C.: The deepfake detection challenge (dfdc) dataset. arXiv preprint arXiv:2006.07397 (2020)
6. Goodfellow, I., et al.: Generative adversarial networks. Commun. ACM. ACM **63**(11), 139–144 (2020)
7. He, K., Zhang, X., Ren, S., Sun, J.: Deep residual learning for image recognition. In: Proceedings of the IEEE Conference on Computer Vision and Pattern Recognition, pp. 770–778 (2016)
8. Hu, J., Shen, L., Sun, G.: Squeeze-and-excitation networks. In: Proceedings of the IEEE Conference on Computer Vision and Pattern Recognition, pp. 7132–7141 (2018)
9. Karras, T., Laine, S., Aila, T.: A style-based generator architecture for generative adversarial networks. In: Proceedings of the IEEE/CVF Conference on Computer Vision and Pattern Recognition, pp. 4401–4410 (2019)
10. Kumar, M., Babaeizadeh, M., Erhan, D., Finn, C., Levine, S., Dinh, L., Kingma, D.: Videoflow: a flow-based generative model for video. arXiv preprint arXiv:1903.01434 2(5), 3 (2019)
11. Masood, M., Nawaz, M., Malik, K.M., Javed, A., Irtaza, A., Malik, H.: Deepfakes generation and detection: state-of-the-art, open challenges, countermeasures, and way forward. Applied Intelligence, pp. 1–53 (2022)
12. Matern, F., Riess, C., Stamminger, M.: Exploiting visual artifacts to expose deepfakes and face manipulations. In: 2019 IEEE Winter Applications of Computer Vision Workshops (WACVW), pp. 83–92. IEEE (2019)
13. Nguyen, H.H., Yamagishi, J., Echizen, I.: Capsule-forensics: Using capsule networks to detect forged images and videos. In: IEEE International Conference on Acoustics, Speech and Signal Processing, ICASSP 2019, Brighton, United Kingdom, May 12–17, 2019, pp. 2307–2311. IEEE (2019). https://doi.org/10.1109/ICASSP.2019.8682602
14. Rössler, A., Cozzolino, D., Verdoliva, L., Riess, C., Thies, J., Nießner, M.: Faceforensics++: learning to detect manipulated facial images. In: 2019 IEEE/CVF International Conference on Computer Vision, ICCV 2019, Seoul, Korea (South), October 27 - November 2, 2019, pp. 1–11. IEEE (2019). https://doi.org/10.1109/ICCV.2019.00009
15. Shiohara, K., Yamasaki, T.: Detecting deepfakes with self-blended images. In: Proceedings of the IEEE/CVF Conference on Computer Vision and Pattern Recognition, pp. 18720–18729 (2022)

16. Smith, L.N., Topin, N.: Super-convergence: Very fast training of neural networks using large learning rates. In: Artificial Intelligence and Machine Learning for Multi-domain Operations Applications, vol. 11006, pp. 369–386. SPIE (2019)
17. Tan, M., Le, Q.: Efficientnet: rethinking model scaling for convolutional neural networks. In: International Conference on Machine Learning, pp. 6105–6114. PMLR (2019)
18. Tariq, S., Abuadbba, A., Moore, K.: Deepfake in the metaverse: security implications for virtual gaming, meetings, and offices. arXiv preprint arXiv:2303.14612 (2023)
19. Tolosana, R., Vera-Rodriguez, R., Fierrez, J., Morales, A., Ortega-Garcia, J.: Deepfakes and beyond: a survey of face manipulation and fake detection. Inf. Fusion **64**, 131–148 (2020)
20. Tran, D., Wang, H., Torresani, L., Ray, J., LeCun, Y., Paluri, M.: A closer look at spatiotemporal convolutions for action recognition. In: Proceedings of the IEEE Conference on Computer Vision and Pattern Recognition, pp. 6450–6459 (2018)
21. Wang, R., Juefei-Xu, F., Ma, L., Xie, X., Huang, Y., Wang, J., Liu, Y.: Fakespotter: a simple yet robust baseline for spotting ai-synthesized fake faces. In: Bessiere, C. (ed.) Proceedings of the Twenty-Ninth International Joint Conference on Artificial Intelligence, IJCAI 2020, pp. 3444–3451. ijcai.org (2020). https://doi.org/10.24963/ijcai.2020/476
22. Weiss, K., Khoshgoftaar, T.M., Wang, D.: A survey of transfer learning. J. Big Data **3**(1), 1–40 (2016)
23. Xie, S., Girshick, R., Dollár, P., Tu, Z., He, K.: Aggregated residual transformations for deep neural networks. In: Proceedings of the IEEE Conference on Computer Vision and Pattern Recognition, pp. 1492–1500 (2017)
24. Yang, X., Li, Y., Lyu, S.: Exposing deep fakes using inconsistent head poses. In: ICASSP 2019–2019 IEEE International Conference on Acoustics, Speech and Signal Processing (ICASSP), pp. 8261–8265. IEEE (2019)
25. Yun, S., Han, D., Oh, S.J., Chun, S., Choe, J., Yoo, Y.: Cutmix: regularization strategy to train strong classifiers with localizable features. In: Proceedings of the IEEE/CVF International Conference on Computer Vision, pp. 6023–6032 (2019)
26. Zhang, H., Cisse, M., Dauphin, Y.N., Lopez-Paz, D.: mixup: beyond empirical risk minimization. arXiv preprint arXiv:1710.09412 (2017)
27. Zhang, K., Zhang, Z., Li, Z., Qiao, Y.: Joint face detection and alignment using multitask cascaded convolutional networks. IEEE Signal Process. Lett. **23**(10), 1499–1503 (2016)
28. Zoss, G., Chandran, P., Sifakis, E., Gross, M., Gotardo, P., Bradley, D.: Production-ready face re-aging for visual effects. ACM Trans. Graph. (TOG) **41**(6), 1–12 (2022)

Prediction Intervals of Principal Component Regression with Applications to Molecular Descriptors Datasets

Yuling Fu[1,2], Zixin Bin[1], Ligong Wei[3(✉)], and Youwu Lin[1,2(✉)]

[1] School of Mathematics and Computational Sciences, Guangxi Colleges and Universities Key Laboratory of Data Analysis and Computation, Guilin University of Electronic Technology, Guilin 541002, Guangxi, China
lyw015813@126.com

[2] Center for Applied Mathematics of Guangxi (GUET), Guilin 541002, People's Republic of China

[3] Guangxi Academy of Sciences High-Tech Group Co., LTD., Nanning 530007, Guangxi, China
22216304@qq.com

Abstract. The prediction interval of the Principal Component Regression (PCR) model is usually based on some strong distributional assumptions. To overcome this drawback, this study extends a prediction interval estimation method by considering a condition of distribution-free. Six different prediction interval estimation methods are developed for constructing prediction intervals for PCR models. The simulated and real data experiment results show that the developed methods perform better than some state-of-the-art methods. This study can enrich the tools of PCR model prediction and statistical inference and is significant for data analysis.

Keywords: Principal component regression · Prediction intervals · Distribution-free · Jackknife

1 Introduction

Data analysis has undergone significant changes in recent years with the rapid development of data collection technology and computing technology. Data with large dimensions and relatively small sample sizes is often obtained. Let n represent the sample size, and p represent the corresponding feature number of the sample. Data with "large p small n" or "large n large p" characteristics is called high-dimensional data. This type of data is generated in many fields, such as bioinformatics, finance, weather forecasting and satellite imagery. High dimensional data often features high dimensionality and high correlation between data, making traditional machine learning or statistical learning methods no longer applicable. Dimensionality reduction is a commonly used method to analyze high-dimensional data.

Principal Component Regression (PCR) is a commonly used method for dimensionality reduction in machine learning. Although the PCR model is widely utilized for point estimation (prediction) of unknown response variables, a particular error exists between the predicted value and the real value due to the presence of a large amount of noise, and the established model merely approximates the real model. Measuring the closeness between real and predicted values has always been significant in data analysis. The prediction interval is an effective analysis method for measuring the degree of approximation between predicted and real values.

Many scholars have proposed a variety of methods for constructing prediction intervals. For example, Patel [1] reviewed methods for constructing prediction intervals in 1989, and many important methods for constructing prediction intervals have been proposed since then. For example, the pivotal methods proposed by Cox [2] and Barndorff-Nielsen et al. [3].

While numerous methods have been proposed to construct prediction intervals, most are based on strong distribution assumptions. Additionally, because the statistical properties of parameter estimation for PCR models have not been fully studied, the construction of prediction intervals for PCR models based on distribution hypothesis has yet to be thoroughly investigated. At the same time, the statistical inference method based on strong distribution assumptions needs revision to verify whether the assumed distribution is suitable when analyzing real data. All these issues make it more challenging to estimate the prediction interval of the PCR model.

In recent years, some renowned statisticians have investigated methods for constructing prediction intervals under distribution-free conditions. For instance, Stine [4] and Efron [5] employed the Jackknife method to estimate the interval of response variables, demonstrating its efficacy in reducing estimation errors. However, the Jackknife method may lose its predictive coverage when the modeling method is unstable. Therefore, Barber et al. [6] improved the Jackknife method and introduced the Jackknife + method, which incorporates the leave-one-out method. When the model is stable, the Jackknife + method can achieve higher coverage with similar lengths to the Jackknife method. Notably, the Jackknife + method exhibits greater stability when the model is unstable, employing a leave-one-out regression model instead of a full one. Additionally, Barber et al. [6] proposed the Cross-validation + (CV+) method, which partitions the training dataset into L subsets to obtain L leave-one-out block models. When L equals the number of samples n, , the Jackknife + can be considered a special case of the CV + method. Kim et al. [7] combined the advantages of ensemble learning and the Jackknife + method, introducing the Jackknife + after-bootstrap (J + aB) method, which utilizes available bootstrapped samples. Lin et al. [8] proposed the leave-one-bag-out + (LOBO+) method, which extracts a certain proportion of samples by equally returning them. The LOBO + method is stable and can generate more residuals, enhancing the accuracy of the model. However, the estimation of prediction intervals for the PCR model under distribution-free conditions has yet to be explored. Motivated by the aforementioned studies, this paper introduces several prediction interval construction methods under distribution-free conditions into the PCR model. The proposed method is applied to analyze medical data, and its advantages are demonstrated by comparing numerical results with several different prediction interval construction methods. The results show that the proposed method has been well applied, enriching the statistical inference tools for PCR model prediction.

The main results of this paper can be summarized as follows: (1) Our method offers a means to construct PCR prediction intervals without assuming data distribution; (2) Our method proves to be suitable for diverse types of data, including high- and low-dimensional data, and is applicable in real data analysis; (3) Our method has better performance, with higher coverage and shorter interval length. Furthermore, our method can be extended to more types of real data analysis.

2　Notation

$\mu(x) = E(Y|X = x), x \in R^p$ is the regression function. $\hat{\mu}(x)$ is the estimate of $\mu(x)$ obtained by the sample. $\mu_{-D}(x)$ represents the regression function after the subsample D is subtracted, and $\hat{\mu}_{-D}(x)$ represents the estimate of the regression function obtained by subtracting the sample from the subset D. $R_i = |Y_i - \hat{\mu}(x_i)|, i = 1, 2, ..., n$ is the fitting residual. $R_{-i} = |Y_i - \hat{\mu}_{-D}(x_i)|$ represents the residual obtained after removing the subset D. When the number of samples in the subset is 1, it is called the leave-one-out method. R_{-i} represents the residual obtained after subtracting the i sample. $Q_{n,\alpha}\{R_i\}$ and $Q_{n,1-\alpha}\{R_i\}$ are the α-quantile and $1 - \alpha$-quantile of sequence $R_1, R_2, ..., R_n$, where α is the mis-coverage level.

3　Theory and Algorithm

Consider the linear regression model

$$Y = X\beta + \varepsilon,$$

where X is a $n \times p$ data matrix, Y is a single response vector of $n \times 1$, β is the regression coefficient, and ε is a random error vector of size $n \times 1$ with $N(0, \sigma^2)$.

To estimate β, the PCR method is used in this paper, and the estimation formula of the PCR regression coefficient is as follows:

$$\hat{\beta}_{k,PCR} = V(V^T \sum_x V)^{-1}V^T \sigma_{xy},$$

where $\hat{\beta}_{k,PCR}$ is a $p \times k$ matrix, $V = (\xi_1, ..., \xi_k)$ is the eigenvector of the x covariance matrix, and k is the number of components.

The following subsections will introduce the theory and algorithm of several prediction interval estimation methods.

3.1　The Jackknife Method

The Jackknife method is proposed to avoid the overfitting problem. Algorithm 1 introduces the leave-one-out method, which consists of three main steps.

Algorithm 1 The Jackknife method

Step1. The PCR model is established using all the training dataset samples, and
$\hat{\mu}(x)$ is obtained. In addition, one sample point is removed each time, and
n residual models $\hat{\mu}_{-1}(x),\ldots,\hat{\mu}_{-n}(x)$ are obtained;

Step2. Calculate the predicted value $\hat{\mu}(X_{n+1})$ of the new sample point X_{n+1}, and
the fitted residual R_i of the leave-one-out method;

Step3. Calculate the prediction interval by
$$\hat{L}_{Jackknife,\alpha}\left(X_{n+1}\right)=\left[\hat{\mu}\left(X_{n+1}\right)-\hat{Q}_{(n,1-\alpha)}\left\{R_{-i}\right\},\hat{\mu}\left(X_{n+1}\right)+\hat{Q}_{(n,1-\alpha)}\left\{R_{-i}\right\}\right].$$

Jackknife uses the quantile of residual sequence obtained by the leave-one-out method instead of the naive method to solve the problem of overfitting, which can improve estimation accuracy and achieve higher coverage. However, the defect of the Jackknife method is that the coverage of calculations could be lower when the modeling method is unstable.

3.2 The Jackknife + method

As shown in Algorithm 2, Barber et al.[6] proposed the Jackknife + method based on the Jackknife method.

Algorithm 2 The Jackknife+ method

Step1. The PCR model is established using all the training dataset samples, and
$\hat{\mu}(x)$ is obtained. In addition, one sample point is removed each time, and
n residual models $\hat{\mu}_{-1}(x),\ldots,\hat{\mu}_{-n}(x)$ are obtained;

Step2. Calculate the predicted value $\hat{\mu}(X_{n+1})$ of the new sample point X_{n+1} and
fitted residual $R_{-i}=\left|Y_{-i}-\hat{\mu}_{-i}\left(X_{-i}\right)\right|,i=1,2,\ldots,n$ of the leave-one-out
method, and the fitted the α-quantile $\hat{Q}_{(n,\alpha)}\{R_{-i}\}$ and $1-\alpha$-quantile
$\hat{Q}_{(n,1-\alpha)}\{R_{-i}\}$ of the residual sequence $\{R_i\}$;

Step3. Calculate the prediction interval by
$$\hat{L}_{Jackknife+,\alpha}\left(X_{n+1}\right)=\left[\hat{Q}_{(n,\alpha)}\left\{\hat{\mu}_{-i}\left(X_{n+1}\right)-R_{-i}\right\},\hat{Q}_{(n,1-\alpha)}\left\{\hat{\mu}_{-i}\left(X_{n+1}\right)+R_{-i}\right\}\right].$$

Jackknife + is an improved version of Jackknife. Compared with the Jackknife method, the Jackknife + method uses $\hat{\mu}_{-i}(X)$ obtained by the leave-one-out method instead of estimating the predicted value $\hat{\mu}(X_{n+1})$ of the sample point X_{n+1}, which overcomes the defect of the Jackknife method.

3.3 The Cross-Validation + (CV+) Method

The CV + method is a generalization of the Jackknife + method. The main steps of the CV + method are shown in Algorithm 3.

Algorithm 3 The CV+ method

Step1.   The training dataset is divided into L disjoint subsets $D_1,\ldots,D_L$, each containing $m = n/L$ samples. According to the divided training dataset, one subset is removed each time to fit the PCR model by the leave-one-out method. In this way, L leave-one-out models $\hat{\mu}_{-D_1},\ldots,\hat{\mu}_{-D_L}$ are obtained;

Step2.   Calculate L different predicted values $\hat{\mu}_{-D_1}(X_{n+1}),\ldots,\hat{\mu}_{-D_L}(X_{n+1})$ for the new sample point X_{n+1}, and the residual

$$R_{-D_l(i)} = \left| Y_i - \hat{\mu}_{-D_l}(X_i) \right|, i \in D_l ;$$

Step3.   Calculate the prediction interval:

$$\hat{L}_{CV+,\alpha}(X_{n+1}) = \left[\hat{Q}_{(n,\alpha)} \left\{ \hat{\mu}_{-D_l}(X_{n+1}) - R_{-D_l(i)} \right\}, \hat{Q}_{(n,1-\alpha)} \left\{ \hat{\mu}_{-D_l}(X_{n+1}) + R_{-D_l(i)} \right\} \right].$$

Compared with the Jackknife + method, one advantage of the CV + method is that the computational cost is much lower than that of the Jackknife + method when $L << n$. However, the CV + method has a longer prediction interval because it removes a larger subset and uses fewer sample points in the training dataset when establishing the PCR model.

3.4 The Jackknife + -After-Bootstrap(J + aB) Method

The J + aB method proposed by Kim et al. [7] combines the advantages of the Jackknife method and ensemble learning, and its algorithm consists of four steps.

Algorithm 4 The J+aB method

Step1.   A bootstrapped sample can be obtained by randomly sampling m samples each time, the sampling can be put back or not put back, and repeated sampling B times to obtain B bootstrapped samples $D_b = \left(i_{b,1}, \ldots, i_{b,m}\right), b = 1, \ldots, B$, requiring $m \le n$;

Step2.   PCR models of B bootstrapped samples are fitted respectively, and B sub-models $\hat{\mu}_1, \ldots, \hat{\mu}_B$ are obtained;

Step3.   Calculate $\hat{\mu}_{f,-i} = f(\{\hat{\mu}_b : b = 1, \ldots, B, i \notin D_b\})$ through the sub-models and find the residual $R_{f,i} = \left| Y_i - \hat{\mu}_{f,-i}(X_i) \right|, i = 1, \ldots, n$, where f is a function of μ_b, which can be a mean function or some other functions;

Step4.   Calculate the prediction interval:

$$\hat{L}_{J+aB,\alpha}\left(X_{n+1}\right) = \left[\hat{Q}_{(n,\alpha)}\left\{\hat{\mu}_{f,-i}\left(X_{n+1}\right) - R_{f,i}\right\}, \hat{Q}_{(n,1-\alpha)}\left\{\hat{\mu}_{f,-i}\left(X_{n+1}\right) - R_{f,i}\right\}\right].$$

The J + aB method provides a framework for quantifying the uncertainty in set prediction. It also provides prediction intervals with limited sample coverage.

3.5   The Leave-One-Bag-Out + (LOBO +) Method

Lin et al. [8] proposed the LOBO + method based on the empirical distribution of the leave-one-bag-out prediction error, which has four steps shown as Algorithm 5.

Algorithm 5 The LOBO+ method

Step1. Samples with a fixed proportion of A are extracted from the training dataset using the method of equal probability without replacing. The sample size taken each time is $N_1 = n \times a$, which is taken as the training dataset D_{N1}, and the remaining samples in the training dataset are taken as the test dataset D_{N2}, including $N_2 = n - N_1$ samples;

Step2. A PCR model is established with the training dataset D_{N1}. Calculate $\hat{\mu}_{-D_{N2}}$, the residual $R_{-D_{N2(i)}} = |Y_i - \hat{\mu}_{-D_{N2}}(X_i)|, i \in N_2$ of the PCR model on the test dataset D_{N2};

Step3. Step 1 and Step 2 are repeated B times to obtain B models with corresponding residuals $R_{-D_{N2(i),1}}, \ldots, R_{-D_{N2(i),B}}$, and the model $\hat{\mu}_f = f(\{\hat{\mu}_{-D_{N2},1}, \ldots, \hat{\mu}_{-D_{N2},B}\})$ is integrated by function f;

Step4. Calculate the prediction interval:

$$\hat{L}_{LOBO+,\alpha}(X_{n+1}) = [\hat{\mu}_f(X_{n+1}) - \hat{Q}_{(B*N_2,1-\alpha)}\{R_{B*N_2}\}, \hat{\mu}_f(X_{n+1}) + \hat{Q}_{(B*N_2,1-\alpha)}\{R_{B*N_2}\}],$$

where R_{B*N_2} is the set containing $R_{-D_{N2(i),1}}, \ldots, R_{-D_{N2(i),B}}$ and the cardinality is $B*N_2$.

The model prediction results obtained by the LOBO + method are more accurate and stable. With the increase in sample size and repetition times, the prediction error information obtained is more accurate, and the prediction error estimation is more accurate than other similar methods.

3.6 Evaluation Index

Two evaluation indexes are used to evaluate the performance of different prediction interval construction methods: coverage rate and interval length. The coverage rate refers to the probability that the true value of the test dataset falls into the prediction interval. The higher the coverage rate, the more the true response value falls into the prediction interval. However, the prediction interval cannot be infinitely long. Otherwise, the true value will fall into the prediction interval. So, interval length is used to solve this problem, which is the value of the right endpoint of the forecast interval minus the left endpoint.

The coverage rate is calculated by

$$P(Y_{n+1} \in \hat{C}(X_{n+1})) \geq 1 - \alpha,$$

where $\hat{C}(X_{n+1}) \subseteq R$ is the mapping of X_{n+1} to a prediction interval containing the true response value Y_{n+1}. The coverage rate is based on the training dataset $(X_1, Y_1), ..., (X_n, Y_n)$ and the test dataset (X_{n+1}, Y_{n+1}).

The interval length is calculated by

$$L_n = b_n - a_n,$$

where a_n and b_n are the left and right endpoints of the interval.

4   Analysis of Simulation and Real Experiments

In this section, the performance of six different prediction interval construction methods will be compared using simulation experiments and real-world data. Each experiment will be repeated 40 times, and the results will be averaged to ensure the accuracy of the experimental results. Figure 1 shows the flow of the work done in this article, and the calculation formulas take the naive method as an example.

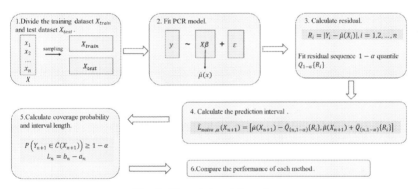

Fig. 1. Experimental flow chart

4.1   Simulation Experiments

4.1.1   Simulation Data

The experimental process of the simulated data is as follows: firstly, samples with sample size n and dimension p are randomly generated as training datasets to establish PCR models, and 100 samples with the same distribution are regenerated as test datasets to calculate coverage rate and interval length. The number of sample n and dimension p affect the high or low dimensionality of the data. The population covariance matrix has an autoregressive structure $\rho^{|i-j|}$, where ρ is the correlation coefficient. These three parameters will affect the results of the experiment, so the different values of these three factors are considered as follows:

1. $n = 50$;

2. $p = 300$;
3. $\rho = 0$(uncorrelated), 0.9(highly correlated).

There are 2 different simulation settings, and it is generally accepted that the coverage level of $1 - \alpha = 0.9$ is better.

4.1.2 Analysis of Simulation Experiments

The results of the simulation experiments are shown in Fig. 2. The vertical coordinate is the coverage rate or interval length values of different methods, and the horizontal coordinate is the component number. All results are based on 40 simulation experiments and expressed as the mean $\pm$ standard deviation of 40 experiments.

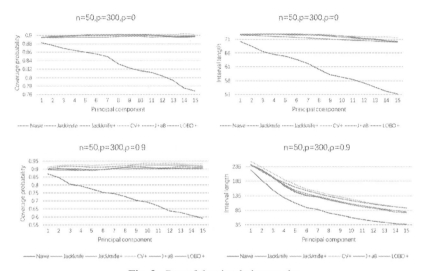

Fig. 2. Part of the simulation results

According to the simulation experiments, we can get the following conclusions:

1) For two simulation experiments, the performance of the naive method is the worst, and the coverage rate is the lowest, which does not reach the 0.9 level.
2) The coverage rate and interval length of Jackknife, Jackknife +, and the J + aB methods are similar, reaching or close to the level of 0.9, in which the curves of Jackknife and Jackknife + almost coincide in some cases.
3) The coverage rate of the CV + method is the highest, basically reaching 0.9 level, and the interval length generated is also the longest, indicating that the CV + method is more conservative than the other methods;
4) The coverage level of the LOBO + method is higher than other methods except the CV + method.

4.2   Real Experiments

In this section, the performance of different prediction interval methods is compared by three real-world datasets, and all results are expressed as the mean ± standard deviation of 16 experiments.

4.2.1   Real-World Datasets

Dataset 1: COX-2 Activity Dataset. Each independent variable of the sample point corresponds to a descriptor, and the dependent variable is the enzyme's activity. This dataset consists of 462 sample points, each with 225 variables. In this experiment, two hundred thirty-one samples are used as the training dataset, and 231 samples are used as the test dataset.

Dataset 2: Fathead Minnow Acute Aquatic Toxicity Dataset. It contains several data frames and consists of 322 compounds that are experimentally evaluated for toxicity. The experiment used one of the AquaticTox_Lcalc data frames; each sample represents a compound, each independent variable corresponds to a descriptor, and the dependent variable is the toxicity result. This data frame consists of 332 sample points, each with 23 variables. In this experiment, 166 samples are used as the training dataset and 166 samples are used as the test dataset.

4.2.2   Analysis of Real Experiments

According to the results of Fig. 3, it can be seen that the naive method yields the lowest coverage rate and the shortest interval length, exhibiting a decreasing trend with the increase in the number of components, accompanied by significant standard deviations. In contrast, the remaining five methods demonstrate relatively good coverage performance, all approaching 0.9. Generally, the LOBO + method outperforms others, achieving a coverage rate of 0.9 with favorable interval length and standard deviation, indicating superior stability. The Jackknife method also attains a higher coverage rate, reaching the 0.9 level, but with a higher standard deviation, indicating poorer stability. Notably, when the number of components is large, the coverage rates and interval lengths of Jackknife, Jackknife+, and the J + aB methods are very similar, with the curves of Jackknife and Jackknife + nearly overlapping.

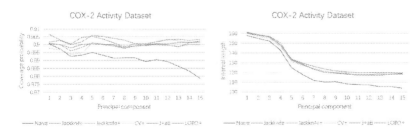

Fig. 3. Coverage probability and the interval length of the COX-2 Activity dataset

In Fig. 4, it is evident that the naive method exhibits the lowest coverage rate, failing to reach the 0.9 level. The CV + method attains a maximum coverage rate of 0.9, with the

longest interval length. The performance of Jackknife, Jackknife+, J + aB, and LOBO +
is comparable. Specifically, the coverage rates of J + aB and LOBO + reach 0.9, while
certain components of Jackknife and Jackknife + fall short of this threshold. Notably,
the interval length of J + aB is lower than that of the other three methods.

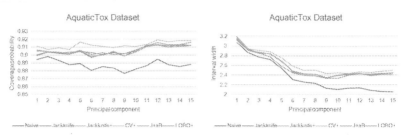

Fig. 4. Coverage probability and the interval length of the AquaticTox dataset

5   Conclusion

In this paper, we introduce a distribution-free prediction interval construction method
to enhance the PCR model. The performance of various prediction interval construction
methods is assessed through simulations and real experiments. The experimental results
show that the naïve method exhibits the poorest performance, with prediction accuracy
significantly below the standard. Notably, among all methods, both the CV + method and
the LOBO + method demonstrate superior performance, nearly reaching the specified
level. The CV + method boasts a higher coverage rate, while the LOBO + method
exhibits greater stability.

6   Declaration of Competing Interest

The authors declare that they have no known competing financial interests or personal
relationships that could have appeared to influence the work reported in this paper.

Acknowledgment. We are grateful to the anonymous reviewers and editors for their constructive
comments that help us improve quality of this manuscript. This work is financially supported by
the National Innovation Project of University Students (202110595046).

Data Availability. Data will be made available on request.

References

1. Patel, J.K.: Prediction intervals-a review. Commun. Stat. Theory Methods **18**(7), 2393–2465
(1989)

2. Cox, D.R.: Prediction intervals and empirical bayes confidence intervals. J. Appl. Probab.Probab. **12**(S1), 47–55 (1975)
3. Barndorff-Nielsen, O.E., Cox, D.R.: Prediction and asymptotics. Bernoulli **2**(4), 319–340 (1996)
4. Stine, R.A.: Bootstrap prediction intervals for regression. J. Am. Stat. Assoc. **80**(392), 1026–1031 (1985)
5. Efron, B.: Bootstrap methods: another look at the Jackknife. Ann. Stat. **7**(1), 1–26 (1979)
6. Barber, R.F., Candes, E.J., Ramdas, A., et al.: Predictive inference with the Jackknife+. Ann. Stat. **49**(1), 486–507 (2021)
7. Kim, B., Xu, C., Barber, R.: Predictive inference is free with the Jackknife+-after-bootstrap. Adv. Neural. Inf. Process. Syst. **33**, 4138–4149 (2020)
8. Lin, Y., Xu, C., Zhou, Z., et al.: Distribution-free predictive inference for partial least squares regression with applications to molecular descriptors datasets. J. Chemom.Chemom. **36**(12), 34–57 (2022)

Tourist Attraction Recommendation System Based on Django and Collaborative Filtering

Yufeng Jiang[1,2,3], Yushu Zhang[4], Zhujun Li[5], Wendong Yu[6], Hongwei Wei[6], and Lin Yuan[1,2,3]($\boxtimes$)

[1] Key Laboratory of Computing Power Network and Information Security, Ministry of Education, Shandong Computer Science Center, Qilu University of Technology (Shandong Academy of Sciences), Jinan 250353, Shandong, China
yuanl@qlu.edu.cn

[2] Shandong Engineering Research Center of Big Data Applied Technology, Faculty of Computer Science and Technology, Qilu University of Technology (Shandong Academy of Sciences), Jinan 250353, Shandong, China

[3] Shandong Provincial Key Laboratory of Computer Networks, Shandong Fundamental Research Center for Computer Science, Jinan 250353, Shandong, China

[4] School of Art, Nantong University, Nantong 226019, Jiangsu, China

[5] Jinan Springs Patent and Trademark Office, Jinan 250014, Shandong, China

[6] Shandong Tianyi Information Technology Co., Ltd., Jinan, Shandong 250101, China

Abstract. Recently, with the development of social and economic levels and people's pursuit of quality of life, tourism has become the first choice for more and more people. However, the traditional travel agency-based tourism method has also begun to expose some problems. However, with the development of machine learning and big data technology, personalized recommendation systems have become more and more important, which also provides us with new ideas to solve this problem. This paper therefore explores the application of spatial clustering algorithm and collaborative filtering algorithm in tourist attraction recommendation. This algorithm can not only mine the geographical location information of tourist attractions, but also analyze and study the data connections between users and attractions, and use these data to make personalized recommendations, providing more considerate and convenient services for travel planning.

Keywords: personalized recommendation · clustering algorithm · collaborative filtering · tourism

1 Introduction

Smart tourism research has recently highlighted the importance of recommender systems. The vast amounts of transaction data generated by the tourism industry provide critical insight into tourist consumption behaviors and patterns [1]. Tourism has a significant multiplier effect on the economy, making it one of the industries with the highest impact [2]. Tourism is a vast and thriving industry in our modern era. In 2022, the United Nations' World Tourism Organization reported that over 900 million tourists embarked

D.-S. Huang et al. (Eds.): ICAI 2023, CCIS 2015, pp. 226–235, 2024.
https://doi.org/10.1007/978-981-97-0827-7_20

on international travel [3, 4]. The revenue from the tourism industry exceeded 4 trillion yuan, and it is still in a stage of steady development. The proliferation of tourism websites and social networks has led to a massive influx of data and comments, potentially resulting in information overload [5, 6]. Personalized recommendation systems effectively alleviate the issue of information overload [7, 8]. Recommender systems are widely utilized to offer personalized recommendations to users based on their preferences, interests, and observed behavior towards items [9–11]. Sometimes tourists may not know if an attraction will meet their expectations or preferences before they visit it. If the attraction does not align with what they were hoping for, they may feel disappointed or dissatisfied. Tourists can rely on system-adjusted tourist attraction recommendations, taking into account ratings from previous visitors, to avoid disappointment and simplify their decision-making process when choosing tourist destinations [12, 13].

Geographic information refers to data represented spatially or geographically, and the systems used to store, retrieve, analyze, and display such data have various definitions [14, 15]. It is important to highlight that, thanks to the development of positioning technology, geographical location information has become ubiquitous in social network services. The advancement of wearable sensors connected to smart devices enables the collection of diverse information about users [16, 17]. For example, from its inception in 2004 to January 2018, Flickr's database has stored a total of 6.47 billion photos. These photos usually contain information such as geographical location, shooting time, and tag text, which can be used to reconstruct tourists' movement trajectories and analyze the travel of different tourists, model and recommend tourist attractions that tourists are interested in. Current travel recommendation methods include two types: general recommendation and personalized recommendation. Modern recommender systems now offer suggestions across multiple media formats, encompassing audio, text, visual (images), and videos [18, 19], using media data for personalized travel recommendations has become a new method of travel recommendation services and has great application prospects.

2 System-Related Theories

2.1 Introduction of Django Framework

Django is a Python-based open-source web application framework that adheres to the Model-View-Controller (MVC) design pattern, enabling developers to quickly build high-quality web applications. Django supports multiple database backends including MySQL, PostgreSQL, SQLite, and more. Its advantages include ease of learning, extensibility, maintainability, and high security. Django enhances website and application security by preventing various attacks like XSS, CSRF, SQL injection, and clickjacking [20]. As a result, Django has become one of the most popular frameworks for Python web development.

2.2 MySQL Database

MySQL is a platform-independent, secure, reliable, and highly efficient database. It is developed and implemented by MySQL AB [21]. Due to its small size, fast speed, and the cost-saving benefits of open-source code, MySQL has been widely adopted by small to mid-sized companies.

2.3 System Architecture

The overall structure of a recommendation system is illustrated in Fig. 1.

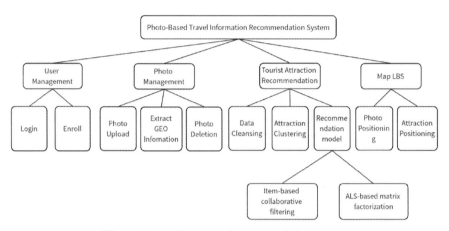

Fig. 1. The architecture of recommendation system

2.4 Collaborative Filtering

Collaborative filtering algorithm utilizes users' historical behavior data to analyze their interests and preferences, accordingly recommend relevant items to the users. Collaborative filtering typically models using a matrix consisting of m items and m users. However, this matrix is often sparse because not all users have rating data for all items. In such cases, we need to predict the rating relationships between blank items and data based on the existing sparse data. Subsequently, the system recommends items with the highest predicted ratings to users.

2.5 Map LBS

In the context of Location-Based Services (LBS), map information, GIS services, and infrastructures play a crucial role as essential support services [22]. Based on the EXIF information stored in the photos, the GEO information of each scenic spot photo is obtained first, as shown in Fig. 2. Then, by combining the latitude and longitude information with the API interface provided by Baidu Maps, it is easy to find the location of the scenic spot on the map and query its specific information, as shown in Fig. 3.

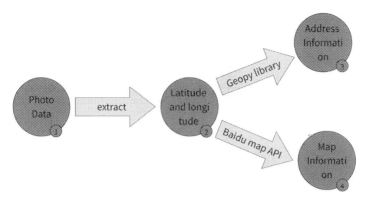

Fig. 2. Data Processing

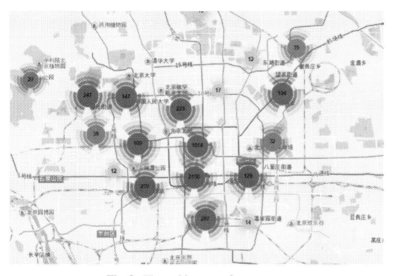

Fig. 3. The architecture of our system

3 Methods

3.1 DBSCAN Clustering Algorithm

The DBSCAN algorithm, as a prominent density-based clustering method, exhibits excellent clustering outcomes in various applications [23]. The DBSCAN algorithm employs a straightforward approach where it randomly selects an unclassified core object as a seed, identifies all density-reachable sample sets from this core object, and forms a clustering cluster. This process is repeated by selecting another unclassified core object to find density-reachable sample sets, resulting in additional clustering clusters. This process continues until all core objects have a category, as shown in Table 1 and Fig. 4:

Table 1. The DBSCAN Clustering Algorithm

Table 1. The DBSCAN Clustering Algorithm

Algorithm 1: DBSCAN Clustering Algorithm

Input:	Dataset $D = (x_1, x_2, ..., x_m)$, neighborhood parameters (ϵ, MinPts), sample distance metric Output: Clustering partition C
Output:	Clustering partition C
1	Initialize the set of core objects $\Omega = \emptyset$, initialize the number of clusters k = 0, initialize the set of unvisited samples $\Gamma = D$, set clustering partition $C = \emptyset$.
2	For j = 1, 2, ..., m, perform the following steps to find all core objects: • Use the distance metric to find the ϵ-neighborhood subset $N \in (x_j)$ for sample x_j • If the number of samples in the subset satisfies $|N \in (x_j)| \geq$ MinPts, add sample x_j to the set of core objects: $\Omega = \Omega \cup \{x_j\}$.
3	If the set of core objects $\Omega = \emptyset$, then terminate the algorithm. Otherwise, proceed to step 4.
4	Select a random core object o from the set of core objects Ω. Initialize the current cluster core object queue $\Omega_{cur} = \{o\}$, set category index k = k + 1, initialize the current cluster sample set $C_k = \{o\}$,, update the set of unvisited samples $\Gamma = \Gamma - \{o\}$.
5	If the current cluster core object queue $\Omega_{cur} = \emptyset$, then the current cluster C_k is complete. Update the clustering partition $\{C_1, C_2, ..., C_k\}$. Update the set of core objects $\Omega = \Omega - C_k$. Go to step 3.
6	Select a core object o' from the cluster's core object queue Ω_{cur}. Identify all ϵ-neighborhood subsets $N \in o'$ based on the specified distance threshold ϵ. Allow $\Delta = N \in (o') \cap \Gamma$. Update the current cluster sample set $C_k = C_k \cup \Delta$. Update the set of unvisited samples $\Gamma = \Gamma - \Delta$. Update $\Omega_{cur} = \Omega_{cur} \cup (N \in (o') \cap \Omega$. Go to step 5.
7	Output the result: Clustering partition $C = \{C_1, C_2, ..., C_k\}$.

3.2   ALS Matrix Factorization

Collaborative filtering introduces a matrix factorization technique that supports incomplete rating matrices without the need for imputation. This approach achieves excellent recommendation accuracy. In Spark MLlib, there is an implementation of collaborative

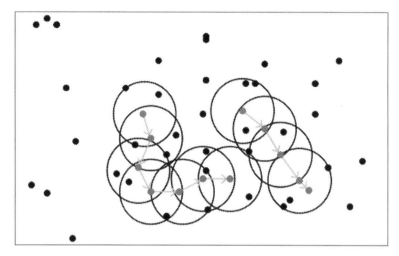

Fig. 4. Description of the performance of DBSCAN

filtering based on ALS (Alternating Least Squares) matrix factorization. ALS (Alternating Least Squares) is a powerful matrix that captures the latent features linking users and items in their interactions [24].

The core assumption of ALS is that the rating matrix A is approximately low-rank, which means that a rating matrix $A_{m \times n}$ can be approximated by the product of two smaller matrices $U_{m \times k}$ and $V_{n \times k}$:

$$A_{m \times n} \approx U_{m \times k} V_{n \times k}^T, \ k \ll m, n \tag{1}$$

If we interpret ratings as similarity measures, then the rating matrix $A_{m \times n}$ can be approximated by the product of user preference feature matrix $U_{m \times k}$ and product feature matrix $V_{n \times k}$.

We use the feature vector u_i of the i-th user in the user preference feature matrix $U_{m \times k}$ and the feature vector v_j of the j-th product in the product feature matrix $V_{n \times k}$ to predict the rating a_{ij} in the rating matrix $A_{m \times n}$. The loss function for the matrix factorization model can be expressed as follows:

$$C(U, V) = \sum_{(i,j) \in R} [\left(a_{ij} - u_i v_j^T\right)^2 + \lambda(u_i^2 + v_j^2)] \tag{2}$$

Once we have the loss function, we can discuss the optimization methods. Then we use Alternating Least Squares (ALS) to optimize the loss function.

4 Results and Presentation

4.1 DBSCAN Clustering Effectiveness

Taking a dataset of 38,293 photos in Beijing as an example, the DBSCAN clustering results with different parameters are shown in the Table 2.

Table 2. DBSCAN Clustering Results with Different Parameters

ϵ	minPts	clusters	noise rate
0.0004	15	422	37.20
0.0006	15	380	31.45
0.001	15	320	26.12
0.002	15	244	15.30
0.004	15	120	7.02

When $\epsilon = 0.001$ and minPts $= 15$, the clustering results for each province are as shown in Table 3, and the clustering results are projected on Baidu Maps as shown in Fig. 5.

Table 3. Clustering Results for Different Provinces

province	data	cluster	noise rate	province	data	cluster	noise rate
Macao	5954	37	11.79	Guizhou	946	15	51.37
Hong Kong	65953	381	19.27	Zhejiang	5894	80	51.55
Taiwan	272237	2462	20.32	Anhui	1742	22	53.45
Shanghai	31561	249	23.11	Chongqing	2368	28	53.76
Beijing	38228	333	25.16	Chongqing	1988	16	54.1
shaanxi	5450	77	31.69	Heilongjiang	886	13	54.95
Hainan	1461	15	35.25	Sichuan	5954	59	55.06
Jiangxi	1129	16	40.06	Hubei	2138	26	56.8
Tianjin	1347	20	42.57	Nei Monggol	1181	13	59.2
Fujian	3028	46	43.07	Henan	977	15	61.29
Guangdong	12462	155	46.01	Hunan	1510	19	62.47
Tibet	4543	56	47.23	Shandong	2507	23	62.73
Jiangsu	6562	87	47.71	Gansu	1207	15	66.44
Jilin	748	9	47.94	Gansu	756	9	70.64
Jilin	1335	22	49.03	Xinjiang	1967	21	70.96
Yunnan	6326	76	49.94	Hebei	1354	12	74.33
GZAR	4147	54	50.27	Ningxia	48	0	100

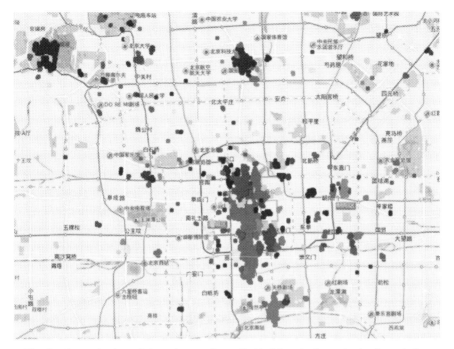

Fig. 5. The projection of clustering results on Baidu Maps is shown in the figure below

4.2   System Navigation Interface

Our system's final functional interface is shown in Fig. 6. By clicking on the "Upload Photos" option in the navigation bar at the top of the system, users can select and upload their own travel photos to their album. The uploaded photos can also be deleted if needed.

Based on the GEO information extracted from the user's uploaded photos, we perform clustering analysis and utilize a collaborative filtering algorithm built upon ALS matrix factorization to provide personalized recommendations for attractions. For users who haven't uploaded any travel photos, we can still offer high-quality attraction recommendations to effectively address the cold start problem and enhance user experience.

Furthermore, we integrate the API of Baidu Maps to facilitate route planning for our travel. This feature allows us to provide services such as restaurant and hotel recommendations as well.

Fig. 6. System Navigation Layout

5   Conclusion

Throughout this paper, we have developed a user-oriented tourism attraction recommendation system. In this system, we use the YFCC 100M dataset as our source photo dataset and store the data using MySQL database. The system is developed using the Django framework to ensure ease of use and operability. Personalized attraction recommendations are provided to users through DBSCAN clustering algorithm and collaborative filtering algorithm based on ALS matrix factorization. With this system, we aim to provide a more thoughtful, convenient, and efficient way for travel planners to solve problems related to travel planning and satisfy people's expectations and needs for travel.

Acknowledgement. This work was supported by the Natural Science Foundation of Shandong Province, China (No. ZR2020QF038), the Ability Improvement Project of Science and Technology SMES in Shandong Province (No. 2023TSGC0279), the Youth Innovation Team of Colleges and Universities in Shandong Province (2023KJ329), and the Qilu University of Technology (Shandong Academy of Sciences) Talent Scientific Research Project (No. 2023RCKY128).

References

1. Hong, M., Chung, N., Koo, C., et al.: TPEDTR: temporal preference embedding-based deep tourism recommendation with card transaction data. Int. J. Data Sci. Anal. **16**, 147–162 (2023)
2. Gamidullaeva, L., Finogeev, A., Kataev, M., et al.: A Design Concept for a Tourism Recommender System for Regional Development, Algorithms 16 (2023)
3. Chalkiadakis, G., Ziogas, I., Koutsmanis, M., et al.: A Novel Hybrid Recommender System for the Tourism Domain, Algorithms 16 (2023)
4. Yuan, L., Zhu, L., Guo, W.-L., et al.: Nonconvex penalty based low-rank representation and sparse regression for eQTL mapping. IEEE/ACM Trans. Comput. Biol. Bioinf.Bioinf. **14**, 1154–1164 (2016)

5. Abbasi-Moud, Z., Vahdat-Nejad, H., Sadri, J.: Tourism recommendation system based on semantic clustering and sentiment analysis. Expert systems with applications, p. 167 (2021)
6. Yuan, L., Guo, L.-H., Yuan, C.-A., et al.: Integration of multi-omics data for gene regulatory network inference and application to breast cancer. IEEE/ACM Trans. Comput. Biol. Bioinf.Bioinf. **16**, 782–791 (2018)
7. Zheng, X., Han, B., Ni, Z.: Tourism route recommendation based on a multi-objective evolutionary algorithm using two-stage decomposition and pareto layering. IEEE/CAA J. Automatica Sinica **10**, 486–500 (2023)
8. Yuan, L., Sun, T., Zhao, J., et al.: A Novel Computational Framework to Predict Disease-Related Copy Number Variations by Integrating Multiple Data Sources, Frontiers in genetics 12 (2021)
9. Yu, H.-F., Hsieh, C.-J., Si, S., et al.: Scalable coordinate descent approaches to parallel matrix factorization for recommender systems. In: 2012 IEEE 12th International Conference on Data Mining, 765–774 (2012)
10. Fayyaz, Z., Ebrahimian, M., Nawara, D., et al.: Recommendation Systems: Algorithms, Challenges, Metrics, and Business Opportunities, Applied Sciences, 10 (2020)
11. Yuan, L., Zhao, J., Sun, T., et al.: A machine learning framework that integrates multi-omics data predicts cancer-related LncRNAs. BMC Bioinform. **22**, 1–18 (2021)
12. Pratama, Y., Sianturi, R.A., Chandra, D., et al.: Restricted Boltzmann machine and matrix factorization-alternating square algorithm for development tourist recommendation system. J. Phys. Conf. Ser. 2022;2394
13. Shen, Z., Shao, Y.L., Liu, W., et al.: Prediction of Back-splicing sites for CircRNA formation based on convolutional neural networks. BMC Genomics **23**, 581 (2022)
14. Lü, G., Batty, M., Strobl, J., et al.: Reflections and speculations on the progress in Geographic Information Systems (GIS): a geographic perspective. Int. J. Geogr. Inf. Sci. **33**, 346–367 (2018)
15. Yuan, L., Lai, J., Zhao, J., et al.: Path-ATT-CNN: A Novel Deep Neural Network Method for Key Pathway Identification of Lung Cancer, Frontiers in genetics 2022;13
16. Ko, H., Lee, S., Park, Y., et al.: A survey of recommendation systems: recommendation models, techniques, and application fields. Electronics 11 (2022)
17. Yuan, L., Yang, Z., Zhao, J., et al.: Pan-Cancer Bioinformatics Analysis of Gene UBE2C. Front. Genetics 13 (2022)
18. Deldjoo, Y., Schedl, M., Cremonesi, P., et al.: Recommender Systems Leveraging Multimedia Content. ACM Comput. Surv.Surv. **53**, 1–38 (2020)
19. Yuan, L., Zhao, J., Shen, Z., et al.: ICircDA-NEAE: Accelerated attribute network embedding and dynamic convolutional autoencoder for circRNA-disease associations prediction. PLoS Comput. Biol. Comput. Biol. **19**, e1011344 (2023)
20. Shyam, A., Mukesh, N.: A Django Based Educational Resource Sharing Website: Shreic. J. Sci. Res. **64**, 138–152 (2020)
21. Combaudon, S.: MySQL 5.7: administración y optimización. Ediciones Eni (2018)
22. Zipf, A.: User-Adaptive Maps for Location-Based Services (LBS) for Tourism. Inf. Commun. Technol. Tourism **2002**, 329–338 (2002)
23. Deng, D.: DBSCAN clustering algorithm based on density. In: 2020 7th International Forum on Electrical Engineering and Automation (IFEEA), pp. 949–953 (2020)
24. Hartatik, H., Sejati, B.P., Nur Fitrianto, A., et al.: A comparison study of model based collaborative filtering using alternating least square and singular value decomposition. In: 2021 3rd International Conference on Electronics Representation and Algorithm (ICERA), pp. 185–190 (2021)

Effective Rules for a Rule-Based SIEM System in Detecting DoS Attacks: An Association Rule Mining Approach

Federica Uccello[1]([⊠]) (ID), Marek Pawlicki[2], Salvatore D'Antonio[1] (ID), Rafał Kozik[2], and Michał Choraś[2]

[1] University of Naples 'Parthenope' Centro Direzionale, Isola C4, 80133 Napoli, Italy
`federica.uccello@assegnista.uniparthenope.it`,
`salvatore.dantonio@uniparthenope.it`
[2] Bydgoszcz University of Science and Technology, PBS, Bydgoszcz, Poland
`chorasm@pbs.edu.pl`

Abstract. In today's interconnected digital landscape, Security Information and Event Management (SIEM) systems play a vital role as the frontline defense against cyber threats, providing prompt detection of the most common cyber-threats. As Denial of Service (DoS) and Distributed Denial of Service (DDoS) attacks remain among the most challenging hazards for organizations worldwide, their quick and effective detection is a major concern. This research paper explores innovative methods to enhance the effectiveness of rule-based SIEM systems in detecting DoS and DDoS attacks. The SIEM rule sets are augmented leveraging Association Rule Mining (ARM), a data mining technique for uncovering hidden relationships within dataset's features. By identifying and applying association rules to network traffic data, our methodology aims to strengthen SIEM rules, ultimately leading to more accurate DDoS attack detection.

Keywords: SIEM · DoS Attacks · Association Rule Mining · Benchmark Datasets · Rule-based Systems

1 Introduction

In an increasingly interconnected digital landscape, the importance of Security Information and Event Management (SIEM) systems cannot be overstated. SIEM systems serve as the first line of defense against cybercrime, as they are in charge of monitoring networks and systems for signs of intrusion and malicious activity [1–3]. Among the threats that organizations face, Denial of Service (DoS) and Distributed Denial of Service (DDoS) attacks represent a persistent and disruptive menace [4, 5]. These attacks flood target systems with malicious traffic, rendering them unresponsive and causing significant downtime, financial losses, and reputational damage [6].

The sheer scale and sophistication of contemporary cyber threats have necessitated the evolution of SIEM systems. While traditional rule-based SIEM systems have been instrumental in detecting known patterns of attacks, their adaptability and accuracy in

D.-S. Huang et al. (Eds.): ICAI 2023, CCIS 2015, pp. 236–246, 2024.
https://doi.org/10.1007/978-981-97-0827-7_21

handling evolving threats, such as DoS and DDoS attacks, have been called into question [7, 8]. In particular, as shown by the latest ENISA Threat Landscape (ETL), there has been a "significant rise on attacks against availability, particularly DDoS, with the ongoing war being the main reason behind such attacks" [9]. Effective DDoS attack detection is, therefore, paramount for the security and operational continuity of modern enterprises. For the aforementioned reasons, this research paper delves into the realm of enhancing the effectiveness of rule-based SIEM systems in DDoS detection. To fortify the rule sets used by SIEM systems, an innovative approach leveraging Association Rule Mining (ARM), a data mining technique that seeks to discover hidden relationships and patterns within datasets, is proposed. By identifying and applying association rules to network traffic data, the proposed methodology aims to bolster the discriminatory power of SIEM rules, leading to more accurate DDoS detection. The present research revolves around the need for SIEM systems to evolve and adapt in the face of evolving cyber threats [10]. While rule-based systems have served as dependable supporter in network security, the dynamic nature of DoS attacks demands a more adaptive and nuanced approach [3]. The fundamental issue lies in the ability to discern between legitimate network traffic and malicious DoS or DDoS attacks in real-time [11], a challenge exacerbated by the ever-evolving tactics of cyber adversaries [12]. The proposed solution combines Machine Learning (ML) and ARM to improve traditional rule-based SIEM, supporting the definition of reliable and accurate detection rules.

The main contribution can be summarized as follows.

– Integration of ARM: The concept of integrating ARM into the rule creation process of SIEM systems is introduced. By discovering hidden associations and correlations within network traffic data, the accuracy of rule-based DDoS attack detection is increased.
– Enhanced Discrimination Criteria: The present approach results in the identification of more precise discrimination criteria, enabling SIEM systems to differentiate between benign and malicious traffic with increased accuracy.
– Experimental Validation: Comprehensive experimental results are presented to validate the effectiveness of the proposed methodology. Through rigorous testing on real-world datasets, significant improvements in DDoS attack detection accuracy are demonstrated in comparison to traditional static rules.

The remainder of this paper is organized as follows. Section 2 provides an overview of related work. Section 3, showcases the materials and methods used in our research. Section 4 details the experimental setup, including datasets, parameter settings, and results, while Sect. 5 provides a comprehensive discussion of the obtained results and the implications of the proposed approach, including limitations and areas for future research. Finally, Sect. 6 offers a conclusion summarizing the key findings and contributions of the research.

2   Related Works

SIEM systems have been the subject of discussion and research in recent years [13–15], highlighting the importance of SIEM solutions in cybersecurity, especially in critical infrastructures and corporations, and remarking the need for early detection and response to cyberattacks. Multiple research works have proposed the usage of ML to enhance DoS and DDoS detection [16–18], for the benefits brought in terms of pattern recognition, real-time detection, and continuous learning.

Among the related works, [19–21] discuss the usage of rule-based systems in various application domains of cybersecurity. As shown in the literature, ARM offers a valuable toolkit for enhancing cyber incident detection, optimizing alert management, and reducing false positives. In [22], the authors investigate the integration of associative rules into SIEM systems to bolster the detection of cyber incidents. Their approach incorporates fuzzy sets theory and data mining techniques to construct associative rules based on an analysis of cyber incidents. In [23], the focus shifts towards the utilization of data mining, specifically sequential rule mining, in the analysis of intrusion detection alerts. The primary objective is to reduce the volume of alerts received by security professionals. The work presented in [24] addresses the common issue of false positive alerts generated by intrusion detection systems. The study proposes a novel network security Intrusion Detection System (IDS) framework that leverages Modified Frequent Pattern (MFP-Tree) with the K-means algorithm in detecting various attacks, including DoS. The authors of [25] present a correlation approach to SQL Injection Attack detection, improving the ability to spot those attacks. The authors of [26] introduce a method for mining association rules from multi-source logs to identify various intrusion behaviors in cloud computing platforms.

In comparison to existing literature, the proposed approach aims to provide support to traditional rule-based SIEM systems in the definition of static discrimination criteria. While the usage of ARM and rule-based SIEM is widespread across different contexts, the present research introduces a novel integration of ARM into the context of rule-based SIEM systems. Major novelties are the definition of a methodology for the definition of highly accurate detection rules, backed by experimental evaluation tailored, but not limited, to DDoS detection. In comparison with rules defined based on correlation analysis, the ones defined after ARM show a significant improvement in terms of Precision, Recall, F1-Score, and Accuracy. The present research also opens the door to future developments in real-time and adaptive security systems.

3   Materials and Methods

This section details the materials and methods employed in the present research, including the application of ARM and the fundamentals of SIEM systems, with a specific emphasis on rule-based systems. Finally, the applied method along with the improved SIEM rules are presented.

3.1   Association Rule Mining

ARM is a powerful data mining technique widely utilized in various domains, including cybersecurity, for discovering hidden patterns and relationships within large datasets. In the context of this research, ARM plays a pivotal role in enhancing the discriminatory criteria of traditional rule-based SIEM systems. The application of ARM can be summarized as follows.

– Dataset Selection: The first step has consisted in the selection of appropriate network traffic datasets for analysis. The choice of datasets, such as CICIDS2017 [27] and CICFlowMeter Dataset [28], was made based on their relevance to the focus on DDoS attack detection.

– Data Preprocessing: Before applying ARM, essential data preprocessing steps have been conducted to ensure the quality and suitability of the datasets, and consistency among them. This included handling missing values, removing duplicates, and addressing class imbalance issues through techniques like Synthetic Minority Over-sampling Technique (SMOTE) [29].

– Feature Selection: To streamline the analysis and reduce dimensionality, the SelectKBest feature selection method from the scikit-learn library has been employed. SelectKBest utilizes univariate statistical tests, such as ANOVA F-value, to identify the most relevant features for our analysis.

– ARM Algorithm: For ARM, the APRIORI algorithm has been employed.

– APRIORI is a classic algorithm for association rule discovery, that efficiently identifies correlations between attributes in the dataset.

– Association Rule Discovery: the APRIORI algorithm has been applied to the preprocessed datasets, resulting in the identification of association rules that describe relationships between different features. These rules provide insights into the behaviors and patterns associated with network traffic.

3.2   SIEM and Rule-Based Systems

SIEM systems are central to the present research, serving as the foundational framework for the development and testing of the proposed rule-based approach for DDoS attack detection.

A SIEM system typically consists of several essential components, including data collection agents, event correlation engines, a database or storage repository, and a user interface for monitoring and reporting. These components work together to collect, process, and analyze security events and logs from various sources within an organization's network. Event correlation is a crucial function of SIEM systems. It involves the analysis of incoming security events to identify patterns or sequences that may indicate potential threats. In rule-based SIEM systems, correlation rules play a central role in such a process.

Rule-based systems are responsible for defining the logic that determines how security events should be processed and classified. These rules are typically crafted based on domain knowledge, known attack patterns, and heuristic analysis. The present research aims to enhance the effectiveness of these correlation rules through the integration of association rules derived from ARM. In particular, the SIEM presented in [30] is enhanced, extending the authors' previous research.

3.3   Proposed Innovative Methodology for Rule-Based DDoS Detection with Rule Discovery

The applied method is formalized as follows. The experiments focus on DDoS attacks detection for a proof-of-concept approach, but the same methodology can serve a wider scope, as the core concept is the analysis of hidden discrimination criteria within traffic. Let X represent the set of features in the generic original dataset D_i, and let $y = \{DDoS, BENIGN\}$ be the corresponding labels. The aim is to identify x_{rule} and use it as discrimination criteria between DDoS and BENIGN samples. ARM is applied to D_i to generate a set of detection rules. The application of such rules leads to the creation of a set of SIEM-labelled datasets D_{si}.

As shown in Eq. 1, to minimize the number of mislabelled samples, the difference between the original labels y and the SIEM labels y_{Dsi} must be minimized.

$$xrule \in X : minxrule|yDsi \backslash y| \tag{1}$$

To test the proposed methodology, the CICIDS2017 dataset [27] (D_1) and the CICFlowMeter Dataset [28] (D_2) have been employed. In particular, the APRIORI algorithm has been applied to D_1 and D_2 to mine the most correlated features with the target variable (DDoS), and discover hidden relationships between features. For D_1, an interesting rule has been found between the following features: *BwdPacketLengthMean* (mean packet length observed in the backward direction), *FwdPacketLengthMean* (average length of packets in the forward direction), and *InitWinBytesForward* (record of the initial window size in bytes for forward traffic flows at the beginning of network communication sessions). For D_2, the rule mining returned a hidden correlation between the following features: *SrcPort* (the source port number), *ActiveMean* (mean duration of active communication session), and *FwdIATTot* (cumulative total of inter-arrival times for forward traffic flow). Two different sets of SIEM-labelled datasets, D_{s1} and D_{s2}, have been created according to the refined rules defined after the ARM analysis. The logic behind the creation of each of the SIEM-labelled datasets is formalized in the following paragraphs. To test the impact of considering the association rules in the definition of static correlation rules for the SIEM, three different versions of D_{s1} have been considered. The three versions are the results of the correlation rules shown in Eq. 2, Eq. 3, and Eq. 4. In the equations, the features *BwdPacketLengthMean* and *FwdPacketLengthMean* are represented as *BPLengthMean* and *FPLengthMean* respectively, for the sake of brevity. Among the rules, Eq. 2 has been the best rule defined throughout past experiments, without the usage of data rule mining, and it has been used as the control. The feature *BwdPacketLengthMean* has been chosen as a key feature for the rule as it was the most correlated one with the DDoS label. The parameters *Threshold*$_1$ (T_1) and *ThresholdMultiplier* (*TM*) constitute threshold values for the considered features, and their definition is explained in detail in Sect. 4.

$$yDs1 = \begin{cases} DDoSifBPLengthMean \geq T_1 \\ BENIGN \quad otherwise \end{cases} \tag{2}$$

$$yDs1 = \begin{cases} DDoSifBPLengthMean \geq TMxFPLengthMean \\ BENIGN \quad otherwise \end{cases} \tag{3}$$

$$yDs1 = \begin{cases} DDoSifInitWinBytesForward = 256 \\ \quad or \\ BPLengthMean \geq TMxFPLengthMean \\ BENIGN \quad otherwise \end{cases} \tag{4}$$

Having obtained the association rules, Eq. 3 and Eq. 4 have been defined. It has been observed that, for D_1, only DDoS instances had a specific value for *InitWinBytes-Forward*. As suggested by the association rules, this has been a key feature for highly accurate detection. To validate the proposed approach, a second set of datasets D_{s2} has been created. The correlation rules employed in the generation of D_{s2} follow the same approach described previously, and are shown in Eq. 5, Eq. 6, and Eq. 7.

$$yDs1 = \begin{cases} DDoSifFwdIATToT \geq T_2 \\ BENIGN \quad otherwise \end{cases} \tag{5}$$

$$yDs1 = \begin{cases} DDoSifFwdIATToT \geq T_2 \\ \quad or \\ ActiveMean > \varepsilon \\ BENIGN \quad otherwise \end{cases} \tag{6}$$

$$yDs1 = \begin{cases} BENIGNifSrcPort \in Whitelist \\ DDoSifSrcPort \in Blacklist \\ \quad or \\ FwdIATToT \geq T_2 \\ \quad or \\ ActiveMean > \varepsilon \\ BENIGN \quad otherwise \end{cases} \tag{7}$$

where the *Whitelist* and *Blacklist* sets have been defined by identifying all the ports that were exclusive for BENIGN and DDoS instances respectively. The parameter *Threshold*$_2$ (T_2) and ε constitute threshold values for the considered features, and their definition is explained in detail in Sect. 4.

4 Experimental Setup and Results

This section shows the details of the experiments conducted, including the Datasets employed, parameters setting, and results.

4.1 Datasets

To test the proposed approach, the CICIDS2017 dataset [27] and the CICFlowMeter Dataset [28] have been employed. The CICIDS2017 is a well-recognized benchmark dataset designed for evaluating IDSs and Intrusion Prevention Systems (IPSs) against sophisticated and evolving network attacks. The choice of this dataset was motivated by its careful addressing of the shortcomings of other datasets, which suffer from outdated information, lack of diversity, and anonymization of critical data. Both datasets provide a realistic representation of real-world network traffic captured from Packet Capture (PCAPs) data, as they include a variety of benign background traffic and various attack scenarios. In line with the topic of this research, only the DDoS attack subset along with the benign traffic captured in that scenario have been utilized for this experiment.

Given the imbalanced nature of the original datasets, an initial pre-processing phase was implemented, leveraging the SMOTE Technique as an effective strategy for ameliorating data disproportionality [29]. Records containing missing values and duplicates were discarded. In order to facilitate an accurate comparison of identical samples across different analyses, an auxiliary attribute comprising a distinct index for each data instance was incorporated into the datasets. It is essential to note that this index attribute was excluded from the feature set, recognizing that its role was solely for identification purposes and it bore no relevance to the process of inferential analysis.

4.2 Parameter Settings

To conduct the experiments and implement the proposed methodology effectively, several parameters needed to be defined. The present section outlines the key parameter settings used throughout the research.

Feature Selection Parameters. For the dimensionality reduction, the datasets were processed with scikit-learn's SelectKBest, and the individual scores for each feature were analyzed to identify the most relevant features. SelectKBest performs a univariate statistical test using the Analysis of variance (ANOVA) F-value between the labels and the features. The top 14 features were selected. This value was chosen after observing a significant drop in feature scores beyond this point, indicating that these 14 features were the most relevant. The correlation between features has also been studied to avoid redundant features and only select the most representative ones.

Association Rule Mining Parameters. The APRIORI algorithm has been employed for discovering association rules in the considered datasets. The Minimum Support Threshold controls the minimum frequency or occurrence of an itemset in the dataset for it to be considered in the rule mining process. The authors experimented with different support thresholds, including 0.1, 0.05, and 0.01, to observe the impact of varying support levels. The Minimum Confidence Threshold determines the minimum level of confidence required for an association rule to be considered relevant. Different confidence thresholds have been tested, such as 0.7, 0.8, and 0.9, to assess the impact on rule discovery.

SIEM Parameters. For defining SIEM correlation rules, specific threshold values were set based on the association rules. These threshold values were derived from the statistical properties of the selected features and the discovered association rules. In particular, in Eq. 2, the T_1 parameter has been set equal to the mean value of *BwdPacketLengthMean* for all the DDoS samples of the original dataset. To define a suitable value for the *TM* parameter shown in Eqs. 3 and 4, multiple approaches have been considered. Ultimately, the optimal value has been obtained by calculating the average *BwdPacketLengthMean* to *FwdPacketLengthMean* ratio for DDoS instances. This value was set to optimize the discrimination between DDoS and benign traffic. In Eq. 5, the *Threshold$_2$ T_2* parameter has been set equal to the mean value of *FwdIATTot* for all the DDoS samples of the original dataset. In Eqs. 6 and 7, the ε parameter is defined as shown below (Eq. 8):

$$\varepsilon \to 0 \tag{8}$$

This value has been selected as threshold for the *ActiveMean* parameter as it has been observed that only benign flow had an average duration of the session close to zero.

4.3   Results

This section shows how association rules have provided significant support in the definition of more accurate correlation rules for the considered SIEM system. The results are summarized in Table 1 and Table 2. In particular, a comparison is provided between the static rules defined prior to the usage of ARM, and the ones defined using the proposed methodology.

Table 1. Summary of classification reports for SIEM-labelled dataset D_{s1} implemented using different correlation rules. Equations 2 has been defined prior to the ARM application, based on correlation analysis. Equation 3 and 4 have been derived from ARM.

Correlation Rules	Precision BENIGN	DDoS	Recall BENIGN	DDoS	F1-Score BENIGN	DDoS	Accuracy
Equation 2 (before ARM)	0.73	0.96	0.97	0.64	0.83	0.76	0.8
Equation 3	0.73	0.99	1	0.63	0.84	0.77	0.81
Equation 4	1	0.99	0.99	1	0.99	0.99	0.99

Table 2. Summary of classification reports for SIEM-labelled dataset D_{s2} implemented using different correlation rules. Equation 5 has been defin7ed prior to the ARM application, based on correlation analysis. Equation 6 and have been derived from ARM.

Correlation Rules	Precision BENIGN	DDoS	Recall BENIGN	DDoS	F1-Score BENIGN	DDoS	Accuracy
Equation 5 (before ARM)	0.73	0.88	0.91	0.66	0.81	0.75	0.79
Equation 6	0.76	0.88	0.91	0.71	0.83	0.79	0.81
Equation 7	0.87	0.93	0.93	0.86	0.90	0.89	0.90

5   Discussion

The primary objective of this research was to enhance the effectiveness of rule based SIEM systems in DDoS attacks by integrating ARM. The presented findings suggest that this approach has yielded substantial improvements in the accuracy of DDoS attack detection. The association rules discovered through ARM provided valuable insights into the hidden relationships and patterns within network traffic data. These rules served as a foundation for creating more precise correlation rules in the SIEM system. The results, as demonstrated in Tables 1 and 2, showcase a remarkable enhancement in the system's ability to distinguish between benign and malicious traffic in comparison to static SIEM rules, defined without the knowledge brought by ARM.

Notably, the correlation rules derived from association rules, particularly Eq. 4 and Eq. 7, achieved near-perfect precision and recall values. This suggests that by incorporating association rules, SIEM systems can significantly reduce false positives and false negatives, thereby enhancing overall effectiveness in detecting DDoS attacks.

5.1 Limitations

While the present research shows promising results, it is essential to acknowledge certain limitations that should be considered. In particular, dataset specificity could constitute a threat to validity, as the effectiveness of the proposed approach may be influenced by the choice of datasets. The authors utilized the CICIDS2017 and CICFlowMeter datasets, which, while comprehensive, may not cover all possible network scenarios. Additionally, it is possible that different DoS and DDoS attacks would lead to other network patterns, affecting the correctness of the defined rules. Further evaluation on a broader range of datasets is warranted. A further limitation is constituted by parameter tuning. The determination of optimal parameters, such as threshold values in correlation rules, relies on domain knowledge and empirical observations. More advanced techniques for automatic parameter tuning could be explored to further optimize the performance. Additionally, as network traffic data volumes continue to grow, the scalability of the proposed approach may become a concern. Efficient methods for handling large-scale network datasets should be investigated.

5.2 Future Work

Building upon the foundations laid in this research, several avenues for future work emerge, such as real-time implementation. Extending the present approach to operate in real-time environments is essential. Developing a real-time SIEM system that integrates association rules for on-the-fly DDoS attack detection would be a valuable advancement. Also, Investigating methods for dynamically adapting correlation rules based on evolving network conditions and threat landscapes is crucial. An adaptive rule-based SIEM system that self-adjusts its rules in response to emerging threats would be highly desirable. Finally, expanding the scope of research to address multi-class classification challenges beyond DDoS attacks could provide a more comprehensive security solution. Identifying and distinguishing between various types of attacks would be a valuable addition.

6   Conclusion

The present research has explored and demonstrated the potential of ARM as a valuable tool for enhancing the accuracy of rule-based SIEM systems in the context of DoS/DDoS attacks detection. By leveraging association rules derived from network traffic data, the authors refined the correlation rules used in their SIEM system, resulting in significantly improved precision, recall, and F1-score values. Notably, some correlation rules achieved near-perfect accuracy in distinguishing between benign and malicious traffic, highlighting the effectiveness of the proposed approach. While certain limitations exist, such as dataset specificity and scalability concerns, the experimental results provide insights for future advancements in real-time, adaptive rule-based SIEM systems. The integration of ARM into SIEM represents a promising avenue for strengthening network security and mitigating the impact of evolving cyber threats.

Acknowledgement. This research has received funding from the European Union's Horizon 2020 Research and Innovation Programme under Grant Agreement No 101020560 CyberSEAS, and from the Industrial Cyber Shield (ICS) No C83C22001460001 project funded by INAIL. The content of this publication reflects the opinion of its authors and does not, in any way, represent opinions of the funders. The European Commission and INAIL are not responsible for any use that may be made of the information that this publication contains.

References

1. Tariq, A., Manzoor, J., Aziz, M.A., Tariq, Z.U.A., Masood, A.: Open source siem solutions for an enterprise. Inform. Comput. Secur. **31**(1), 88–107 (2022)
2. Fakiha, B.S.: Effectiveness of security incident event management (siem) system for cyber security situation awareness. Indian J. Forensic Med. Toxicol. **14**(4) (2020)
3. Sheeraz, M., et al.: Effective security monitoring using efficient siem architecture. Hum.-Centric Comput. Inf. Sci. **13**, 1–18 (2023)
4. Khader, R., Eleyan, D.: Survey of dos/ddos attacks in iot. Sustainable Eng. Innov. **3**(1), 23–28 (2021)
5. Priyanka, S., Vijay Bhanu, S.: A survey on variants of dos attacks: Issues and defense mechanisms. J. Appl. Res. Technol. **21**(1), 12–16 (2023)
6. Pawlicka, A., Chora´s, M., Pawlicki, M.: The stray sheep of cyberspace aka the actors who claim they break the law for the greater good. Personal Ubiquitous Comput. **25**(5), 843–852 (2021)
7. Cucu, C., Cazacu, M.: Current technologies and trends in cybersecurity and the impact of artificial intelligence. In: The International Scientific Conference eLearning and Software for Education, vol. 2, pp. 208–214, Carol I" National Defence University (2019)
8. Campfield, M.: The problem with (most) network detection and response. Netw. Secur.. Secur. **2020**(9), 6–9 (2020)
9. Ardagna, C., Corbiaux, S., Impe, K.V., Sfakianaki, A.: Enisa threat landscape (2022)
10. Pawlicki, M., Pawlicka, A., Kozik, R., Chora´s, M.: The survey and meta-analysis of the attacks, transgressions, countermeasures and security aspects common to the cloud, edge and iot. Neurocomputing, p. 126533 (2023)
11. Awan, M.J., et al.: Real-time ddos attack detection system using big data approach. Sustainability **13**(19), 10743 (2021)
12. Adedeji, K.B., Abu-Mahfouz, A.M., Kurien, A.M.: Ddos attack and detection methods in internet-enabled networks: Concept, research perspectives, and challenges. J. Sens. Actuator Netw.Netw. **12**(4), 51 (2023)
13. Podzins, O., Romanovs, A.: Why siem is irreplaceable in a secure it environment? In: 2019 Open Conference of Electrical, Electronic and Information Sciences (eStream), pp. 1–5 (2019)
14. Gonz´alez-Granadillo, G., Gonza´lez-Zarzosa, S., Diaz, R.: Security information and event management (siem): analysis, trends, and usage in critical infrastructures. Sensors **21**(14), 4759 (2021)
15. Rosenberg, M., Schneider, B., Scherb, C., Asprion, P.M.: An adaptable approach for successful siem adoption in companies, arXiv preprint arXiv:2308.01065 (2023)

16. d. Lima Filho, F.S., Silveira, F.A., de Medeiros Brito Junior, A., Vargas-Solar, G., Silveira, L.F.: Smart detection: an online approach for dos/ddos attack detection using machine learning. Secur. Commun. Networks **2019**, 1–15 (2019)

17. Alsulaiman, L., Al-Ahmadi, S.: Performance evaluation of machine learning techniques for dos detection in wireless sensor network, arXiv preprint arXiv:2104.01963 (2021)

18. Al-Shareeda, M.A., Manickam, S., Ali, M.: Ddos attacks detection using machine learning and deep learning techniques: analysis and comparison. Bull. Electr. Eng. Inform. **12**(2), 930–939 (2023)

19. Holkoviˇc, M., Ryˇsavy', O., Dudek, J.: Automating network security analysis at packet-level by using rule-based engine. In: Proceedings of the 6th Conference on the Engineering of Computer Based Systems, pp. 1–8 (2019)

20. Nespoli, P., D´ıaz-Lo´pez, D., Ma´rmol, F.G.: Cyberprotection in iot environments: a dynamic rule-based solution to defend smart devices. J. Inf. Secur. Appl. **60**, 102878 (2021)

21. Siddabathula, K.S., Panneerselvam, R.K., Vasana, V., Vejendla, J., Rafi, M., Gummadi, S.B.: Yaracapper–yara rule-based automated system to detect and alert network attacks. In: Research Advances in Network Technologies, pp. 25–47. CRC Press (2023)

22. Subach, I., Mykytiuk, A.: Methodology of formation of fuzzy associative rules with weighted attributes from siem database for detection of cyber incidents in special information and communication systems. Inf. Technol. Secur. **11**(1) (20) (2023)

23. Husa´k, M., Bajtoˇs, T., Kaˇspar, J., Bou-Harb, E., Celeda, P.: Predictive cyberˇ situational awareness and personalized blacklisting: a sequential rule mining approach. ACM Trans. Manage. Inf. Syst. (TMIS) **11**(4), 1–16 (2020)

24. Sivanantham, S., Mohanraj, V., Suresh, Y., Senthilkumar, J.: Association rule mining frequent-pattern-based intrusion detection in network. Comput. Syst. Sci. Eng. **44**(2) (2023)

25. Chora´s, M., Kozik, R., Puchalski, D., Ho lubowicz, W.: Correlation approach for sql injection attacks detection. In: International Joint Conference CISIS'12ICEUTE´12-SOCO´12 Special Sessions, pp. 177–185. Springer (2013)

26. Lou, P., Lu, G., Jiang, X., Xiao, Z., Hu, J., Yan, J.: Cyber intrusion detection through association rule mining on multi-source logs. Appl. Intell.Intell. **51**, 4043–4057 (2021)

27. Sharafaldin, I., Lashkari, A.H., Ghorbani, A.A.: Toward generating a new intrusion detection dataset and intrusion traffic characterization. ICISSp **1**, 108–116 (2018)

28. Sarhan, M., Layeghy, S., Portmann, M.: Evaluating standard feature sets towards increased generalisability and explainability of ml-based network intrusion detection. Big Data Res. **30**, 100359 (2022)

29. Chawla, N.V., Bowyer, K.W., Hall, L.O., Kegelmeyer, W.P.: Smote: synthetic minority over-sampling technique. J. Artif. Intell. Res.Artif. Intell. Res. **16**, 321–357 (2002)

30. Coppolino, L., D'Antonio, S., Uccello, F., Lyratzis, A., Bakalis, C., Touloumtzi, S., Papoutsis, I.: Detection of radio frequency interference in satellite ground segments. In: 2023 IEEE International Conference on Cyber Security and Resilience (CSR), pp. 648–653 (2023)

Beibu Gulf Marine Ranch: Utilizing BeiDou Grid Code and Multi-system Integration for Modernized Management and Monitoring

Guilin Xu[1], Hengtong Qiu[1], Xiaomin Yan[1], Man Wu[2], Jing Guo[1], Zhaoyong Huang[1], and Wenlong Huang[1(✉)]

[1] Nanning Normal University Key Laboratory of Environment Change and Resources Use in Beibu Gulf, Ministry of Education, Nanning 530100, Guangxi, China
wenlonghuang999@gmail.com

[2] Guangxi Key Laboratory of Marine Environmental Science, Guangxi Academy of Marine Sciences, Guangxi Academy of Sciences, Nanning 530007, China

Abstract. This paper delineates the conception, cultivation, and application of the Geospatial-Temporal Data Grid Smart Service Platform for the Beibu Gulf Marine Ranch. This platform amalgamates a multitude of pivotal data elements and both software and hardware components. It employs BeiDou grid code technology to orchestrate geospatial-temporal data, refining retrieval algorithms to elevate data indexing efficiency and augment the discernment and governance of diverse, heterogeneous data sources within the marine ranch. The platform encompasses a multitude of modules, spanning marine ecological milieu surveillance, buoy data scrutiny and analysis, as well as radar oversight systems, thereby enabling a comprehensive scrutiny of marine ranch production operations and ecological integrity. Against the backdrop of rapid technological advancement, this platform serves to satiate the informatization exigencies of marine data management. Through the edifice of a contemporary marine ranch monitoring framework, this platform provides substantive underpinning for the elevation of the marine industry, amplifying its cachet and fiscal utility. In summation, this platform furnishes the Beibu Gulf Marine Ranch with tangible, efficacious information services and decision-making support, thus galvanizing the robust evolution of the marine economy.

Keywords: Geospatial-temporal data · BeiDou grid code · Marine ecological monitoring · Data governance · Marine industry

1 Introduction

In recent years, China's coastal fisheries resources have witnessed a severe decline. Ocean ranching, as a pivotal means for rehabilitating marine fisheries resources and aquatic ecosystems, plays a crucial role in coordinating the strategic spatial planning of marine fisheries and achieving sustainable development [1]. Compared to developed marine fishing nations like Japan and Norway, China's marine ranching initiatives started relatively late, commencing in the late 1970s [2]. After more than four decades of development, China's marine ranching is currently in an accelerated construction phase [3]. The first batch of 20 national-level marine ranching demonstration zones was approved for construction in 2015, with plans to establish as many as 178 national-level marine ranching demonstration zones by 2025 [4]. However, in recent years, seasonal hypoxia events have been recurring in the marine waters of Muping, Yantai, Shandong, leading to the significant mortality of marine organisms. In the summer of 2018, Liaoning experienced an unprecedented and sustained high-temperature period, resulting in a large-scale die-off of cultured sea cucumbers and causing substantial losses to aquaculture farmers. An increasing number of occurrences indicate that localized anomalies in the marine ecological environment are a significant factor contributing to the mass mortality of marine organisms. Therefore, the establishment of disaster response and early warning mechanisms, through real-time monitoring of marine environmental information and efficient information transmission and processing methods, has become crucial. This can mitigate the systemic risks associated with marine ranching development.

In response to the inadequacies of traditional marine ranching, such as the lack of comprehensive three-dimensional monitoring and insufficient levels of automation, Shandong Province initiated the establishment of the "Marine Ranch Observation Network" in 2015. This endeavor achieved real-time and online monitoring of crucial ecological elements on the seabed, including multiple water quality parameters, high-definition underwater video capture, and seawater profile flow velocity and direction [4]. Jia Wenjuan and her colleagues explored the development of oceanic environmental data collection devices that not only satisfy the long-distance transmission demands of massive on-site observations in marine ranching but also exhibit cost-effectiveness, low power consumption, and robust stability [5]. As marine ranching data continue to grow in richness, encompassing various aspects such as marine biology, ecology, economics, and physics, the collection, processing, and storage formats of different types of data may vary, posing a significant obstacle to the integration and sharing of marine data. However, the unified storage, retrieval, and management of marine big data within a grid framework represent an effective approach for efficiently handling heterogeneous data from diverse sources in the marine domain.

The Beibu Gulf, situated at the southwestern end of China's mainland coastline and the northwest of the South China Sea, adjacent to ASEAN and the Greater Bay Area of Guangdong, Hong Kong, and Macau, serves as the maritime entry point for the new land-sea corridor in western China, highlighting its strategic significance. However, the Beibu Gulf region is marked by pronounced natural disasters, with frequent occurrences of tropical cyclones [6] and algal blooms [7, 8], posing significant harm to marine life. Furthermore, the springtime mortality of the flagship industry's Qinzhou oysters in Guangxi is linked to persistently high environmental salinity [9]. Therefore, there is an urgent need to establish an all-encompassing, intelligent marine ranching service platform that integrates monitoring, early warning, traceability, disaster prevention, ecological leisure, and online transactions across the entire industry chain. This platform will provide essential technological support for the ecological monitoring of Beibu Gulf marine ranch and the development of marine fisheries.

2 Platform System Design and Development

In the wake of advancements in marine environmental sensing and information transmission technologies, the wealth of maritime spatiotemporal data resources has grown significantly. This has imposed heightened demands on the organization and management of oceanic data. Leveraging a grid-based model, the spatiotemporal data organization method provides an avenue for the efficient management of vast, diverse, and dynamic oceanic spatiotemporal data on a large scale [10, 11]. Furthermore, in accordance with the Beidou grid positioning code (GB/T 39409-2020), the temporal and spatial attributes of all entities can be assigned and defined within the national Earth-centric spatiotemporal framework. The establishment of a fused spatiotemporal data Beidou grid code index promises a substantial enhancement in the efficiency of querying, organizing, and managing marine spatial information [12].

2.1 Platform System Design Principles

The technical principles underpinning this service platform encompass three fundamental aspects: communication mechanism design, spatiotemporal big data preprocessing, and service platform development. These components collectively ensure the acquisition, transmission, fusion, integration, and presentation of data within the Beibu Gulf Marine Ranch demonstration area. The project has established an intelligent service area for the Beibu Gulf Marine Ranch, comprising a hall (exhibition hall), a chamber (monitoring room), a center (Beibu Gulf Marine Ranch oyster trading center), a bay (Beibu Gulf geospatial spatiotemporal data set), a zone (core demonstration area), three sets

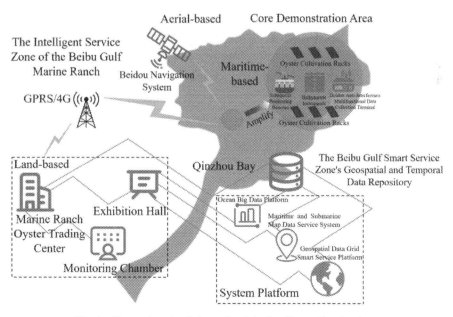

Fig. 1. Comprehensive Schematic of the Intelligent Service Area.

of equipment (beacons, depth sounders, multifunctional devices), and three platforms (marine big data platform, offshore and underwater map service system, and geospatial spatiotemporal data grid smart service platform). The overall schematic of the intelligent service area is depicted in Fig. 1.

The service platform not only necessitates the automated collection, remote transmission, processing, and storage of marine geospatial data to meet the requirements of application demonstrations, but also requires the remote modification of telemetry station operating parameters and communication parameters based on this foundation. The communication mechanism is designed as follows: ① Establishing an integrated information system with the control center at its core; ② Providing each sub-center with an automated platform to control the collection of data from both above and below the sea within its jurisdiction, automatically collecting data from all telemetry stations and sending it to the center according to the configured operational mode through communication monitoring front-end machines; ③ Enabling the setting of operational parameters for all running telemetry stations and controlling the operation of collection stations; ④ Remotely and simultaneously extracting historical solid-state data from all telemetry stations through mobile technology channels (4G) [13]. Based on mobile 4G technology, this remote data collection technology achieves real-time monitoring and data collection from remote marine ecological sensors.

The management process of spatiotemporal big data is governed by a series of standard constraints. Building upon existing industry standards within the realm of big data, and following a meticulous process of comparison and analysis, we have curated industry standards tailored to the processing of marine ranch big data. By referencing prevailing international standards, we have crafted novel standards for marine ranch data processing, thereby enhancing and complementing the standards framework for marine big data management. According to these standards, structured data (relational databases) are stored in the Hbase database, while semi-structured data (JSON logs) and binary data (videos, audio) are stored in the HDFS files.

The platform data primarily comprises two categories: ① Real-time data and storage, encompassing Beidou spatiotemporal grid data, high-precision navigation and positioning data for marine (surface/underwater) environments, data from the deployment of marine geodetic benchmarks (beacons) and responsive sensing devices, prototype system data for offshore/underwater location services, and operational data from the digital marine integrated business system.② Historical survey data, including information from the "908 Special Project" for comprehensive assessment of China's coastal marine areas, which covers data from fundamental surveys of nearshore waters, coastal zones, islands, and socio-economic conditions in coastal regions. By integrating and managing data that spans multiple levels of databases, formats, and applications, we aim to enhance the level of data integration and management.

2.2 The Development of the Service Platform

Oceanic heterogeneous data from various sources, such as databases, videos, audios, and logs, undergo a series of steps including data cleansing, transformation, and quality control. These data are then classified and written into target repositories like HDFS, Hcatalog, Hbase, Redis, etc. Data extraction and presentation are carried out through web or log modules. ETL tools [14] are employed for data cleansing and integration, ensuring that disparate and chaotic data finds structured storage.

Based on the wide-ranging sources and diverse forms of marine data, this paper introduces the application of the Hadoop 2.2.0 framework [15] in various aspects of marine heterogeneous data storage, extraction, integration, and analysis. In this cluster environment, the marine environmental perception system and database, combined with the JBoss server and integrated with the MapReduce [16] programming model, map data that meets query conditions to the visualization platform. Simultaneously, the platform adopts a J2EE architecture [17], designed based on REST API design principles, with a persistence layer utilizing frameworks such as Hibernate and MyBatis. The platform primarily utilizes the Java programming language and server-side Java technologies, while communication data packets are in a lightweight JSON format [18]. Both the business application system and data integration platform must be built on top of Hadoop 2.2.0 and Oracle 11g (or higher) large-scale databases. The platform integrates at least the following components: HBase, Hive, Impala, Sqoop, and Phoenix. Marine big data is presented using the HTML5 framework and open-source ECharts and webGL. The technical roadmap for the Beibu Gulf Ocean Ranch Geographic Spatiotemporal Data

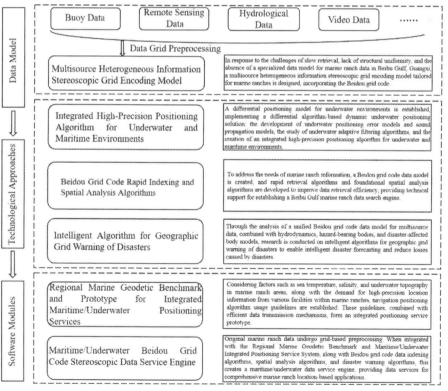

Fig. 2. Technological Roadmap of the Multi-Source Heterogeneous Data Grid Processing Platform.

Grid Smart Service Platform's multi-source heterogeneous data gridding processing platform is depicted in Fig. 2.

To enhance the quality of information services, the platform has employed Web Service client and server mechanisms, enabling remote users to engage in unhindered HTTP communication. Furthermore, it has incorporated the principles of inversion of control and dependency injection to eliminate direct dependencies between components or modules, thereby rendering the software system's development more extensible. Additionally, it has elevated data retrieval techniques to ensure both retrieval performance and speed, ensuring that the extraction of information from vast datasets remains seamless.

3   Platform Functions and Applications

3.1   Platform Main Page

The geospatial-temporal data grid smart service platform for the Beibu Gulf Marine Ranch, as constructed in this study, showcases its web main page (Fig. 3) predominantly in deep blue hues. The access to the site is available at http://47.92.145.187:8081/front/index and is now freely accessible to researchers and aquaculturists. It furnishes comprehensive marine ecological data and disaster alert information for the Beibu Gulf Marine Ranch.

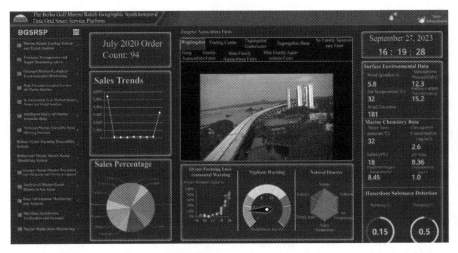

Fig. 3. Web Main Page of the Beibu Gulf Marine Ranch Geospatial-Temporal Data Grid Smart Service Platform.

3.2 Platform's Principal Modules

(1) Marine Ranch High-Precision Positioning Service: Leveraging marine geodesy benchmark beacons and responsive sensing equipment, it furnishes high-precision positioning services to marine ranch production facilities such as surface vessels, sightseeing and fishing platforms, net cages, rafts, oyster racks, underwater fish reefs, AUVs, ROVs, and more.

(2) Analysis of Adjacent Marine Area Geoid Models: Utilizing marine geodesy benchmark beacons and responsive sensing equipment, it computationally derives and displays the gridded geoid models of the marine ranch's adjacent marine areas on the platform.

(3) Demonstration Area Hydrodynamic Numerical Model Analysis: Visualizing the gridded spatial distribution and temporal variations of hydrodynamic elements, it vividly portrays calculated results, offering a scientific basis for studying the laws of waves, tides, and sediment movement in the nearshore region.

(4) Intelligent Query of Marine Scientific Data: Utilizing the Beibu Gulf Economic Zone's intelligent scientific data query system and auxiliary analysis and decision-making system, it conducts data visualization, browsing, viewing, and statistical presentation for climate resource data, geological information, hydrological information, and other related datasets.

(5) Buoy Information Monitoring and Analysis: By collecting oceanographic monitoring data including wind direction, wind speed, surface current speed, surface current direction, average wave height, wave count, air temperature, as well as oceanographic chemical monitoring data like water temperature, conductivity, salinity, dissolved oxygen, chlorophyll concentration, and pH, it enriches the environmental information elements of the marine ranch.

(6) Beibu Gulf Marine Ranch Radar Monitoring System: By radar, it remotely detects the routes of civilian vessels navigating in the Beibu Gulf waters, tracks vessels engaged in theft or accidents that may harm the marine ranch, ensuring safety in production.

(7) Precision Management and Rapid Monitoring and Early Warning: Realizing fine management of the marine industry within the demonstration area, it provides high-precision positioning services and attribute management for marine ranch production facilities and equipment such as surface vessels, sightseeing and fishing platforms, net cages, rafts, oyster racks, underwater fish reefs, AUVs, and ROVs. It enables informed decisions for the management team. Simultaneously, it achieves dynamic monitoring and early warning by performing comparative analysis, calculations, and scoring based on data from the planning and monitoring modules, quickly determining whether there are deviations that warrant early warning. The current status data from planning and monitoring are presented in different ways to assist decision-making.

(8) Oyster Trading System and Trend Analysis: Through the trading system, it tracks, queries, stores, displays, and sells oyster products, records information on stored goods, and efficiently locates and identifies goods stored by various merchants in the same area to maximize storage space utilization. It conducts statistical analysis of all e-commerce in-stock products, inspects the validity period of perishable goods,

and provides alerts in case of impending expiration. E-commerce can also check the shipping status of goods. Through analysis, it showcases oyster ecology and sales trends.

4 Discussion

The development of the Beibu Gulf Marine Ranch Geospatial-Temporal Data Grid Smart Service Platform involves the integration of various key data information and software-hardware components. Building upon the foundational grid algorithm for the regional spatiotemporal Beidou grid code within the demonstration area, it combines environmental data spatiotemporal evolution, multifunctional data collection for Beidou anti-interference, investigation and assessment models for marine vulnerability, core databases and storage standards for land-sea integration, hydrodynamic numerical simulation models, marine geodesy benchmarks (beacons), and responsive sensing equipment. Simultaneously, it enhances grid data indexing efficiency through optimized retrieval algorithms, elevating the perception and management levels of diverse, heterogeneous data sources concerning the marine ranch's ecological environment, biological resources, meteorology, hydrodynamics, and more. This platform serves as a valuable reference for industry professionals.

Located in the oyster main production area of Maowei Sea, Qinzhou, the platform comprises multiple systems, including marine ecological environmental monitoring, buoy information monitoring and analysis, and Beibu Gulf Marine Ranch radar surveillance. By connecting to these systems, it achieves comprehensive monitoring of the oyster rack production and operational status in the marine ranch, as well as potential ecological security threats.

In the current era of rapid technological advancement, the capabilities for marine data perception and transmission continue to improve, leading to an explosive growth in marine data. However, the management of marine ranch data, given its diverse sources and complex processing requirements, requires an enhancement in information technology [19]. Leveraging the spatiotemporal geospatial data organization method based on Beidou grid codes, which are strongly correlated with time and space and exhibit high coupling characteristics, can lay the foundation for realizing multidimensional information monitoring of broad-spectrum ecological environments in modern marine ranches. It provides support for the development of modern marine ranches, even fully integrated marine ecological ranches, and contributes to the efficient management of marine big data. As China's marine ranch strategic space continues to expand and guided by the strategic concept of a new stage for marine ranches characterized by "ecology, precision, intelligence, and integration," the construction of a modern marine ranch multidimensional dynamic stereoscopic monitoring system through means such as scientific site selection, resource nurturing, and security guarantees [20] is essential. This system will create an integrated platform for the entire industry chain of modern marine ranches, incorporating monitoring, early warning, traceability, disaster prevention, ecological leisure, and online transactions. The goal is to enhance brand reputation, increase efficiency, and mitigate losses, thereby promoting industry transformation and upgrading.

5 Conclusion

(1) We have built a smart service platform that integrates online collection and analysis of marine scientific data, monitoring and early warning of marine disasters, and online trading of marine products for global traceability. Realize the transformation, storage, and display of geographic spatiotemporal grid data for marine pastures and their surrounding areas, forming a corresponding collection and analysis system for complex ecological, industrial, and life land and sea regulatory information, and carry out refined management and monitoring and early warning for marine ranches.

(2) The service platform has introduced the Beidou grid code algorithm and high-precision underwater location services for marine pastures in the blind spot of Beidou/GPS signals, which are important features of this service platform. Relying on the global subdivision grid, the traditional polygonal spatial intersection retrieval algorithm is transformed into a grid encoded numerical matching retrieval algorithm, greatly reducing the computational complexity of spatial retrieval and improving the retrieval efficiency of spatial data. Relying on underwater marine geodetic reference beacons and response sensing equipment, and through the solution of high-precision positioning algorithms integrated in the sea and sea, a regional marine geodetic reference and marine/underwater integrated position service system is established to provide high-precision position services for users in the region.

(3) The service platform promotes industrial integration. The platform mainly serves the Qinzhou "Longmen Oyster Bay" aquaculture industry demonstration zone, shifting from marine fishing to multi industry integration in industrial integration, covering the breeding of local advantageous natural oyster seedlings, large oyster cultivation and trading in Qinzhou, Guangxi, promoting the combination of industrial value-added and ecological protection.

On the basis of traditional ocean data management information systems, a grid based intelligent service platform for geographic spatiotemporal data of marine ranches in the Beibu Gulf has been constructed using technical means and methods such as network databases, rapid data retrieval, data network analysis, Beidou grid code, and visualization http://47.92.145.187:8081/front/index. This service platform has the characteristics of user-friendly and beautiful interface, high resolution, and fast retrieval speed. It can provide practical and effective information services and decision-making support for marine ranches in the Beibu Gulf, promoting the healthy development of the marine economy in the Beibu Gulf.

6 Prospects

The informationization construction of marine ranches still faces many challenges. Relying on intelligent management platforms to carry out digital cruising of marine ranches, timely sensing and responding to environmental stress and death disasters of breeding objects, is an important way to improve the survival rate of breeding target organisms. Strengthen the research and development of multi-functional sensing devices such as online monitoring of marine environmental physicochemical factors and real-time monitoring of breeding object behavior, while improving the transmission, storage, processing and analysis capabilities of marine ranch data, and developing key control points for response mechanisms to enhance information service capabilities for marine ranches.

Acknowledgements. This research was financially supported by: Special Initiative on Marine Agriculture and Freshwater Aquaculture Technology Innovation of National Key Research and Development Program (No. 2022YFD2401200); Central Government-Guided Local Science and Technology Development Funding Project of China (No. Guike AD20238054, Guike AD21075012); Key Projects of Regional Innovation Development Joint Fund of the National Natural Science Foundation of China (No. U20A20105); The Opening Foundation of Key Laboratory of Environment Change and Resources Use in Beibu Gulf, Ministry of Education, Nanning Normal University (No. NNNU-KLOP-X1815).

References

1. Hongsheng, Y.: Construction of marine ranching in China: reviews and prospects. J. Fish. China **40**(7), 1133–1140 (2016)
2. Huayong, Q., Yong, C., Xiumei, Z., Shouyu, Z., Guofan, Z.: Modern marine ranching: status and development strategy. Strateg. Study CAE **18**(3), 79–84 (2016)
3. Hongsheng, Y., Shouyu, Z., Xiumei, Z., Pimao, C., Tao, T., Tao, Z.: Strategic thinking on the construction of modern marine ranching in China. J. Fish. China **43**(4), 1255–1262 (2019)
4. Fangguo, Z., Yanzhen, G., Peiliang, L., Liyuan, S., Xin, L., Dong, C., et al.: Construction and development of marine ranch observation network in Shandong Province. Mar. Sci. **44**(12), 93–106 (2020)
5. Wenjuan, J., Xiaowei, Z., Chenyang, Y., Hongzhi, L.: Internet of things technology for online monitoring of marine ranch ecological environments. Mar. Sci. **46**(1), 83–89 (2022)
6. Shouwen, Q., Zhigang, B., Yi, S., Liyang, W., Chao, Z., Hongyuan, S.: Analysis of the variation characteristics of tropical cyclones affecting Beibu gulf between 1980 and 2015. Guangxi Sci. **26**(6), 663–668 (2019)
7. Xingyi, L.: Implementing marine ecological engineering creating ecological bays. Guangxi Econ. **7**, 30 (2015)
8. Pingping, S., Yuzao, Q., Linjian, O.: Phaeocystis globosa in coastal China: taxonomy, distribution, and its blooms. Mar. Sci. **42**(10), 146–162 (2018)
9. Wei, L., et al.: Investigation and analysis of death of Qinzhou oyster in Spring in Guangxi. Aquaculture **6**, 7–13 (2020). (in Chinese)
10. Yi, L.: Research on efficient grid coding method and application for large-scale spatiotemporal data organization. Master Dissertation of Strategic Support Force Information Engineering University, pp. 1–78 (2020)
11. Junning, C., Naijun, H., Qing, L.: Design of ocean data management and visualization platform. China Sci. Technol. Inf. **2**, 113–115 (2015)
12. Hongfang, S., Zhiyu, W., Yinger, Z., Yubin, X.: Refined management system of airspace resources based on BeiDou Grid location code. J. Civ. Aviat. **6**(1), 44–47 (2022)
13. Singh, G., Bansal, N., Rani, P.: 4G mobile network technology (LTE). Int. J. Adv. Res. Comput. Sci. **7**(6), 97–100 (2016)
14. Kherdekar, V.A., Metkewar, P.S.: A technical comprehensive survey of ETL tools. Int. J. Appl. Eng. Res. **11**(4), 2557–2559 (2016)

15. Merceedi, K.J., Sabry, N.A.: A comprehensive survey for hadoop distributed file system. Asian J. Res. Comput. Sci. **11**(2), 46–57 (2021)
16. Dean, J., Ghemawat, S.: MapReduce: simplified data processing on large clusters. Commun. ACM **51**(1), 107–113 (2008)
17. Johnson, R.: J2EE development frameworks. Computer **38**(1), 107–110 (2005)
18. Bourhis, P., Reutter, J.L., Vrgoč, D.: JSON: Data model and query languages. Inf. Syst. **89**, 101478 (2020)
19. Hongsheng, Y., Da, H., Qiang, X.: Views on modern marine ranching. Oceanol. Limnol. Sin. **47**(6), 1069–1074 (2016)
20. Hongsheng, Y., Dewen, D.: Marine ranching version 3.0: history, status and prospects. Bull. Chin. Acad. Sci. **37**(6), 832–839 (2022)

A Member Inference Attack Defense Method Based on Differential Privacy and Data Enhancement

Gaoxiang Cui[1,2], Lina Ge[1,2,3]([✉]), Yiteng Zhao[1,2], and Teng Fang[1,2]

[1] School of Artificial Intelligence, Guangxi Minzu University, Nanning 530000, China
66436539@qq.com
[2] Key Laboratory of Network Communication Engineering, Guangxi Minzu University, Nanning 530000, China
[3] Guangxi Key Laboratory of Hybrid Computation and IC Design Analysis, Nanning 530000, China

Abstract. The development of deep learning has brought about the business model of Machine Learning as a Service (MLaaS). Malicious users can infer whether a member has participated in model training through Membership Inference Attacks (MIA), thereby stealing user privacy. Although various methods have been proposed to defend against membership inference attacks, they are all aimed at defending against a certain type of membership inference attack and cannot defend against various membership inference attacks at the same time. This paper proposes the MEWDP method, which can defend against multiple types of membership inference attacks. Firstly, it uses multi-round MIXUP data augmentation method to process privacy data and adds non-interference noise to the data in the form of data fusion. Then, during the model training stage, Gaussian noise that satisfies differential privacy is added to protect model privacy, and label smoothing method is used to prevent the training model from overfitting. The results show that this defense method can reduce the success rate of metric-based membership inference attacks to 51.2%, and reduce the success rate of model-based membership inference attacks to 50.9%. Compared with other defense methods, the MEWDP defense method has universality and better defense effect. For the CIFAR10 dataset, it can reduce the success rate of member attacks to 50.8%.

Keywords: Differential Privacy · Member Inference attack · Privacy Protection · Data Enhancement

1 Introduction

With the development of science and technology, Artificial Intelligence (AI) has penetrated into all aspects of people's lives. In particular, the usefulness of ChatGPT [1] has been widely recognized and has had a huge impact on people, prompting further attention to the field of artificial intelligence. At the same time, the security of various models has also been paid attention to, because there is no absolutely safe system in the world, machine learning is no exception. Since the birth of machine learning, research on its security has never been interrupted. For each stage of machine learning, there are corresponding attack means, including confidentiality attack [2], integrity attack and availability attack [3]. Common attack methods include data poisoning attack [4] and backdoor attack [5]. Common attacks on model data include data theft attacks, attribute inference attacks and member inference attacks. In Membership Inference Attacks (MIA) [6], the attacker can infer whether a piece of information has participated in the training of the model. This attack is covert and easy to implement. Compared with other attacks, Membership Inference Attacks have lower attack difficulty, fewer conditions involved in the attack, and are easier to implement. They can leak sensitive user data and have become one of the major threats to current machine learning.

In recent years, a commercial model of machine learning, the Machine Learning as a Service (MLaas) model [7], has become a major cause and environment for privacy risks. The working principle of this model is that each manufacturer publishes its own trained model interface for users to query for free or for a fee. This model is convenient for users but also brings the risk of privacy disclosure. Because the training data used in some models contains some sensitive information, privacy information will be leaked. The harm caused by personal privacy disclosure is self-evident, and lawbreakers can use illegal means to obtain information to commit fraud, blackmail and other acts that endanger social stability.

This paper proposes the MIXUP Enhance With Differential privacy (MEWDP) defense method to defend against member inference attacks. This method uses differential privacy and label smoothing to train neural network models, and performs MIXUP data enhancement processing on the training data to defend against member inference attacks. This method can effectively resist member inference attacks and has less interference with model accuracy.

The rest of this study is organized as follows: Sect. 2 introduces related work and background knowledge, including currently common membership inference attack methods and member inference attack defense methods. Section 3 describes the method proposed in this article, including multi-round hybrid data enhancement technology, differential privacy noise-adding mechanism and privacy budget selection, and label smoothing to prevent over-fitting technology. Section 4 presents the experimental results and analysis, and Sect. 5 summarizes the conclusions and prospects.

2   Related Work

2.1   Member Inference Attack

Shokri et al. [6] proposed the concept of member inference attack for the first time in 2017. Due to the easy deployment and implementation of member inference attack, it has been widely studied and the methods of member inference attack have been developed and expanded [8]. Subsequently, research on member inference attack in all directions of deep learning has been carried out, such as image recognition [6]. Speech recognition [9], positioning information [10], etc. The purpose of member inference is to infer whether a sample data participates in the training of this model, and to steal the privacy information of users participating in the training model by constructing an attack model. Existing member inference attacks are mainly carried out in two ways: Model-based Attack and Metric-based Attack.

Model-Based Attack

Shokri et al. [6] obtains training data with the same distribution as the target model, trains multiple shadow models that mimic the target model, and obtains prediction data as a training dataset to train a binary classification attack model. The model-based Membership Inference Attack is shown in Fig. 1.

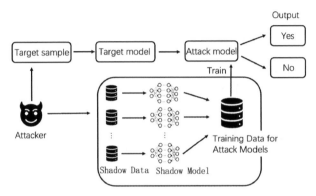

Fig. 1. Model-based Membership Inference Attack

As can be seen from the figure, there are three main stages of model-based member inference attacks.

The first stage is data preparation. In order to train the shadow model, we need to prepare the shadow data set $D_1^{shadow} \ldots D_n^{shadow}$, the second stage needs to establish the shadow model $M_1^{shadow} \ldots M_n^{shadow}$, the shadow model to close to the target model M_{train}, according to the different prior knowledge of the attacker. In the third stage, the attack model M_{attack} is trained. A binary classification model is trained by combining the output of the shadow model with the raw data as the input of the attack model. After the training is completed, a complete attack model is obtained. Input a sample P_{test} to be tested, and the output of the model indicates whether the test sample is member data.

Metrics-Based Attacks

Model-based member inference attacks require the preparation of shadow data and association attack models, and the attack cost is high. Therefore, model-based member inference attacks, also known as metrics-based member inference attacks, are proposed and applied. Since there is no need to train the attack model, only some output parameters of the target model are required, and the attack cost is low and the success rate is high. In 2021, song et al. [11] proposed a metrics-based attack, which is simpler and easier to use. The main attack methods include sample label threshold based attack, loss threshold based attack and model output confidence based attack.

In the attack based on sample label threshold, the attack idea is that the attacker determines whether test sample x is the member data of the target model by whether the label output by the target model is the real label of test sample x. In the attack based on loss threshold, the attack mainly uses the deep learning training process to always minimize the loss, so the loss obtained by the training sample input model should be smaller than that of the non-member sample, so a threshold value is set. If the loss is less than this value, the sample can be identified as the member sample. A model-based output confidence attack that uses deep learning to output a prediction vector that describes the probability that the input data x belongs to a class. The goal of deep learning training is to maximize the probability of true labels in the prediction vector as much as possible, so the probability of true labels in the prediction vector output by the model should be greater than that of non-member samples as input.

2.2 Defense Against Member Inference Attacks

The harm caused by the member inference attack cannot be underestimated, so the defense methods are endless. Shokri et al. [6] believe that member inference attacks are caused by overfitting during model training, and propose that regularization can resist member inference attacks. The classic Memguard method proposed by Jia et al. [12] adds noise to the model prediction to disturb the attacker. Nasr et al. [13] proposed an adversarial regularization method to defend against member inference attacks, which introduced a new training method to optimize the defense model by simultaneously training the attack model and the defense model. Differential privacy, as a privacy protection method with strict mathematical proof, was first proposed by Dwork et al. [14, 15]. Differential privacy has a good protective effect on private data, which is widely used in various data publishing scenarios and subsequently applied to machine learning [16, 17]. At present, differential privacy is also used to defend against member inference attacks [18].

Because differential privacy has strict mathematical proof, it is widely used in privacy protection. Hagestedt et al. [19] proposed a differential privacy mechanism that can effectively prevent the risk of privacy leakage of biomedical data members. This defense method effectively defends against model-based member inference attacks. Chen et al. [20] proposed a model training method that satisfies differential privacy for machine learning models of genetic inference, which can defend against model-based member inference attacks. Chen et al. [21] proposed the EMT method to defend against metric-based membership inference attacks.

However, most of the existing defense methods are aimed at one kind of member inference attack method. Member inference attack methods are becoming more and more diversified. Based on the use of differential privacy, this paper applies MIXUP data enhancement to member inference attack defense and proposes MEWDP defense method, which uses the MIXUP data enhancement method combined with differential privacy to protect user privacy data, introduces label smoothing in model training to prevent model overfitting, and can resist model-based member inference attacks and metric-based member inference attacks.

3   Our MEWDP Defense Method

In this paper, MEWDP defense method is proposed, which can simultaneously defend against model-based and metric-based member inference attacks. In this method, noise conforming to differential privacy is added during model training, so as to protect members' private data during model training. Before the training starts, multiple rounds of MIXUP are carried out to enhance the private data. Compared with adding useless noise to the training data to improve the defense effect, this operation enhances the availability of noise and reduces the impact on the model accuracy. In addition, the label smoothing technique is used to prevent the model from overfitting, thus defending against various metry-based member inference attacks. The defense method architecture is shown in Fig. 2.

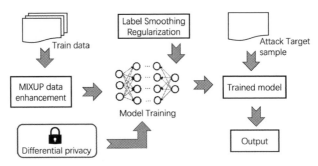

Fig. 2. Defense approach architecture

3.1   Multi-rounds of MIXUP Mixing Enhancement

The MEWDP defense method uses multiple rounds of MIXUP for data enhancement. Different from traditional data enhancement methods, MIXUP integrates two types of data in a certain proportion to create new data. The core idea of MIXUP is to do fusion, randomly pick two training samples of vectors and labels. The new vectors and labels are generated by linear interpolation, as shown in Eq. 1.

$$\tilde{x} = \lambda x_i + (1 - \lambda)x_j; \tilde{y} = \lambda y_i + (1 - \lambda)y_j \tag{1}$$

where $\tilde{x}$ is a new vector generated by linear fusion of two class vectors x_i and x_j, and $\tilde{y}$ is a new label of the same fusion. The data after MIXUP enhancement processing can significantly reduce the overfitting degree of the model and make the classification boundary smoother.

In the training stage of the deep learning model, batch is used as the minimum unit for gradient update. Batch refers to the training data according to the batch size. All the training data is packaged according to the batch size, which is set according to the specific data amount. As the minimum input unit in each round, MIXUP mixing is based on a batch of data. For each batch of data, n new batches are copied, the sample order in the copied batch is randomly disrupted, and then fused n times to generate n new batch data based on MIXUP enhanced fusion. The process of multi-round MIXUP hybrid enhancement method is shown in Fig. 3.

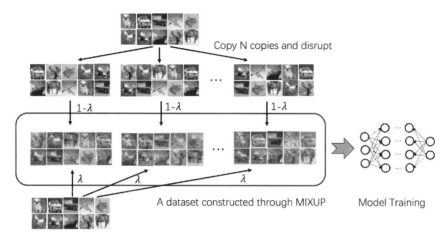

Fig. 3. Multi-round MIXUP process of hybrid enhancement method

As can be seen from the figure, for each batch that will be input into the model for training, a new n new batch will be copied, and then the initial batch will be mixed up for data enhancement fusion. After n rounds of fusion, the new batch data enhanced with data will be obtained and input into the model for training as input data. After multiple rounds of replication and generation, the number of samples of training data is increased and the data set is expanded, which can make the model training more accurate.

3.2 DPSGD Algorithm

When the MEWDP method trains the model on the training data after data enhancement, it uses the DPSGD algorithm to add noise that is consistent with differential privacy. Based on experimental verification, it sets an appropriate privacy budget value to protect users in a query-based machine learning as a service environment. Private data.

DPSGD Algorithm Description
The Stochastic Gradient Descent with Differential Privacy (DPSGD) algorithm applied to deep learning is proposed by Abadi et al. The core of the algorithm is to make the neural network model conform to the definition of differential privacy by adding appropriate amount of noise in the gradient of the training process of the neural network model.

Privacy Budget
Deep learning training is achieved through continuous iteration, and differential privacy requires the privacy budget to be set for the entire model. In order to solve the problem of adding privacy budget in each iteration training, this algorithm proposes a privacy budget accumulation algorithm, which means that the privacy budget generated by each iteration

during the model training process will be superimposed. Therefore, privacy budget is an important parameter that needs to be adjusted. If the privacy budget is small, it will generate too much noise in each iteration, which will seriously affect the model training effect and the success rate of prediction and recognition of the final model. If the privacy budget is large, it will not exert a constraint on the model and it is difficult to protect the model privacy.

In this paper, 15 privacy budget values are selected from 0–10 to verify their impact on the model accuracy and protection effect on the model, and select appropriate privacy budget values to minimize the impact on the model accuracy and effectively defend against member inference attacks.

DPSGD Algorithm Description

The DPSGD algorithm first calculates the gradient according to the data sample and model architecture, and then introduces the gradient norm boundary C to clipping the gradient. The core of gradient clipping is to calculate the L2 norm of the sample gradient and compare the parameter C to select the smaller one to achieve the clipping effect. Then, Gaussian noise is added to the processed gradient, that is, the gradient value is added to the disturbance following Gaussian distribution. The cumulative privacy budget is recorded, followed by gradient descent to update the model parameters. The detailed description of the DPSGD algorithm is shown in Algorithm 1.

Algorithm 1: Differential privacy SGD

Hyper-parameters: Learning rate η_t; model parameter θ; Model boundary C

While: circuit training

1: Obtain training data$\{x_1, x_2, x_3, \dots, x_n\}$ and labels$\{y_1, y_2, y_3, \dots, y_n\}$

2: Calculate gradient: $g_t(x_i) \leftarrow \nabla_{\theta_i} L(\theta_t, x_i)$

3: Crop Gradient: $\bar{g}_t(x_i) \leftarrow g_t(x_i)/max(1, \frac{\|g_t(x_i)\|_2}{C})$

4: Adding noise: $\tilde{g}_t \leftarrow \frac{1}{L_t}(\sum_i \bar{g}_t(x_i) + N(0, \sigma^2 C^2 I))$

5: Gradient descent: $\theta_{t+1} \leftarrow \theta_t - \eta_t \widetilde{g_t}$

End while

Privacy Loss Measurement of DPSGD

To measure the Privacy loss of deep learning training process, this paper uses the Zero-Concentrated Differential Privacy (zCDP) privacy loss measurement mechanism, which is defined as follows:

Definition: For adjacent data sets D and D 'that differ by at most one piece of data, if centralized difference privacy is obeyed, the stochastic algorithm A can satisfy the following formula:

$$D_\alpha(A(D)\|A(D')) \triangleq \frac{1}{\alpha - 1}\log(\mathbb{E}[e^{(\alpha-1)L^{(o)}}]) \leq \rho\alpha \tag{2}$$

Including $\alpha \in (1, +\infty)$, the smaller the rho said the higher level of privacy protection, and privacy protection degree is lower, entries $D_a\left(A(D)\|A(D')\right)$ said $A(D)$ and $A(D')$

"a - Renyi" distance. The privacy loss of adjacent data sets is calculated according to the results of the stochastic algorithm A, and the calculation method is as follows:

$$L^{(O)}_{\left(A(D)\|A\left(D'\right)\right)} \; log \frac{P_r(A(D) = O)}{P_r\left(A(D') = O\right)} \tag{3}$$

where, O is the calculation result of stochastic algorithm A, according to the sequence combination and parallel combination of differential privacy, that is, If random algorithm $A_i (1 \leq i \leq k)$ satisfies $\rho_i - zCDP$, then combined algorithm $A(1 \leq i \leq k)$ composed of random algorithm A_i satisfies $\left(\Sigma_i^k \rho_i\right) - zCDP$, if the data set D into $D_1, D_2, \ldots, D_k$, then $A(D) = (A_l(D \cap D_l), ..., A_k(D \cap D_k))$ satisfies $\frac{1}{k}\left(\Sigma_{j=1}^k \rho_i\right) - zCDP$. It follows that:

Corollary: If random algorithm A satisfies $\rho - zCDP$, random algorithm A satisfies $\left(\rho + 2\sqrt{\rho log(\frac{1}{\delta})}, \delta\right) - DP$ differential privacy for any $\delta > 0$.

In this paper, high filament noise and Gaussian noise mechanism $N\left(0, \Delta^2 \sigma^2 I\right)$ satisfy $\left(\frac{1}{2\sigma^2}\right) - zCDP$. After one round of convolutional network training, the neural network model satisfies $\frac{1}{k}\left(\Sigma_{j=1}^k \frac{1}{2\sigma^2}\right) - zCDP$. That is $\left(\frac{1}{2\sigma^2}\right) - zCDP$. The neural network needs to train T rounds. After training T rounds, according to the parallel combination, the trained model should satisfy $\left(\frac{T}{2\sigma^2}\right) - zCDP$, that is, the trained model satisfies differential privacy. According to the inference, the model satisfies $\left(\frac{T}{2\sigma^2} + 2\sqrt{\frac{T}{2\sigma^2} log(\frac{1}{\delta})}, \delta\right) - DP$.

3.3 Label Smoothing Regularization

When training the model, we adds label smoothing regularization processing to further improve the model's ability to resist member inference attacks, strengthen the defense effect of MEFDP, and enable it to defend against member inference attacks designed based on model overfitting. Label Smoothing Regularization is proposed in deep learning to train convolutional neural networks to avoid overfitting. Preventing model overfitting can also improve the robustness of the model.

The deep learning training process is constantly approaching the real label, which is infinitely close to 1 on the basis of one-hot coding. The generation of overfitting means that the trained model can accurately identify the training data set, but the recognition success rate for the non-training data set is not high. Therefore, a parameter ε can be set to make the result infinitely approximate to 1-ε, so that the real label will not be excessively approximated, thus inhibiting the generation of overfitting. The essence of label smoothing is to change the distribution of real labels, as shown below:

$$q'(k|x) = (1 - \epsilon)\delta_{k,y} + \epsilon u(k) \tag{4}$$

where $u(k)$ represents the label distribution independent of the training sample x, ε represents the smooth parameter, $q(k|x) = \delta_{k,y}$ is the original label distribution for the sample x and the label y, $q'^{(k|x)}$ is the new label distribution in place of the original distribution, which is the weighted average of the original distribution $q(k|x)$ and the fixed distribution $u(k)$. Using the training results of label smoothing will make the features more clustered and have better generalization effect.

The MEWDP defense method first uses multiple rounds of hybrid data enhancement methods to process data to weaken the privacy features of individual data, and then conducts model training on the enhanced data. During training, differential privacy noise that conforms to DPSGD is added to protect user privacy data, and labels are modified. Label smoothing regularization processing prevents model overfitting, thereby achieving the purpose of defending against member inference attacks.

4   Experimental Results and Analysis

This section deploys the MEWDP defense method for the target model to be protected, verifies the influence of multi-rounds of MIXUP hybrid enhancement method on the model accuracy, the effect of different differential privacy budgets in the DPSGD algorithm on the model accuracy and the defense against member inference attacks, and compares the defense effect of the MEWDP defense method proposed in this paper against different data sets. The defense effect of other defense methods against inference attacks of different members is compared.

In this paper, data sets ADULT [6] and CIFAR10 [22] are used. The ADULT data set contains 48,442 pieces of data, each of which contains 14 personal privacy items such as age, working hours, marital status, race, nationality, and annual salary. CIFAR10 includes aircraft, cars, boats, trucks, birds, frogs, cats, dogs, horses, deer, snakes ten categories of labels, a total of 60,000 32*32 pictures. Model training and prediction were conducted respectively for no and with data enhancement. The results showed that when CIFAR10 data set was not enhanced, the success rate of model test set recognition was 94.5%; when CIFAR10 data set was enhanced and the same model was used, the model recognition rate was 95.1%. The training times of the model are all 100 times, and the rounds of multiple rounds of data enhancement are defined as 3. When the model is trained using data enhancement, the training time is increased because each round of data enhancement requires training, and the success rate of test set recognition is improved. The situation of ADULT data set is similar to that of CIFAR10, specific data are shown in Table 1.

Table 1. The effect of data enhancement on recognition rate

Dataset	Test sets predict success rates		
	Non-enhancement	multi-round MIXUP data enhancement method	
		3 rounds	5 rounds
CIFAR10	94.5%	95.1%	96.2%
ADULT	96.7%	96.8%	97.2%

The results showed that the multi-round MIXUP data enhancement did not affect the recognition rate of the model, and the recognition success rate would be improved when the number of enhancement rounds was modified. The MEWDP defense method adds noise conforming to differential privacy during model training and performs label smoothing for data labels. To protect the user's private data. In the model training stage, the important parameter is the allocation of privacy budget. Although a small privacy budget can provide higher protection for users, it will reduce the recognition success rate of the model, and a high privacy budget will weaken the privacy protection effect. In this paper, several key privacy budget values were selected and the changes in the success rate of model recognition under different privacy budgets were calculated, as shown in Fig. 4.

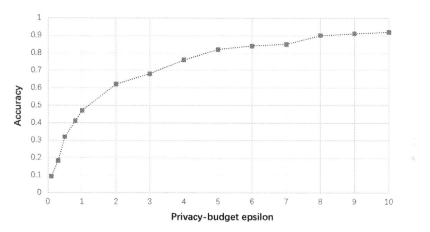

Fig. 4. The change of model recognition success rate with privacy budget

As can be seen from the figure, the size of the privacy budget is positively correlated with the success rate of model recognition. The larger the privacy budget, the higher the success rate of model recognition, and correspondingly, the weaker the protection effect on users. In the experiment, it is found that when the privacy budget is selected near 5, it has less impact on the success rate of model recognition and can provide more protection. In this paper, the privacy budget that has little influence on the accuracy of the model is selected.

In view of the diversified means of member inference attack, MEWDP defense method can resist different member inference attacks, including model-based member inference attack and metric-based member inference attack respectively. For the two data sets selected in this paper, the attack success rates of the two categories of member inference attack are shown in Table 2.

It can be seen from the table that the MEWDP defense method proposed in this paper can deal with multiple member inference attacks, reduce the success rate of the attacker's member inference attacks to the dimension of random guessing, and can effectively defend against multi-level attacks.

MEWDP defense methods compare with Memguard [12], AdvReg [13] and EMT [21] defense methods. Memguard adds noise to the prediction vector of the model to protect against inference attacks by members, while AdvReg defends against inference

Table 2. Defense performance of different attack methods

Dataset	Model-based attack	metrics-based attack		
		label threshold	Loss threshold	Output confidence
CIFAR10	51.8%	49.9%	50.3%	50.9%
ADULT	50.6%	51.4%	51.6%	51.7%

attacks by regularization. The EMT defense method uses data to increase the method of processing private data to resist member inference attacks. The defense effects of different defense methods against different data sets are shown in Table 3.

Table 3. Comparison of different defense methods

Dataset	Attack success rate				
	undefended	Memguard	AdvReg	EMT	MEWDP
CIFAR10	75.3%	51.4%	51.8%	52.3%	50.8%
ADULT	78.8%	52.2%	51.3%	50.6%	51.3%

As can be seen from the table, compared with Mem, AdvReg and EMT, the defense methods proposed in this paper can better defend against member inference attacks. For CIFAR10 data set, the defense effect is better, and the attack success rate is reduced to 50.8%. Both data sets can protect user privacy information and resist member inference attacks.

5   Conclusion

In this paper, a MEWDP defense method is proposed to process data through multiple rounds of MIXUP data enhancement technology to prevent the model from overfitting with little impact on the model accuracy. After that, label smoothing technology and differential privacy are used for model training. Differential privacy is used as a privacy protection algorithm, combined with deep learning model, Gaussian noise is added to the gradient of the model, and appropriate privacy budget is selected to make the model meet the differential privacy. This defense method can effectively reduce multi-dimensional member inference attacks, and reduce model accuracy less. It is an effective and stable member inference attack defense method with availability and generalization ability. In this paper, only two data sets are selected for testing, and the deep learning model has a small number of layers. Now the era of large models has arrived, and more and more extensive data are required for training models. In the face of more complex neural network models, there is still a long way to go to protect users' privacy.

References

1. Brown, T., Mann, B., Ryder, N., et al.: Language models are few-shot learners. Adv. Neural. Inf. Process. Syst. **33**, 1877–1901 (2020)
2. Tramèr, F., Zhang, F., Juels, A., et al.: Stealing machine learning models via prediction {APIs}. In: 25th USENIX Security Symposium (USENIX Security 2016), pp. 601–618 (2016)
3. Shen, S., Tople, S., Saxena, P.: Auror: defending against poisoning attacks in collaborative deep learning systems. In: Proceedings of the 32nd Annual Conference on Computer Security Applications, pp. 508–519 (2016)
4. Barreno, M., Nelson, B., Sears, R., et al.: Can machine learning be secure? Proceedings of the 2006 ACM Symposium on Information, Computer and Communications Security, pp. 16–25 (2006)
5. Gu, T., Dolan-Gavitt, B., Garg, S.: BadNets: Identifying Vulnerabilities in the Machine Learning Model Supply Chain (2017). arXiv preprint arXiv:1708.06733
6. Shokri, R., Stronati, M., Song, C., et al.: Membership inference attacks against machine learning models. In: 2017 IEEE Symposium on Security and Privacy (SP), pp. 3–18. IEEE (2017)
7. Zhao, L., Wang, Q., Wang, C., et al.: Veriml: enabling integrity assurances and fair payments for machine learning as a service. IEEE Trans. Parallel Distrib. Syst. **32**(10), 2524–2540 (2021)
8. Li, Z., Zhang, Y.: Membership leakage in label-only exposures. In: Proceedings of the 2021 ACM SIGSAC Conference on Computer and Communications Security, pp. 880–895 (2021)
9. Hui, B., Yang, Y., Yuan, H., et al.: Practical blind membership inference attack via differential comparisons. arXiv preprint arXiv:2101.01341 (2021)
10. Park, Y., Kang, M.: Membership inference attacks against object detection models. arXiv preprint arXiv:2001.04011 (2020)
11. Song, L., Mittal, P.: Systematic evaluation of privacy risks of machine learning models. In: 30th USENIX Security Symposium (USENIX Security 2021), pp. 2615–2632 (2021)
12. Jia, J., Salem, A., Backes, M., et al.: Memguard: defending against black-box membership inference attacks via adversarial examples. In: Proceedings of the 2019 ACM SIGSAC Conference on Computer and Communications Security, pp. 259–274 (2019)
13. Nasr, M., Shokri, R., Houmansadr, A.: Machine learning with membership privacy using adversarial regularization. In: Proceedings of the 2018 ACM SIGSAC Conference on Computer and Communications Security, pp. 634–646 (2018)
14. Dwork, C.: Differential privacy: a survey of results. In: Agrawal, M., Du, D., Duan, Z., Li, A. (eds.) TAMC 2008. LNCS, vol. 4978, pp. 1–19. Springer, Heidelberg (2008). https://doi.org/10.1007/978-3-540-79228-4_1
15. Dwork, C., Roth, A.: The algorithmic foundations of differential privacy. Found. Trends® Theor. Comput. Sci. **9**(3–4), 211–407 (2014)
16. Han, R., Li, D., Ouyang, J., et al.: Accurate differentially private deep learning on the edge. IEEE Trans. Parallel Distrib. Syst. **32**(9), 2231–2247 (2021)
17. Owusu-Agyemeng, K., Qin, Z., Xiong, H., et al.: MSDP: multi-scheme privacy-preserving deep learning via differential privacy. Pers. Ubiquitous Comput. 1–13 (2021)
18. Rahman, M.A., Rahman, T., Laganière, R., et al.: Membership inference attack against differentially private deep learning model. Trans. Data Priv. **11**(1), 61–79 (2018)
19. Hagestedt, I., Zhang, Y., Humbert, M., et al.: MBeacon: privacy-preserving beacons for DNA methylation data (2019)
20. Chen, J., Wang, W.H., Shi, X.: Differential privacy protection against membership inference attack on machine learning for genomic data. In: BIOCOMPUTING 2021: Proceedings of the Pacific Symposium, pp. 26–37 (2020)

21. Chen, Z., Li, H., Hao, M., Xu, G.: Enhanced mixup training: a defense method against membership inference attack. In: Deng, R., Bao, F., Wang, G., Shen, J., Ryan, M., Meng, W., Wang, D. (eds.) ISPEC 2021. LNCS, vol. 13107, pp. 32–45. Springer, Cham (2021). https://doi.org/10.1007/978-3-030-93206-0_3
22. Krizhevsky, A., Hinton, G.: Learning multiple layers of features from tiny images (2009)

Air Defense Deployment of Anti-reconnaissance Based on Immune Optimization Algorithm with Nested Double Particle Swarm

Yexin Song[1], Yanjie Wu[2], Chunsheng Gao[1], and Yongkai Liu[1($\boxtimes$)]

[1] Department of Basic Courses, Naval University of Engineering, Wuhan, China
kaikai0220@163.com
[2] Department of Operational Research and Programming, Naval University of Engineering, Wuhan, China

Abstract. The position deployment and firepower allocation of weapons are very important in military conflict. To the anti-reconnaissance aircraft cluster problem of multi-type air defense weapons, considering the location and fire distribution of different types of air defense weapons simultaneously, a multi-objective mixed deployment model is established taking the overall effectiveness and cost of air defense as the optimization objectives. An immune optimization algorithm with nested double particle swarm is designed to solve the model. Through a series of comparative analysis of algorithm parameters and performance, the proposed algorithm is verified, and the final simulation results illustrate the feasibility and effectiveness of the model and algorithm.

Keywords: Air defense weapons · Fire distribution · Location deployment · Nested algorithm

1 Introduction

In modern air defense operations, the deployment of air defense weapons is particularly important. In recent years, research on this issue has received increasing attention [1–4]. Özdemir and Ic [2] focus on the optimal deployment and positioning of missile systems to protect a region against possible threats using the layered air defense system, a multi-objective optimization model is proposed to maximize the probability of killing and minimize the total cost. Farlik et al. [3] provided an autonomous air defense effectors deployment algorithm for modeling and simulation purposes. Zhong et al. [5, 6] comprehensively analyzed the main factors affecting air defense deployment, designed an air defense deployment model based on the adaptive nested particle swarm optimization algorithm. Yan et al. [7] established an air defense deployment optimization model based on dynamic programming-genetic algorithm. Aiming at the problem of coordinated deployment of various types of weapon systems for anti-UAV cluster, a multi-objective hybrid air defense deployment model was established, and a multi-layer dynamic genetic algorithm was provided to obtain the optimal deployment plan by Xue

D.-S. Huang et al. (Eds.): ICAI 2023, CCIS 2015, pp. 271–282, 2024.
https://doi.org/10.1007/978-981-97-0827-7_24

et al. [8]. Li et al. [9] studied an optimization problem for deploying air defense systems against reconnaissance drone swarms, the location of each air defense system is determined in a predetermined region, such that the cost for enemy drones to pass through the region would be maximized by using an exact iterative search algorithm and an evolutionary framework. In addition, the queuing theory model is adopted to optimize the deployment of air defense weapons taking the total number of air defense weapons and defense cost as the optimization objectives in literatures [10, 11] and [12].

The above research either just discusses a single type of air defense weapon, or just studies the fire distribution or location deployment of air defense weapons. In this paper, to the problem of multi-type air defense weapons coordinated anti-reconnaissance aircraft cluster, taking into account the location and fire distribution of different types of air defense weapons simultaneously, a multi-objective mixed deployment planning model is established, and an immune optimization algorithm with nested double particle swarm (IONDPS) is designed to solve the model.

2 Air Defense Deployment Model of Anti-reconnaissance Aircraft Cluster

2.1 Description of the Problem

Air defense weapon anti reconnaissance aircraft cluster operations generally contain air defense (AD) side and reconnaissance aircraft (RA) side. The RA side mainly uses various reconnaissance methods to conduct reconnaissance on military facilities in AD side, and the AD side is composed of various air defense weapons, mainly performing regional air defense tasks.

Assuming that the RA side has n different reconnaissance aircraft, and the AD side has h different types of air defense weapons and m alternative positions of air defense fire points. It is stipulated that each alternative position can only set up one type of air defense weapons at most. Now it is necessary to study the location deployment and fire distribution of different types of air defense weapons.

2.2 Constraint Analysis

In the operation of the above-mentioned air defense weapons against reconnaissance aircraft cluster, the position deployment and firepower allocation of different types of air defense weapons must be considered simultaneously.

Suppose that the decision variables w_{ij} is denoted as

$$w_{ij} = \begin{cases} 1, & The\ i-th\ position\ deploys\ the\ j-type\ air\ defense\ weapon \\ 0, & otherwise \end{cases}$$

The decision variable x_{ijk} represents the ammunition consumption of the j-type air defense weapon deployed at the i alternative location to attack the k RA.

Obviously, if d_{ik} is the distance between the i alternative position and the k RA, and r_{ij} is the farthest range for deploying the j-type air defense weapon at the i alternative location, then when $w_{ij} = 0$ or $r_{ij} < d_{ik}$, $x_{ijk} = 0$, and when $w_{ij} = 1$ and $r_{ij} \geq d_{ik}$, $x_{ijk} \geq 0$.

Since each RA must have at least one fire point to intercept it, the air defense deployment must meet the following constraint conditions.

$$\sum_{i=1}^{m} \sum_{j=1}^{h} x_{ijk} > 0, \quad k = 1, 2, 3, \cdots, n \tag{1}$$

Due to the limitation of operational resources, the number of each type of air defense weapons and their total ammunition quantity are limited, these constraint conditions can be expressed as follows.

$$\sum_{i=1}^{m} w_{ij} \leq g_j, \quad j = 1, 2, 3, \cdots, h \tag{2}$$

$$\sum_{i=1}^{m} \sum_{k=1}^{n} x_{ijk} \leq q_j, \quad j = 1, 2, 3, \cdots, h \tag{3}$$

where, g_j represents the number of the j-type air defense weapons, and q_j represents the total ammunition inventory of the j-type air defense weapons.

In addition, if each deployment site can deploy at most one type of air defense weapon, it can be expressed as follows.

$$\sum_{j=1}^{h} w_{ij} \leq 1, \quad i = 1, 2, 3, \cdots, m \tag{4}$$

2.3 Optimization Objective Analysis

The optimization objective of anti-reconnaissance aircraft cluster is to maximize the effectiveness of air defense and minimize the loss of its own air defense [13].

1. Air defense effectiveness objective

The effectiveness of air defense can be measured by the total residual value of reconnaissance aircraft cluster. In order to maximize air defense effectiveness, it is necessary to minimize the total residual value. Denote p_k^R is the survival probability of the k RA at the end of the battle, so

$$p_k^R = \prod_{i=1}^{m} (1 - \sum_{j=1}^{h} p_{jk}^W w_{ij})^{\sum_{j=1}^{h} x_{ijk}}, \quad k = 1, 2, 3, \cdots, n \tag{5}$$

where, p_{jk}^W is the single-shot damage probability of the j-type weapon attacking the k

RA, and $(1 - \sum_{j=1}^{h} p_{jk}^W w_{ij})^{\sum_{j=1}^{h} x_{ijk}}$ is the survival probability of the k RA after being attacked

$\sum_{j=1}^{h} x_{ijk}$ times by air defense weapon deployed at the i alternative location.

If the value of the k RA is u_k^R, the overall residual value of the RA cluster can be measured as follows.

$$W_1 = \sum_{k=1}^{n} p_k^R u_k^R \tag{6}$$

2. Air defense cost objective

The air defense cost of AD side can be calculated by their consumption of various weapons and ammunition [13]. If c_j is the value of single ammunition of the j-type weapon, the total air defense cost of AD side can be measured as follows.

$$W_2 = \sum_{i=1}^{m} \sum_{j=1}^{h} \sum_{k=1}^{n} c_j x_{ijk} \tag{7}$$

2.4   Multi-objective Deployment Model of Air Defense Weapons

Through the above analysis, the multi-objective mixed deployment planning model of anti-reconnaissance aircraft cluster of air defense weapons can be obtained as follows.

$$\min W_1 = \sum_{k=1}^{n} p_k^R u_k^R = \sum_{k=1}^{n} \left[\prod_{i=1}^{m} (1 - \sum_{j=1}^{h} p_{jk}^W w_{ij})^{\sum_{j=1}^{h} x_{ijk}} \right] u_k^R$$

$$\min W_2 = \sum_{i=1}^{m} \sum_{j=1}^{h} \sum_{k=1}^{n} c_j x_{ijk}$$

$$s.t. \quad \sum_{i=1}^{m} \sum_{j=1}^{h} x_{ijk} > 0, \ k = 1, 2, 3, \cdots, n$$

$$\sum_{i=1}^{m} \sum_{k=1}^{n} x_{ijk} \leq q_j, j = 1, 2, 3, \cdots, h$$

$$\sum_{j=1}^{h} w_{ij} \leq 1, \ i = 1, 2, 3, \cdots, m$$

$$\sum_{i=1}^{m} w_{ij} \leq g_j, j = 1, 2, 3, \cdots, h$$

$$w_{ij} = 0 \ or \ 1, \quad x_{ijk} \in N$$

$$x_{ijk} \geq 0, \quad if \ w_{ij} = 1 \ and \ r_{ij} \geq d_{ik}$$
$$x_{ijk} = 0, \quad if \ w_{ij} = 0 \ or \ r_{ij} < d_{ik}$$

In the above model, the optimization variables w_{ij} can determine the optimal positions air defense weapon deployed, and the optimization variable x_{ijk} can determine the optimal ammunition consumption of the j-type air defense weapon deployed at the i alternative location to attack the k RA.

2.5 Single Objective Transformation of Model

Normalize two optimization objectives,

$$S_1 = \sum_{k=1}^{n} p_k^R u_k^R / \sum_{k=1}^{n} u_k^R \tag{8}$$

$$S_2 = (\sum_{i=1}^{m} \sum_{j=1}^{h} \sum_{k=1}^{n} c_j x_{ijk}) / (\sum_{j=1}^{h} c_j q_j) \tag{9}$$

If α and β $(0 \leq \alpha, \beta \leq 1, \ \alpha + \beta = 1)$ are the weights of the above two objectives respectively, measured by the experience and psychological expectations of decision makers, the model objectives can be transformed into following form.

$$\min \ D = \alpha S_1 + \beta S_2 \tag{10}$$

3 Model Solution Based on Immune Optimization Algorithm with Nested Double Particle Swarm

The mixed deployment of air defense weapons is a multi-dimensional complex problem with large scale and high difficulty. It needs to design a fast and accurate intelligent algorithm to solve it. When dealing with high-dimensional multi-objective complex optimization problems, particle swarm optimization has a relatively fast speed of approaching the optimal solution [14, 15]. The disadvantage is that it is easy to fall into the local optimal solution. The immune optimization algorithm has the advantages of global optimization, maintaining population diversity, avoiding falling into the local optimal solution, and good robustness [16, 17]. The disadvantage is that its convergence speed is slower. Aiming at the mixed deployment model of air defense weapons, an immune optimization algorithm with nested double particle swarm (IONDPS) is provided in the following.

The IONDPS algorithm is designed as internal and external layers. The outer of the algorithm uses immune optimization to select the optimal air defense positions according to the objective function value fed back by the inner algorithm. The inner algorithm designs the first particle swarm optimization to search the types of air defense weapons to be deployed at each position provided by the outer algorithm, and the second particle swarm to optimize the air defense fire distribution according to the air defense positions and the types of weapons deployed. The algorithm design flowchart is shown in Fig. 1.

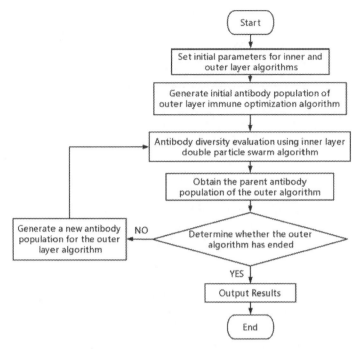

Fig. 1. The design flowchart of the IONDPS algorithm.

The specific solution steps of the IONDPS algorithm are as follows.

Step 1. Initialize the system, set the maximum number of iterations, population size, crossover probability and mutation probability of the inner and outer algorithm.

Step 2. Generate the initial air defense position selection scheme, and use it as the initial antibody group of the outer immune optimization algorithm.

Step 3. Antibody diversity evaluation of the outer immune optimization algorithm.

The optimal objective function value fed back by the inner algorithm is used as the corresponding fitness evaluation index of the outer antibody, and its corresponding optimal weapon deployment and air defense fire distribution scheme are recorded.

Step 4. The outer immune optimization algorithm determines whether the conditions are met. If yes, it ends, and if not, it continues to the next step.

Step 5. The outer immune optimization algorithm generates a new antibody group.

Step 6. Continue to step 3.

4 Simulation and Analysis

4.1 Simulation Background

Assuming that the air defense area of the engagement is a rectangular strip area, the RA cluster composed of 10 large reconnaissance aircraft try to detect the different mission targets of the AD side respectively. Among the 10 large reconnaissance aircrafts, No.1, No.2, and No.3 are type I RA (RA1), No.4, No.5, and No.6 are type II RA (RA2),

No.7, No.8, No.9 and No.10 are type III RA (RA3). The reconnaissance values of these three types of RA are 1.5, 2 and 1.7, respectively. The early warning radar of the AD side can obtain the aerial position of RAs. They must select at most 4 positions from the 10 alternative positions in the linear configuration zone of air defense to deploy the most appropriate air defense weapons and determine their ammunition inventory and air defense tasks.

The AD side is prepared to conFig. two kinds of air defense weapons, AD1 and AD2, respectively. The performance, ammunition inventory, ammunition value and quantity of the two air defense weapons are shown in Table 1. The hit rates of different types of RA hit by different weapons in the best range are shown in Table 2. The alternative location of air defense weapons and the location coordinates of detected reconnaissance aircraft are shown in Table 3. Here, assuming that reconnaissance aircraft operate within a small range in the air during weapon deployment, and their positions are approximately represented by points in the air.

Table 1. Air defense weapon performance, ammunition inventory, ammunition value and weapon quantity

Air defense weapon	AD1	AD2
Air defense radius/km	50	100
Stock/piece	12	10
Single piece value/USD 10000	10	15
Number of weapons/set	3	3

Table 2. Hit rates of various air defense weapons

RA type	AD1	AD2
RA1	0.6	0.5
RA2	0.7	0.6
RA3	0.5	0.8

4.2   Comparative Analysis of Algorithm Parameters

Although the inner layer nested particle swarm has fast convergence speed and high accuracy, the convergence result is relatively easy to fall into the local optimum to a discrete high-dimensional problem. So, the population size and iteration times of inner particle swarm need to be treated specifically.

Table 3. Alternative positions of air defense weapons and RA positions

	AD coordinate/km (x,y,z)	RA coordinate/km (x,y,z)
1	(221, 176, 0)	(50, 51, 3)
2	(167, 59, 0)	(137, 157, 3)
3	(109, 144, 0)	(52, 230, 3)
4	(102, 188, 0)	(8, 103, 1)
5	(133, 49, 0)	(177, 8, 1)
6	(154, 164, 0)	(235, 105, 1)
7	(150, 39, 0)	(204, 270, 0.5)
8	(72, 210, 0)	(137, 24, 0.5)
9	(18, 219, 0)	(232, 218, 0.5)
10	(31, 137, 0)	(207, 171, 0.5)

In the next, let $\alpha = 0.9$, $\beta = 0.1$, and population size and iteration number of outer immune optimization algorithm is 10 and 50, respectively.

1. Comparison analysis of population size of inner particle swarm algorithm

Set the number of inner iteration is 50, and the population size of inner particle swarm is 5, 10, 20, and 30, respectively. The convergence of the IONDPS algorithm is compared as shown in Fig. 2.

From Fig. 2, it is shown that with the continuous expansion of the population size of the inner particle swarm, the convergence curve of the final nested algorithm is gradually stable.

2. Comparison analysis of iteration times of inner particle swarm algorithm

Set the inner population size is 50, and the number of inner iteration is 10, 20, 50 and 100, respectively. The convergence of the IONDPS algorithm is compared as shown in Fig. 3.

From Fig. 3, it is shown that the convergence curve of the final nested algorithm is gradually stable with the increasing number of iteration of the inner particle swarm. After 50 iterations, the convergence curve remains almost unchanged.

So, to make the convergence result of the IONDPS algorithm more stable, and reduce the amount of calculation, we take the number of iteration 50 and the population size 50 as the parameters of the designed particle swarm algorithm embedded in the inner layer.

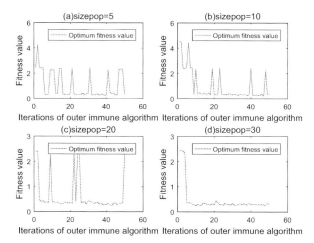

Fig. 2. Comparison of population size of inner layer algorithm.

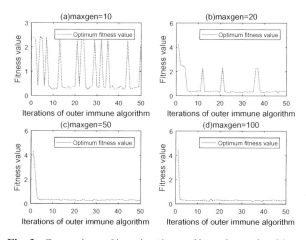

Fig. 3. Comparison of iteration times of inner layer algorithm.

4.3 Comparative Analysis of Algorithm Performance

In the following, we will compare the performance of the IONDPS algorithm with the common immune optimization algorithm nested with random sampling (IONRS).

In the IONRS algorithm, the combination strategy of 2500 kinds of weapons and air defense fire distribution is randomly sampled in the inner layer. The convergence speed and accuracy of two algorithms are shown in Fig. 4. It can be seen that the convergence speed of two algorithms are fast, but the IONDPS algorithm has higher convergence accuracy.

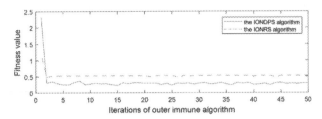

Fig. 4. Comparison of the convergence speed and accuracy of two algorithms.

4.4 Simulation Results and Analysis

For specified $\alpha = 0.9$, $\beta = 0.1$, the simulation results are shown in Table 4. The minimum value of objective function D is 0.3420. For specified $\alpha = 0.6$, $\beta = 0.4$, the simulation results are shown in Table 5. The minimum value of objective function D is 0.4037. With the increase of α, the smaller the objective function D is, and the better the air defense effect is.

Table 4. Simulation results for specified $\alpha = 0.9$, $\beta = 0.1$

the optimal air defense positions	air defense weapon deployed	intercept RAs (ammunition quantity)
1	AD2	2(1), 6(1), 7(1), 9(1),10(1)
5	AD2	1(1), 5(1), 8(1)
8	AD2	2(1), 3(1)
10	AD1	4(1)

Table 5. Simulation results for specified $\alpha = 0.6$, $\beta = 0.4$

the optimal air defense positions	air defense weapon deployed	intercept RAs (ammunition quantity)
1	AD2	2(1), 6(1), 7(1), 9(1), 10(1)
5	AD2	1(1), 5(1), 8(1)
8	AD1	3(1)
10	AD1	4(1)

5   Conclusion

In this paper, aiming at the anti-reconnaissance aircraft cluster problem of multiple types of air defense weapons, considering the location and fire distribution of different types of air defense weapons simultaneously, a multi-objective hybrid deployment planning model is established taking the overall effectiveness and cost of air defense as the optimization objectives. An immune optimization algorithm with nested double particle swarm is designed to solve the model. Through a series of comparative analysis of algorithm parameters and performance, the proposed algorithm is verified, and the final simulation results illustrate the feasibility and effectiveness of the model and algorithm. Due to the complexity and uncertainty of conflict environments, we will further study the deployment of air defense weapons in uncertain conflict environments in the future.

References

1. Wan, J.Q., Wang, P.F., Tang, J.L., et al.: Research on mixed deployment strategy and model of air defense in key areas based on multi-weapon platform. J. Syst. Simul. **33**(8), 1754–1765 (2021)
2. Özdemir, Z., Ic, Y.T.: Multi-objective deployment model for regional air defense. Soft. Comput. **27**, 7321–7335 (2023)
3. Farlik, J., Kratky, M., Simkova, S.: Autonomous air defense effectors deployment algorithms for modeling and simulation purposes. In: Mazal, J. (ed.) MESAS 2018. LNCS, vol. 11472, pp. 577–587. Springer, Cham (2019). https://doi.org/10.1007/978-3-030-14984-0_43
4. Yang, J.W.: Research on the development of foreign military Anti-UAV. Military Abstracts **28**(5), 46–51 (2022)
5. Zhong, W.J., Li, X.B., Chang, H.T., et al.: Design of air defense deployment optimization model based on adaptive nested PSO algorithm. In: 2nd International Conference on Intelligent Design (ICID), Xi'an, China, pp. 172–177. IEEE Press (2021)
6. Zhong, W.J., Li, X.B., Chang, H.T., et al.: Anti-UAV cluster air defense deployment model based on nested PSO algorithm. Electro-Opt. Control **28**(12), 6–10 (2021)
7. Yan, P.Y., Liu, S., Wang, J.: Air defense deployment optimization model based on dynamic programming-genetic algorithm. Syst. Eng. Electron. Technol. **40**(10), 2249–2255 (2018)
8. Xue, N., Ding, D., Ding, J., et al.: Optimization method for coordination deployment of air defense system based on improved genetic algorithm. In: IEEE 4th Advanced Information Management, Communicates, Electronic and Automation Control Conference (IMCEC), Chongqing, China, pp. 1658–1664. IEEE Press (2021)
9. Li, N., Su, Z., Ling, H., Karatas, M., et al.: Optimization of air defense system deployment against reconnaissance drone swarms. Complex Syst. Model. Simul. **3**(2), 102–117 (2023)
10. Cheng, F., Liang, Z.B., Peng, G.L., et al.: An anti-UAV long-term tracking method with hybrid attention mechanism and hierarchical discriminator. Sensors **22**(10), 3701 (2022)
11. Gao, Z.G., Liu, Y.B., Chen, C.Y., et al.: Optimization method of Anti-UAV cluster weapons deployment based on queuing theory. Equip. Environ. Eng. **19**(6), 68–75 (2022)
12. Yang, Y.L., Ding, T.B., Zhao, J.C., et al.: Research on the Anti-UAV cluster combat capability of combination of artillery and air defense weapons. J. Artillery Launch Control **40**(4), 47–50 (2019)
13. Zhang, S., Zhu, Y., Dai, L.T., et al.: Research on the development history, current situation and future application of military UAV technology. Ship Electron. Eng. **41**(6), 9–13 (2021)
14. Li, P., Liu, X., Chen, H., et al.: Optimization of three-dimensional magnetic field in vacuum interrupter using particle swarm optimization algorithm. IEEE Trans. Appl. Supercond. **31**(8), 1–4 (2021)
15. Hernandez, D., Rodriguez, R., Merchán, E., Santiago, A.: Optimal design of a drive shaft with composite materials through particle swarm optimization. IEEE Lat. Am. Trans. **18**(6), 1008–1016 (2020)

16. Fan, X., Liu, J., Goh, H.H., et al.: Aging analysis of transformer insulation at weakest region: dielectric parameters extraction via immune optimization. IEEE Trans. Transp. Electrif. **9**(1), 1579–1589 (2023)
17. Lu, L., Guo, Z., Wang, Z., et al.: Parameter estimation for a capacitive coupling communication channel within a metal cabinet based on a modified artificial immune algorithm. IEEE Access **9**, 75683–75698 (2021)

Relevant Tag Extraction Based on Image Visual Content

Nancy Fazal$^{(\boxtimes)}$ and Pasi Fränti ⓘ

School of Computing, University of Eastern Finland, Joensuu, Finland
{nancyf,pasi.franti}@uef.fi

Abstract. Social media web services like Flickr allow users to share and freely annotate images with textual tags. These tags play a crucial part for text-based social image retrieval and browsing tasks. However, these tags are usually irrelevant and incomplete which limits their effectiveness and use. One fundamental problem is to interpret the relevance of tags with respect to the image visual content. Existing solutions have targeted either image visual content or user contributed tags separately to address the issue. Our proposed method *Tag-tag*, however, combines both aspects and exploit their semantic relationship.

We use state of the art pretrained machine learning models in Tensorflow for object detection and NLP (Natural Language Processing) for semantic analysis. Our experiments on a dataset of 219 randomly collected Flickr images demonstrates the applicability of our method. Images with missing tags were excluded from the experiments. We identified two reasons where relevant tag was not found as: **(1)** relevant tag itself was missing from user contributed tag list, **(2)** relevant tag got skipped because it was either not found in the Wordnet dictionary or has no pretrained vector in word2vec-google-news-300 model. We identify this as one limitation of the proposed method.

Keywords: Social Tagging · Tag Relevance · Flickr · Object Detection · Semantic Similarity

1 Introduction

The popularity of several digital imaging devices and with the advancement of Internet technologies, digital images can be easily created and distributed. Social media tagging, a process where images, videos and text objects are mostly assigned with tags or keywords by common users, is reshaping the way people generate, manage and search multimedia resources [1]. Services like Flickr which cumulates 10 billion images with around 3.5 million new uploads per day are flourishing [33]. Besides their general use, these rich multimedia databases have triggered many innovative research domains such as tag recommendation [2], landmark recognition [3], tag ranking [4], concept similarity measurement [5], automatic image annotation [6] and personalized information delivery [7].

D.-S. Huang et al. (Eds.): ICAI 2023, CCIS 2015, pp. 283–295, 2024.
https://doi.org/10.1007/978-981-97-0827-7_25

The major concern, however, is that the multimedia objects are usually not annotated properly and includes irrelevant, and often incomplete tags. Some objects are left completely unlabeled also. Liu et al. [4] reported that only 50% of the tags are related to the image content, which poses a great challenge for typical web image search approaches. To address this issue, various tag refinement techniques have been reported by the researchers recently, which can improve the quality of tags [1, 5, 8–11]. Tag refinement improves the quality of tags but does not answer the tags which best represent the visual content of an image [12]. Kennedy et al. [13] reports that the tags associated with images contain many noises and not only describe the image contents but a broad spectrum of semantic space such as location, time, and subjective emotion [14, 15]. Liu et al. [3] claims that the automatic detection of tags which are content-related can support more smart use of the social images and tags such as ease the task of browsing, retrieval and indexing of large-scale image repositories [16].

Li et al. [1] proposed a voting algorithm to find a relevant tag from the tagging behavior of visual neighbors of that image. They conducted three experiments one for image ranking and two tag ranking experiments for verification purpose. Their experiments on 3.5 million Flickr images improved upon baselines and demonstrated the usability of algorithm for both social image retrieval and image tag recommendation. Liu et al. [4] reports that the tags are almost in random order in terms of their relevance to the associated tags. They propose a tag ranking method to automatically rank tags for a given image, according to their relevance to image visual content. Their experiments on 50,000 Flickr photo collection show that the proposed method is both effective and efficient. They further applied tag ranking into three applications i.e., tag-based image search, tag recommendation, and (3) group recommendation, which demonstrates that the proposed tag ranking approach really boosts the performances of social-tagging related applications.

Zhao et al. [16] proposed PoCR a data driven method to assess the probability of a tag relevancy to its corresponding image. Their experiments on 149,915 Flickr images demonstrated that PoCR achieved best performance by obtaining 59.8%, 26.6%, 29.3%, and 20.4% relative improvements compared to Baseline, LiCR g, LiCR l, and LiCR f respectively. Liu et al. [36] proposed the pixel voting method to choose the visual neighbors for seed image to find a representative tag. Their experiments on MIR Flickr dataset show the effectiveness of the method in tag de-noising and tag ranking. They also stated that the concern of tag relevancy learning could not be resolved completely because of the semantic gap between the images and tags.

Lindstaedt et al. [17] proposed a research prototype *tagr* which makes use of data from Flickr group "fruit & veg" and electronic lexical database WordNet. Techniques from the various fields such as image analysis, social network analysis and statistical text analysis were used to develop the service. The results revealed that the despite of low precision and recall values test users did find the *tagr* useful. Zhuang et al. [12] proposed a novel two-view learning approach for social image tag ranking by exploiting both visual and textual contents. To evaluate the method efficiency, extensive set of experiments on automatic image annotation task and text-based social image retrieval were conducted. Encouraging results were reported as proposed method outperformed the conventional approaches. See Table 1 for the summary of the existing methods.

Table 1. Existing Tag relevance methods comparison.

Reference	Method	Datasets Used	Data Context	Properties
[1]	Tagging behavior of visually similar neighbors	3.5 M	Images and tags	Does not require any model training for any visual concept but visually similar neighbors for every seed image are needed
[4]	Probability density estimation and Random walk	50,000	Tags	Does not require model training
[16]	Data-driven method	149,915	Image and tags	Does not require any model training and limited to 270 popular Flickr tags only
[17]	*Tagr*	14,000	Tags, images, and users	Offline analysis and limited to Flickr group *fruit & veg*
[12]	Data-driven method	1 M	Tags and images	No parametric model relevance between images and tags
[36]	Pixel voting method	MIR Flickr	Images and tags	Visually similar neighbors for every seed image required to focus on the local features of an image

In this paper, we propose a new method for extracting a tag which best describes the image visual content. Instead of relying on the existing tags alone, or use image visual content as such, we apply them jointly. We refer this method as *Tag-tag*, which exploits the semantic relationship between objects identified on a given image and associated user tags. The method is completely independent of visually similar images, their associated tags and is not restricted to a certain set of tag groups. It further relies on state of the art pretrained models, thus does not require building anything from the scratch.

2   Representative Tag Extraction Using Tag-Tag

We define the representative tag as the one which best describes the objective aspects of the visual content of an image (see Fig. 1). Our method to select a representative tag for a photo comprise of two approaches i.e., computer vision techniques and natural language processing. We use TensorFlow object detection, a computer vision technique that detects, locates, and traces an object from a still image or video and NLTK (Natural Language Toolkit) to perform semantic analysis of textual tags.

The proposed method consists of three steps as 1) Object detection using Tensor-Flow, 2) Semantic analysis of textual tags using NLTK (WordNet) and 3) Deriving representative tag (Wu & Palmer similarity and Word2Vec). The thorough processing of the *Tag-tag* method is shown in Algorithm.

Photo	User-provided tags	Representative tag
	London, Paddington, UK, Sir Simon Milton, statue, sculpture, deputy mayor, official	sculpture / statue

Fig. 1. Representative tag extraction based on the image visual content.

ALGORITHM: REPRESENTATIVE TAG SELECTION USING WORDNET SIMILARITY AND WORD2VEC SIMILARITY

Input: Image i and its user contributed list of tags t;

Output: A representative tag $rTag$ which best describes the visual content of an image;

1. Obtain unique object detections d using Tensorflow object detection model *FasterRCNN+InceptionResNet V2*

2. Extract set of tags t and object detections d present in the Wordnet dictionary and word2vec-google-news-300 model using *wordnet.synsets(word)* and *word not in google_news_vectors.key_to_index* respectively

3. Perform Part of Speech (POS) tagging using nltk.pos_tag(t)

4. Filter named entities using NLTK and extract tags with NN, NNS, NNP and NNPS, POS labels

5. **for** detection in d **do**

6.   **for** tag in t **do**

7.   WORD2VECSIMILARITY(d, t)

        // *For representative tag extraction using Wordnet Similarity call* **WORDNETSIMILARITY**(d, t)

8.   *scores* ← append(*similarityScore)*

9.   *matrix* ← append(scores)

10. *maxScore* ← [sum(col) for col in zip(*$matrix$[1:])]

11. *rTag* ← *matrix*[0][*maxScore*.index(max(*maxScore*))]

12. **return** *rTag*

13. function WORD2VECSIMILARITY (d, t)

14.   *similarityScore* ← google_news_vectors.similarity(d, t)

15.   **return** *similarityScore*

16. function WORDNETSIMILARITY (d, t)

17. $w1$ ← wordnet.synsets(t)[0]

18. $w2$ ← wordnet.synsets(d)[0]

19.   *similarityScore* ← $w1$.wup_similarity ($w2$)

20.   **return** *similarityScore*

2.1 Object Detection Using Tensorflow

The object detection framework in TensorFlow works on trained models, so it does not require building anything from scratch. Prioritizing accuracy over speed, we chose FasterRCNN + InceptionResNet V2 [19] module for object detection task. This model is trained on Open Images V4 with ImageNet pre-trained Inception Resnet V2 as image feature extractor. The model is further publicly available as part of the TensorFlow object detection API.

The Inception ResNetV2 feature extractor was trained on ImageNet and fine-tuned with FasterRCNN head on OpenImages V4 dataset containing 600 classes. Open Images is a dataset of ~9M images annotated with image-level labels, object bounding boxes and visual relationships. The training set of Open Images V4 [20] contains 14.6 million bounding boxes for 600 object classes on 1.74 million images, making it the largest existing dataset with object location annotations. The images used are diverse and often contain complex scenes with several objects. The hierarchy for 600 boxable classes can be viewed and download as JSON file [21, 34].

For a given image, we first perform the object detection and extract all the identified objects. Next, we apply 2 level filtering to find unique detections made, and extracting only those which are present in the Wordnet dictionary [22] and pretrained vectors of the word2vec-google-news-300 [23] model for deriving representative tag (see Fig. 2).

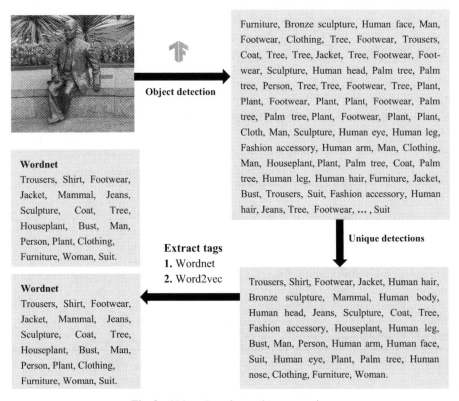

Fig. 2. Object detection and tags grouping

2.2 Semantic Analysis of Textual Tags

Natural language processing (NLP) is a subfield of Artificial Intelligence (AI). This is a widely used technology which deals with the interaction between computers and humans in natural language. It processes and analyses the natural language data such as text and speech with the goal of understanding the meaning behind the language. Some common techniques used in NLP includes tokenization, part-of-speech tagging, named entity recognition, sentiment analysis, machine translation and text classification. In this paper, we have used the Natural Language Toolkit (NLTK) a python package to work with NLP for carrying out the semantic analysis of textual tags. NLTK acts as a toolbox for NLP algorithms and provides easy-to-use interfaces to over 50 corpora and lexical resources such as Wordnet [24].

In this step, the user-provided tags of the photo are refined before computing the semantic similarity. Firstly, we extract two sets of tags i.e., tags which are present in the Wordnet dictionary and the ones which are present in the word2vec-google-news-300 model. Secondly, we apply Part-of-Speech (POS) tagging, a process where each word in a text is labeled with its corresponding part of speech. This may include nouns, pronouns, verbs, adverbs, adjectives, and other grammatical categories. Thirdly, we filter the named entities [35] followed by extracting tags which are identified as Singular Common Nouns (**NN**), Plural Common Nouns (**NNS**), Singular Proper Nouns (**NNP**), and Plural Proper Nouns (**NNPS**) (see Fig. 3).

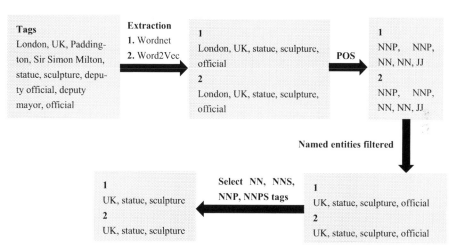

Fig. 3. Semantic analysis and tags refinement

2.3 Extracting Representative Tag

Wu & Palmer Similarity. To select a representative tag for a given photo, we compute the semantic similarity between a unique set of object detections (Fig. 2) and refined tags (Fig. 3). For this purpose, we have considered two NLP techniques i.e., Wu & Palmer similarity [25] and Word2Vec [26, 27]. Wu & Palmer similarity returns a score denoting how similar two-word senses are, by considering depths of two synsets (groups of synonymous words expressing the same concept) in the WordNet taxonomies, along with the depth of the LCS (Least Common Subsumer).

$$Wu - Palmer = 2 * \frac{\text{depth}(\text{lcs}(s1, s2))}{(depth(s1) + depth(s2))} \quad (1)$$

For a given tag and object detected, we first retrieve the list of available synsets. Some words may have only one synset and some may have several. We, however, use the first available synset. The list of synsets can be retrieved using wordnet.synsets(word). For computing Wu & Palmer similarity using NLTK, we use wup_similarity function as follows:

$$synset1.wup\_similarity(synset2) \tag{2}$$

Next, we derive a matrix of $d \times t$ dimensions. Where d represents the number of objects detected and t represents the number of tags. Each element in the matrix represents a wup_similarity score between d and t. Finally, we perform the summation of matrix columns and a tag with maximum column sum is acknowledged as representative tag for a photo (see Fig. 4).

Word embedding – Word2Vec. Word Embedding in NLP is an important aspect which connects a human language to that of a machine. This technique transforms the words into a numerical representation of words (vectors). These vectors try to capture the different characteristics of words regarding the overall text, including semantic relationships of words, definitions, and context etc. [28]. One Hot Encoding, TF-IDF (Term Frequency-Inverse Document Frequency), Bag of Words, Word2Vec, FastText and GloVe (Global Vectors for Word Representations) are frequently used Word Embedding methods. Word Embedding find its applications in music/video recommendation systems, analyzing survey responses, verbatim comments, and others [29].

We chose Word2Vec method to find the semantic similarity between user contributed tags and objects detected. It was developed by Thomas Mikolov in 2013 at Google. It is a popular word embedding technique which embed words in a lower-dimensional vector space using shallow neural network. This results in a set of word vectors where vectors close together in a vector space are semantically related and word vectors distant in vector space have different meanings. For example, *clean* and *tidy* would be close together as compared to the *clean* and *season* [29]. Embeddings learned through Word2Vec has proven to be efficient with learning high-quality vector representations and capturing semantic and syntactic information [30]. It has two neural network-based variants: Continuous Bag of Words (CBOW) and Skip-gram.

While one can train their own Word2Vec embeddings, we take advantage of the pre-trained vectors on part of Google News dataset of about 100 billion words. This model (**word2vec-google-news-300**) contains 300-dimensional vectors for 3 million words and phrases. The model is available through *Gensim*, which is a free open-source python library for unsupervised topic modeling and natural language processing [29, 31]. For a given list of tags and object detections, we compute the cosine similarity using built-in *similarity* function as shown below: [32]

$$google\_news\_vectors.similarity('Statue', 'Clothing') \tag{3}$$

Next, we derive a matrix of $d \times t$ dimensions as discussed above. Where, d represents the number of objects detected and t represents the number of tags (see Fig. 4). Each element in the matrix represents a cosine similarity score between d and t. Finally, we perform the summation of matrix columns and a tag with maximum column sum is acknowledged as representative tag for a photo (see Fig. 4).

	UK	Statue	Sculpture			UK	Statue	Sculpture
Shirt	0.32	0.53	0.55	Shirt	0.07	0.05	0.09
Jeans	0.30	0.50	0.53	Jeans	0.11	0.06	0.17
Tree	0.30	0.40	0.42	Tree	0.03	0.12	0.22
Coat	0.30	0.50	0.53	Coat	0.07	0.15	0.09
Jacket	0.29	0.48	0.50	Jacket	0.09	0.12	0.10
Plant	0.33	0.55	0.59	Plant	0.05	0.14	0.13
Suit	0.32	0.53	0.55	Suit	0.07	0.09	0.01
Footwear	0.33	0.55	0.58	Footwear	0.06	0.20	0.19
Person	0.36	0.47	0.50	Person	−0.02	0.39	0.71
Bust	0.12	0.27	0.29	Bust	0.04	0.05	0.04
Mammal	0.30	0.40	0.42	Mammal	0.09	0.09	0.14
Trousers	0.32	0.53	0.55	Trousers	0.13	0.09	0.07
Woman	0.33	0.42	0.44	Woman	0.09	0.15	0.17
Houseplant	0.33	0.44	0.48	Houseplant	0.01	0.09	0.09
Clothing	0.33	0.59	0.63	Clothing	0.08	0.19	0.21
Furniture	0.33	0.55	0.59	Furniture	0.08	0.03	0.11
Man	0.33	0.42	0.44	Man	0.07	0.07	0.04
Sculpture	0.32	0.84	0.88	Sculpture	0.07	0.07	0.10
Sum	**5.56**	**8.97**	**9.47**	**Sum**	**1.19**	**2.15**	**2.68**

Fig. 4. Representative tag calculation using Wu & Palmer similarity (left) and Word2Vec (right).

3 Experiments

We collected a random sample of 169 Flickr images against 6 different locations around the world for experimental purpose (see Table 2). Flickr standardizes the user-provided tags by removing the space between words and converting the letters into lowercase. A user tag "My Helsinki" would become "myhelsinki" [18]. However, for research purpose we rely on the raw tags. Images with missing tags were excluded. It is worth mention that the authors collected the ground truth data by hand. On average, each photo had 6 to 7 tags ranging from name of the place, content of the photo, weather details, camera information and time information. Our results show that representative tag derivation using Wu & Palmer similarity (Wordnet similarity) outperforms cosine similarity measures in Word2Vec by 11%.

For images, where representative tag was not correctly identified, we observed the following two main reasons:

1. For a given list of user tags, representative tag itself was missing (see Fig. 5)
2. Representative tag existed in the user tags but got filtered because it was either not found in the Wordnet dictionary or has no pretrained vector in the word2vec-google-news-300 model (see Fig. 6)

Table 2. Representative tag selection results.

Locations	Images Inspected	Average number of tags	Images without tags	Wu & Palmer correct prediction	Word2Vec correct prediction
Helsinki Cathedral	27	7	14	38%	23%
Stonehenge	41	6	11	40%	3%
Leaning Tower of Pisa	101	13	22	41%	29%
Koli	1	9	0	100%	100%
Hyde Park	42	7	14	46%	32%
Mont des Arts Garden	7	6	0	14%	28%

Photos		

User-provided Tags		
финляндия, q, finland, helsinki	stonehenge, world heritage site, photographie, landscapes, wbayer.com	stonhenge, stone, circle, neolithic, wiltshire, uk
Representative tags Wordnet / Word2Vec		
helsinki / helsinki	Stonehenge / landscapes	stone / stone
Ground truth Tag		
Pohjola	Graffitied Stone	Sign

Fig. 5. Representative tag itself was missing in user's contributed list of tags

Photos		

User-provided Tags		
London, Camden, Regent's Park, London Parks, Frieze Sculpture, Frieze Sculpture 2022, Public Sculpture, Tim Etchells	Stonehenge, Wiltshire, Neolithic Village	tuscany, pisa, giardino scotto
Representative tags Wordnet / Word2Vec		
Camden / -	Stonehenge / Stonehenge	tuscany / -
Ground truth Tag		
Frieze Sculpture	Neolithic Village	Giardino scotto

Fig. 6. Representative tag was neither present in the Wordnet dictionary nor in Word2Vec model

4 Conclusions and Future Work

Social tagging is subjective, orderless and noisy which restrict the use of tags in many related applications. In this paper, we propose a method named *Tag-tag* to derive a representative tag which best describes the visual content of a given image. For this purpose, we exploit the semantic relationship between visual content of a given image and its tags. Our method is completely independent of visually similar images, their associated tags and is not limited by any set of Flickr tag groups. We further use the existing state of the art pretrained machine learning models for object detection, thus does not require building anything from the scratch.

Our experiments on a set of 169 Flickr images, demonstrate the efficiency of proposed method. We identify two possible reasons where representative tag was not found as: 1) Representative tag itself was missing in the user given list of tags 2) Representative tag got filtered because it was not found in Wordnet dictionary or had no pretrained vector in word2vec-google-news-300 model, which we recognize as the one limitation of our method. For future work, we aim to extend our dataset of images, conduct comprehensive comparison with existing methods and check the applicability of proposed method with other social tagging services.

References

1. Li, X., Snoek, C.G., Worring, M.: Learning social tag relevance by neighbor voting. IEEE Trans. Multimedia **11**(7), 1310–1322 (2009)

2. Sigurbjörnsson, B., Van Zwol, R.: Flickr tag recommendation based on collective knowledge. In: 17th International Conference on World Wide Web, pp. 327–336 (2008)

3. Kennedy, L., Naaman, M., Ahern, S., Nair, R., Rattenbury, T.: How flickr helps us make sense of the world: context and content in community-contributed media collections. In: ACM International Conference on Multimedia, pp. 631–640 (2007)

4. Liu, D., Hua, X.S., Yang, L., Wang, M., Zhang, H.J.: Tag ranking. In: 18th International Conference on World Wide Web, pp. 351–360 (2009)

5. Wu, L., Hua, X.S., Yu, N., Ma, W.Y., Li, S.: Flickr distance. In: 16th ACM International Conference on Multimedia, pp. 31–40 (2008)

6. Torralba, A., Fergus, R., Freeman, W.T.: 80 million tiny images: a large data set for non-parametric object and scene recognition. IEEE Trans. Pattern Anal. Mach. Intell. **30**(11), 1958–1970 (2008)

7. Shamma, D.A., Shaw, R., Shafton, P.L., Liu, Y.: Watch what I watch: using community activity to understand content. In: International Workshop on Multimedia Information Retrieval, pp. 275–284 (2007)

8. Jin, Y., Khan, L., Wang, L., Awad, M.: Image annotations by combining multiple evidence & wordnet. In: 13th Annual ACM International Conference on Multimedia, pp. 706–715 (2005)

9. Wang, C., Jing, F., Zhang, L., Zhang, H.J.: Image annotation refinement using random walk with restarts. In: ACM International Conference on Multimedia, pp. 647–650 (2006)

10. Wang, C., Jing, F., Zhang, L., Zhang, H.J.: Content-based image annotation refinement. In: IEEE Conference on Computer Vision and Pattern Recognition, pp. 1–8 (2007)

11. Wu, L., Yang, L., Yu, N., Hua, X.S.: Learning to tag. In: 18th International Conference on World Wide Web, pp. 361–370 (2009)

12. Zhuang, J., Hoi, S.C.: A two-view learning approach for image tag ranking. In: Fourth ACM International Conference on Web Search and Data Mining, pp. 625–634 (2011)

13. Kennedy, L.S., Chang, S.F., Kozintsev, I.V.: To search or to label? Predicting the performance of search-based automatic image classifiers. In: ACM International Workshop on Multimedia Information Retrieval, pp. 249–258 (2006)

14. Ames, M., Naaman, M.: Why we tag: motivations for annotation in mobile and online media. In: SIGCHI Conference on Human Factors in Computing Systems, pp. 971–980 (2007)

15. Lindstaedt, S., Mörzinger, R., Sorschag, R., Pammer, V., Thallinger, G.: Automatic image annotation using visual content and folksonomies. Multimedia Tools Appl. **42**, 97–113 (2009)

16. Zhao, Y., Zha, Z.-J., Li, S., Wu, X.: Which tags are related to visual content? In: Boll, S., Tian, Qi., Zhang, L., Zhang, Z., Chen, Y.-P. (eds.) MMM 2010. LNCS, vol. 5916, pp. 669–675. Springer, Heidelberg (2010). https://doi.org/10.1007/978-3-642-11301-7_67

17. Lindstaedt, S., Pammer, V., Mörzinger, R., Kern, R., Mülner, H., Wagner, C.: Recommending tags for pictures based on text, visual content and user context. In: IEEE International Conference on Internet and Web Applications and Services, pp. 506–511 (2008)

18. Hu, Y., Gao, S., Janowicz, K., Yu, B., Li, W., Prasad, S.: Extracting and understanding urban areas of interest using geotagged photos. Comput. Environ. Urban Syst. **54**, 240–254 (2015)

19. https://tfhub.dev/google/faster_rcnn/openimages_v4/inception_resnet_v2/1. Accessed 14 Oct 2023

20. https://www.tensorflow.org/datasets/catalog/open_images_v4. Accessed 10 Oct 2023

21. https://storage.googleapis.com/openimages/2018_04/bbox_labels_600_hierarchy_visual izer/circle.html. Accessed 10 Oct 2023

22. https://wordnet.princeton.edu/. Accessed 10 Oct 2023

23. https://radimrehurek.com/gensim/models/word2vec.html. Accessed 10 Oct 2023

24. https://www.nltk.org/. Accessed 10 Oct 2023

25. https://www.geeksforgeeks.org/nlp-wupalmer-wordnet-similarity/. Accessed 10 Oct 2023

26. https://en.wikipedia.org/wiki/Word2vec#:~:text=Word2vec%20is%20a%20technique%20f or,words%20for%20a%20partial%20sentence. Accessed 10 Oct 2023

27. https://towardsdatascience.com/word2vec-explained-49c52b4ccb71. Accessed 10 Oct 2023
28. https://www.turing.com/kb/guide-on-word-embeddings-in-nlp. Accessed 10 Oct 2023
29. https://medium.com/nlplanet/text-similarity-with-the-next-generation-of-word-embeddings-in-gensim-466fdafa4423. Accessed 10 Oct 2023
30. https://www.tensorflow.org/tutorials/text/word2vec. Accessed 10 Oct 2023
31. https://radimrehurek.com/gensim/intro.html. Accessed 10 Oct 2023
32. https://tedboy.github.io/nlps/generated/generated/gensim.models.Word2Vec.similarity.html. Accessed 10 Oct 2023
33. https://phototutorial.com/flickr-statistics/. Accessed 10 Oct 2023
34. https://storage.googleapis.com/openimages/2018_04/bbox_labels_600_hierarchy.json. Accessed 10 Oct 2023
35. https://en.wikipedia.org/wiki/Named-entity_recognition. Accessed 10 Oct 2023
36. Liu, W., Ruan, Y., Cai, X., Chen, H.: Social image tag relevance learning based on pixel voting. In: International Conference on Computer Science and Application Engineering (CSAE) (2017). ISBN: 978-1-60595-505-6

Natural Language Processing and Computational Linguistics

STTS: A Novel Span-Based Approach for Topic-Aware Text Segmentation

Yide Cai, Yuzhe Zhang, and Zhouwang Yang[✉]

School of Data Science, University of Science and Technology of China, Hefei 230026, Anhui, People's Republic of China
{sa21229001,zyz2020}@mail.ustc.edu.cn, yangzw@ustc.edu.cn

Abstract. Topic-aware text segmentation (TATS) involves dividing text into cohesive segments and assigning a corresponding topic label to each segment. The TATS of a document has become increasingly significant for business researchers to obtain comprehensive insights into the behavior of enterprises. However, current models either cannot balance accuracy and generalization or are unable to handle the topic nesting problem, leading to low efficiency in practical needs. This paper proposes a novel **S**pan-based approach for **T**opic-aware **T**ext **S**egmentation called **STTS**, which consists of two components including a sliding window encoder and a span-based NER module. First, we utilize the sliding window encoder to transform the input document into text spans, which are then represented in their embeddings using pre-trained language models. Second, we obtain the coherent segments and assign a topic label to each segment based on the span-based NER method called Global Pointer. Experiments on four real-world business datasets demonstrate that STTS achieves state-of-the-art performance on the flat and nested TATS tasks. Consequently, our model provides an effective solution to TATS tasks with lengthy texts and nested topics, which indicates that our solution is highly suitable for large-scale text processing in practice.

Keywords: Text Segmentation · Span-based NER Task · Overlapping Sliding Window

1 Introduction

Topic-aware text segmentation (TATS) segments text into coherent parts and assign to each segment a topic. The TATS of a document enables business researchers to quickly obtain the desired information by searching for the content under predefined subtopics, better understanding the document, and affording a more convenient application for downstream tasks.

TATS tasks can utilize either supervised or unsupervised methods. The unsupervised statistical models exploit the strong correlation between the topic and lexical usage to determine the segmentation points between paragraphs [2, 4, 6, 14]. Accordingly, supervised deep learning models decompose TATS tasks into two sub-tasks: text segmentation and segment topic labeling. With the advent of pre-trained language models (PreLMs),

D.-S. Huang et al. (Eds.): ICAI 2023, CCIS 2015, pp. 299–309, 2024.
https://doi.org/10.1007/978-981-97-0827-7_26

many approaches exploit PreLMs, which are properly adapted for semi-structured documents, to obtain domain-specific models exhibiting great performance [3, 5, 12, 19, 21].

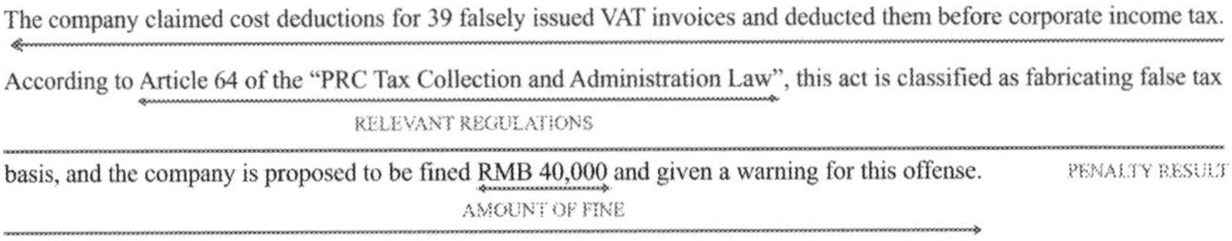

The company claimed cost deductions for 39 falsely issued VAT invoices and deducted them before corporate income tax.

According to Article 64 of the "PRC Tax Collection and Administration Law", this act is classified as fabricating false tax

RELEVANT REGULATIONS

basis, and the company is proposed to be fined RMB 40,000 and given a warning for this offense.   PENALTY RESULT

AMOUNT OF FINE

Fig. 1. An example for topic nesting from Enterprise Violation Announcement. This passage has three topics: PENALTY RESULT, RELEVANT REGULATIONS, and AMOUNT OF FINE. AMOUNT OF FINE and RELEVANT REGULATIONS are nested within PENALTY RESULT.

However, the methods mentioned above suffer from two limitations. First, they tend to solve TATS under a sequence labeling strategy, which assumes that each segment in the document can only have one topic label. Nevertheless, the same segment can belong to different subtopics in real-world datasets, denoted as topic nesting for the remainder of this paper. Figure 1 illustrates an example of TATS with topic nesting, where AMOUNT OF FINE and RELEVANT REGULATIONS are nested within PENALTY RESULT. Second, the above methods rely on complex hierarchical feature engineering when dealing with long texts. Furthermore, TATS has not been thoroughly explored for long Chinese multi-topic datasets.

Spurred by the deficiencies above, this paper proposes a novel Span-based approach for Topic-aware Text Segmentation (STTS), which is a generic end-to-end model for the TATS task. First, STTS employs a sliding window encoder to segment long texts into multiple fragments, which are then encoded using a PreLM. Then, a span-based Named Entity Recognition (NER) model is utilized to identify the target fragment's starting and ending positions and its corresponding topic. Our extensive experiments demonstrate that STTS can outperform existing solutions.

Overall, this paper has the following contributions:

1. Solving the TATS task under the span-based NER paradigm and overcoming the flat and nested topic segmentation problems.
2. Designing a sliding window encoder to process long text input and evaluate the impact of window size on the model's performance.
3. Demonstrating the superiority of STTS on real-world datasets and its effectiveness in industrial applications.

To the best of our knowledge, this is the first work successfully solving the problem of the Chinese TATS task with topic nesting. Our model's effectiveness is demonstrated in industrial applications. The developed code will be released after the paper is accepted.

2   Related Work

Unsupervised Statistical Model. Unsupervised methods utilize the correlation between topics and terms to find the segmentation points for text segmentation. For instance, Hearst et al. [6] presented TextTiling, founded on the principle that a high degree of vocabulary intersection between two adjacent text blocks indicates a high level of coherence between them. Choi [4] proposed a probabilistic algorithm for text similarity detection based on overlapping domain-specific terms. Besides, Chen et al. [2] introduced a novel Bayesian topic model to constrain the potential topic allocation and reflect the underlying document structure. Additionally, TopicTiling [14] uses the LDA model and the frequency of the terms to detect topic transitions. However, unsupervised statistical methods are highly complex and ignore the topic label prediction task.

Supervised Deep Learning Model. Contemporary supervised methods are based on deep learning models. Koshorek et al. [10] were the first to apply hierarchical BiLSTM [7] for text segmentation, outperforming unsupervised methods on the large public dataset Wiki-727k [10]. Li et al. [12] presented SEGBOT, a model that employs BiGRU to decide whether a topic-changing boundary exists and assigns the related topics. With the advent of Transformers [17] such as BERT [9], Zhang et al. [21] segmented documents into paragraphs and trained BERT at the paragraph level to extract content elements from regulatory filings and property lease agreements. Yoong et al. [19] created three BERT models for segmenting a document or dialogue into semantically coherent segments. Lo et al. [13] proposed a Transformer-based model with a bi-level architecture that utilizes BERT as sentence encoders to learn semantic coherence between sentences. Chivers et al. [3] built a classifier based on BERT to identify Human Capital Disclosures (HCD). Besides, Gong et al. [5] utilized a neural topic model to infer latent topic distributions of sentences and divided the document into topically coherent segments. Nevertheless, existing studies regard TATS as a sequence labeling task and thus cannot deal with the challenge of topic nesting.

Spurred by the existing literature, this paper refines a span-based NER model for the TATS task, specifically addressing the challenge of topic nesting that existing methods cannot handle.

3   Approach

3.1   Task Definition

Given the set of K topics $\mathcal{C} = \{c_1, \ldots, c_K\}$ and a document $\mathcal{D}$ containing N independent sentences $\mathcal{D} = \{d_1, \ldots, d_N\}$, the output of a TATS system is a list of tuples $Y = \{(I_i^{head}, I_i^{tail}, t_i)\}_{i=1}^m$, each specifying a segment mentioned in document $\mathcal{D}$. Here, I_i^{head} and I_i^{tail} are the head and tail sentence indices of the i th segment, respectively. t_i is a topic predefined in $\mathcal{C}$ and m is the number of segments in $\mathcal{D}$. Note that this definition holds for both flat and nested topic segmentation scenarios. For convenience, we consider the collection of continuous sentences $\{d_i, d_{i+1}, \ldots, d_j\}$ as a fragment of document $\mathcal{D}$, denoted as $\mathcal{D}[i:j]$.

Remark. We aim to illustrate the similarity between TATS and span-based NER tasks. Recall that the target of the span-based NER task is to find the head and tail token indices of all entities with the given entity type. Formally, given a sentence $X = x_1, \ldots, x_K$ with K tokens, the output of an NER system is a list of tuples $Y = \{(I_i^{head}, I_i^{tail}, t_i)\}_{i=1}^{m}$, each specifying an entity mentioned in X, where I_i^{head} and I_i^{tail} are the head and tail token indices of the i th entity, respectively. t_i is the entity type from a predefined type set and m is the number of named entities in X. In this way, the problem of nested entities can be resolved, which motivates us to solve TATS from the perspective of span-based NER tasks.

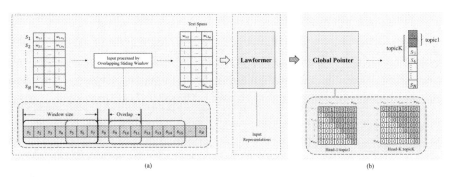

Fig. 2. STTS framework. (a) The STTS model employs a sliding window encoder to acquire the embeddings of multiple input sentences. The sliding window encoder consists of an overlapping sliding window mechanism with a fixed window size that can process the input into text spans, and a pre-trained language model (Lawformer in our method) that can obtain the embedding of the text span. (b) GP conducts the TATS task and outputs the segment boundaries and the topic label of each segment.

3.2 Model Formulation

Figure 2 illustrates the model structure. STTS first transforms the long text document into text spans with a fixed window size using the proposed overlapping sliding window mechanism, and the text spans are encoded into their semantic embeddings by Lawformer [18]. Then, the embeddings of text spans are sent to Global Pointer (GP) [16] to identify the topics.

Sliding Window Encoder. The sliding window encoder consists of the overlapping sliding window mechanism and a PreLM Lawformer.

The overlapping sliding window mechanism segments long texts into smaller units by sliding a fixed-size window over the document with a certain overlap. It reduces the influence of irrelevant information within the sliding window when predicting the segmentation points, considering that a document's segmentation history has a minor impact on predicting the next segmentation point. Overlapping provides a smooth transition of semantic information between two consecutive sliding windows.

Lawformer is a law-specific PreLM based on Longformer [1]. Based on Sliding Window Attention, Dilated Sliding Window, and Global Attention, Lawformer and Longformer reduce the time and memory complexity of the original Transformer from $O(n^2)$ to $O(n \times w)$, where n is the input sequence length, and $w \ll n$ is the fixed sliding window size.

Global Pointer. Given a document $\mathcal{D} = \{d_1, \ldots, d_N\}$ of M sentences, the embedding of each sentence is of dimension v, where v is the hidden layer size of PreLM, and can be represented as:

$$h_1, \ldots, h_M = PreLM(d_1, \ldots, d_M),\tag{1}$$

where $h_i \in \mathbb{R}^v$ is the embedding of sentence d_i, $1 \leq i \leq$ M.

After the sentence representations have been obtained, we calculate the fragment representations depending on each fragment's start and end position. For this purpose, two fully-connected layers are used:

$$q_{i,\alpha} = W_{q,\alpha} h_i + b_{q,\alpha}, \quad k_{i,\alpha} = W_{k,\alpha} h_i + b_{k,\alpha},\tag{2}$$

where $W_{q,\alpha}$ and $W_{k,\alpha}$ are learnable weight matrices, $b_{q,\alpha}$ and $b_{k,\alpha}$ are bias terms, and $q_{i,\alpha} \in \mathbb{R}^d$ and $k_{i,\alpha} \in \mathbb{R}^d$ are the vector representations that are used to identify segments of type α. Then, the score of fragment $\mathcal{D}[i : j]$, which is a segment of type α, is calculated as follows:

$$s_\alpha(i, j) = q_{i,\alpha}^\top k_{j,\alpha}.\tag{3}$$

To avoid GP's insensitivity to the length and span of segments, the scoring function also integrates rotational position encoding (RoPE) [15]. In this way, the scoring function can be written as follows:

$$s_\alpha(i, j) = \left(R_i q_{i,\alpha}\right)^\top \left(R_j k_{j,\alpha}\right) = q_{i,\alpha}^\top R_{j-i} k_{j,\alpha},\tag{4}$$

where RoPE satisfies $R_i^\top R_j = R_{j-i}$.

Training Objective. We adopt the loss function used in GP, which is inspired by circle loss and can alleviate the problem of class imbalance. The loss function can be calculated as follows:

$$\sum_{\alpha \in \mathcal{C}} \left\{ \log\left(1 + \sum_{(q,k) \in P_\alpha} e^{-s_\alpha(q,k)}\right) + \log\left(1 + \sum_{(q,k) \in Q_\alpha} e^{s_\alpha(q,k)}\right)\right\},\tag{5}$$

where q and k represent the start and tail indexes of a fragment, respectively, P_α denotes a collection of fragments with topic type α, whereas Q_α denotes a collection of fragments that are not segments or whose topic type is not α, and $s_\alpha(q, k)$ is the score that a fragment $\mathcal{D}[q, k]$ is of topic type α.

For inference, we determine whether a fragment $\mathcal{D}[q : k]$ belongs to topic type α as a segment by the non-negativity of $s_\alpha(q, k)$.

4   Experiments

4.1   Experimental Setup

Datasets and Baselines. We conduct extensive experiments to demonstrate the effectiveness of STTS on four real-world Chinese datasets, including Legal Documents (LD), Administrative Penalty News (APN), Product Recall News (PRN), and Enterprise Violation Announcements (EVA). LD, APN, and PRN, do not involve topic nesting, whereas, in EVA, about 20% of the samples have nested topics. Table 1 and Table 2 report the corresponding statistics. In the subsequent trials, we challenge our model against PreLM-NER models [21] and PreLM-BiLSTM-CRF models [8, 11] by considering four PreLM types: BERT, Longformer, Lawformer, and BigBird [20].

Table 1. Statistics of the four datasets. *Topics*: the number of topics. *Docs*: the total number of documents. *SLen*: the average sentence length. *S/doc* and *P/doc*: the average number of sentences and paragraphs per document. *S/P*: the average number of sentences per paragraph. *Nesting*: if marked by ✓, topic nesting is present, and otherwise not.

Dataset	Topics	Docs	S/doc	P/doc	S/P	SLen	Nesting
LD	8	24829	41.35	7.05	5.87	45.92	✗
APN	2	13118	54.30	13.50	4.02	20.15	✗
PRN	3	2112	12.04	2.05	5.88	56.09	✗
EVA	8	12752	393.47	18.77	20.96	99.65	✓

Table 2. Topic statistics.

Dataset	Topic(in Chinese)	Topic(in English)
LD	首部, 当事人信息, 诉讼记录, 案件基本情况, 本院认为, 判决结果, 尾部, 法律依据	Header, Party Information, Litigation Records, Basic Case Information, Court's Opinion, Judgment Result, Footer, Legal Basis
APN	违法事实, 处罚决定	Illegal facts, Penalty decision
PRN	产品召回信息概要, 产品信息总览, 产品信息详细	Product Recall Summary, Product Information Overview, Product Information Details
EVA	发文单位, 处罚日期, 文件批号, 相关法规, 违规原因, 处罚结果, 当事人名称, 处罚金额	Issuing authority, Date of penalty, Document number, Relevant regulations, Reason for violation, Penalty result, Name of the party involved, Penalty amount

Evaluation Metrics. We employ the segment-level F_1 score and document-level accuracy as evaluation metrics. For the segment-level F_1 score, a segment is considered correctly recognized only if the detected boundaries and topic are consistent with the manually annotated ground-truth labels. For document-level accuracy, a document is considered correctly segmented only if all segments in the document are correctly recognized.

Parameter Settings. We follow the same model initialization method of GP [16] and tune the models' hyperparameters on each dataset. Table 3 reports the hyperparameters for the best-performing model per dataset.

Table 3. Hyperparameters for the best-performing model.

Dataset	Epochs	LR	Window size	Overlap	Batch size	Optimizer
LD	50	1e−5	400	500	3	BertAdam
APN	50	2e−5	1000	500	5	BertAdam
PRN	20	2e−5	2000	0	10	BertAdam
EVA	50	2e−5	2500	500	5	BertAdam

4.2 Main Results

Table 4 presents the experimental results on the test set of each dataset. Regarding the segment-level F_1 score, our model has an average improvement of 5.6% and 26.51% under flat-topic and nested-topic scenarios compared to Lawformer-BiLSTM-CRF. Regarding the document-level accuracy, our model reaches 94.95% on average on flat-topic datasets and 89.80% on nested-topic dataset, demonstrating an improvement of 17.82 and 32.15 points compared to Lawformer-NER. It should be noted that the proposed method, STTS, performs consistently better regarding the segment-level F_1 and document-level accuracy for TATS with topic nesting. This demonstrates our model's robustness under complex situations. Moreover, our approach achieves state-of-the-art performance due to the overlapping sliding window mechanism and span-based NER method. Due to constraints on the input text length, BERT can only model short texts, while Lawformer, Longformer, and BigBird, although capable of modeling long texts to some extent, still face the incapability of dealing with extremely long texts in real-world datasets. The experimental results demonstrate our model's effectiveness on the TATS task, especially in scenarios with nested topics.

4.3 Overlapping Sliding Window Evaluation

To illustrate the effect of the overlapping sliding window mechanism, we conduct an ablation study on the APN dataset by choosing different sliding window sizes on three PreLMs, including Longformer, BigBird, and Lawformer. Note that we do not take the overlap into account to eliminate its interplay. Moreover, setting the window size to zero means excluding the proposed overlapping sliding window mechanism. Figure 3 depicts the line chart of document-level accuracy versus window size, revealing that choosing an appropriate window size improves the model's performance, demonstrating the effectiveness of our overlapping sliding window.

Table 4. Segment-level F_1 (Segment-F_1) and document-level accuracy (Doc-acc) of different models on four datasets.

Dataset	LD		APN		PRN		EVA	
	Segment-F_1	Doc-acc	Segment-F_1	Doc-acc	Segment-F_1	Doc-acc	Segment-F_1	Doc-acc
BERT-NER	0.6150	0.1429	0.3600	0.1014	0.8527	0.7333	0.1743	0.1225
Longformer-NER	0.9188	0.7177	0.8734	0.7485	0.9291	0.7857	0.6975	0.5586
Lawformer-NER	0.9188	0.7169	0.9015	0.8063	0.9251	0.7905	0.7057	0.5765
BigBird-NER	0.8907	0.5810	0.7301	0.5308	0.9251	0.7857	0.6360	0.5058
BERT-BiLSTM-CRF	0.6365	0.2103	0.5713	0.2320	0.9271	0.7905	0.4027	0.3829
Longformer-BiLSTM-CRF	0.9169	0.7126	0.8620	0.7305	0.9242	0.7810	0.7056	0.5714
Lawformer-BiLSTM-CRF	0.9188	0.7185	0.8962	0.7935	0.9312	0.7905	0.7106	0.5798
BigBird-BiLSTM-CRF	0.9169	0.7044	0.8660	0.7357	0.9251	0.7857	0.7001	0.5674
The proposed model	**0.9857**	**0.9584**	**0.9403**	**0.9230**	**0.9887**	**0.9668**	**0.9757**	**0.8980**

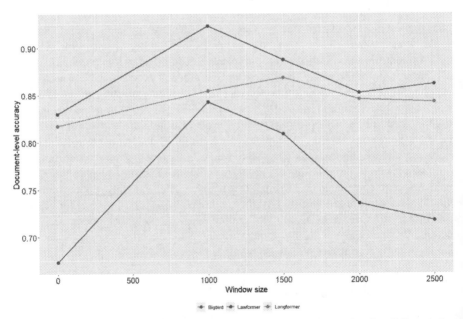

Fig. 3. Comparative evaluation of the document-level accuracy of the overlapping sliding window approach on APN.

4.4 Case Study

In Fig. 4, we present a representative example demonstrating the differing prediction results obtained using STTS and Lawformer-BiLSTM-CRF. In this example, the "penalty amount", "document number" and "name of the party" involved are nested within "penalty result". STTS successfully identifies all segments and assigns appropriate labels to each one, while Lawformer-BiLSTM-CRF correctly identifies the segments nested within the penalty result but fails to identify the segment corresponding to the label "penalty result" itself. This case highlights the advantage of STTS over Lawformer-BiLSTM-CRF in handling nested topic segmentation problems. The example serves as a validation of the effectiveness of STTS in TATS tasks under the span-based NER paradigm.

This case study provides evidence that STTS outperforms Lawformer-BiLSTM-CRF due to its superior syntactic and semantic learning capabilities. Specifically, STTS excels in overcoming both flat and nested topic segmentation challenges, making it more robust and accurate in this context.

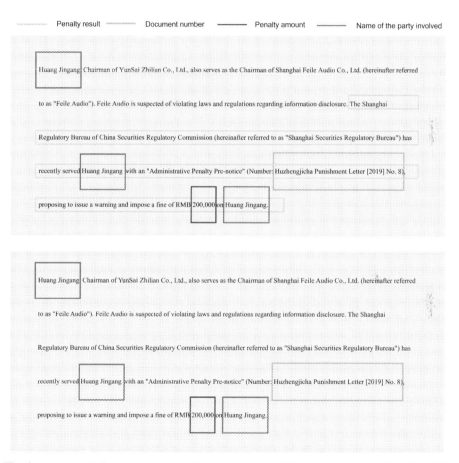

Fig. 4. A representative example with different results by STTS and Lawformer-BiLSTM-CRF.

5   Conclusion

This paper proposes STTS, a framework to address topic-aware text segmentation tasks from the perspective of span-based NER tasks. STTS has two main advantages. On the one hand, it can cope with nested topic segmentation using the span-based NER. On the other hand, STTS enables long document processing via an overlapping sliding window mechanism without losing much contextual information. Extensive experiments demonstrate the superiority of STTS on four real-world business topic segmentation datasets, which indicates that our solution is highly suitable for large-scale text processing in practice.

References

1. Beltagy, I., Peters, M.E., Cohan, A.: Longformer: the long-document transformer. arXiv preprint arXiv:2004.05150 (2020)
2. Chen, H., Branavan, S., Barzilay, R., Karger, D.R.: Global models of document structure using latent permutations. In: Proceedings of Human Language Technologies: The 2009 Annual Conference of the North American Chapter of the Association for Computational Linguistics, pp. 371–379 (2009)
3. Chivers, B., Jiang, M.P., Lee, W., Ng, A., Rapstine, N.I., Storer, A.: ANTS: a framework for retrieval of text segments in unstructured documents. In: Proceedings of the Third Workshop on Deep Learning for Low-Resource Natural Language Processing, pp. 38–47 (2022)
4. Choi, F.Y.: Advances in domain independent linear text segmentation. In: 1st Meeting of the North American Chapter of the Association for Computational Linguistics (2000)
5. Gong, Z., et al.: Tipster: a topic-guided language model for topic-aware text segmentation. In: Bhattacharya, A., et al. (eds.) Database Systems for Advanced Applications: 27th International Conference, DASFAA 2022, Virtual Event, 11–14 April 2022, Proceedings, Part III, pp. 213–221. Springer, Cham (2022). https://doi.org/10.1007/978-3-031-00129-1_14
6. Hearst, M.A.: Text tiling: segmenting text into multi-paragraph subtopic passages. Comput. Linguist. **23**(1), 33–64 (1997)
7. Hochreiter, S., Schmidhuber, J.: Long short-term memory. Neural Comput. **9**(8), 1735–1780 (1997)
8. Ju, M., Miwa, M., Ananiadou, S.: A neural layered model for nested named entity recognition. In: Proceedings of the 2018 Conference of the North American Chapter of the Association for Computational Linguistics: Human Language Technologies, Volume 1 (Long Papers), pp. 1446–1459 (2018)
9. Devlin, J., Chang, M.W., Lee, K., Toutanova, K: BERT: pre-training of deep bidirectional transformers for language understanding. In: Proceedings of NAACL-HLT, pp. 4171–4186 (2019)
10. Koshorek, O., Cohen, A., Mor, N., Rotman, M., Berant, J.: Text segmentation as a supervised learning task. In: Proceedings of the 2018 Conference of the North American Chapter of the Association for Computational Linguistics: Human Language Technologies, Volume 2 (Short Papers), pp. 469–473 (2018)
11. Lample, G., Ballesteros, M., Subramanian, S., Kawakami, K., Dyer, C.: Neural architectures for named entity recognition. arXiv preprint arXiv:1603.01360 (2016)
12. Li, J., Sun, A., Joty, S.: SegBot: a generic neural text segmentation model with pointer network. In: Proceedings of the 27th International Joint Conference on Artificial Intelligence, pp. 4166–4172 (2018)

13. Lo, K., Jin, Y., Tan, W., Liu, M., Du, L., Buntine, W.: Transformer over pre-trained transformer for neural text segmentation with enhanced topic coherence. In: Findings of the Association for Computational Linguistics: EMNLP 2021, pp. 3334–3340 (2021)
14. Riedl, M., Biemann, C.: TopicTiling: a text segmentation algorithm based on LDA. In: Proceedings of ACL 2012 Student Research Workshop, pp. 37–42 (2012)
15. Su, J., Lu, Y., Pan, S., Murtadha, A., Wen, B., Liu, Y.: RoFormer: enhanced transformer with rotary position embedding. arXiv preprint arXiv:2104.09864 (2021)
16. Su, J., et al.: Global pointer: novel efficient span-based approach for named entity recognition. arXiv preprint arXiv:2208.03054 (2022)
17. Vaswani, A., et al.: Attention is all you need. In: Advances in Neural Information Processing Systems, vol. 30 (2017)
18. Xiao, C., Hu, X., Liu, Z., Tu, C., Sun, M.: Lawformer: a pre-trained language model for Chinese legal long documents. AI Open **2**, 79–84 (2021)
19. Yoong, S.Y., Fan, Y.-C., Leu, F.-Y.: On text tiling for documents: a neural-network approach. In: Barolli, L., Takizawa, M., Enokido, T., Chen, H.-C., Matsuo, K. (eds.) Advances on Broad-Band Wireless Computing, Communication and Applications. LNNS, vol. 159, pp. 265–274. Springer, Cham (2021). https://doi.org/10.1007/978-3-030-61108-8_26
20. Zaheer, M., et al.: Big bird: transformers for longer sequences. In: Advances in Neural Information Processing Systems, vol. 33, pp. 17283–17297 (2020)
21. Zhang, R., et al.: Rapid adaptation of BERT for information extraction on domain-specific business documents. arXiv preprint arXiv:2002.01861 (2020)

BERT-KRS: A BERT-Based Model for Knowledge-Grounded Response Selection in Retrieval-Based Chatbots

Jindian Su[1], Shanshan Yu[2(✉)], Xiaobin Ye[3], and Dandan Ma[3]

[1] College of Computer Science and Engineering, South China University of Technology,
Guangzhou 510641, Guangdong, China
`sujd@scut.edu.com`
[2] College of Medical Information Engineering, Guangdong Pharmaceutical University,
Guangzhou 510006, Guangdong, China
`susyu@139.com`
[3] Guangdong Unicomm, Guangzhou 440100, China

Abstract. Knowledge-grounded response selection in retrieval-based chatbots aims to select the proper response given the conversation context and background knowledge. Most of existing BERT-based approaches only consider the relations between the conversation context and response candidates, and often ignore the effect of BERT in domain adaptation with task-specific data and fail to consider the relevance between the context and knowledge. To address these issues, we propose a BERT-based model for Knowledge-grounded Response Selection (BERT-KRS). We first design a domain adaptation task, consistency detection, to yield a domain-adaptive BERT for capturing the consistency relation between conversations and knowledge. Then we take the domain-adaptive BERT as the encoder and propose an attention-based matching network to strengthen the relationship between the context and knowledge before matching with response candidates. Experimental results on two canonical benchmarks indicate that our BERT-KRS model outperforms the strong baseline methods.

Keywords: Response selection · Retrieval-based conversation · Knowledge-grounded conversation · Pre-trained language model

1 Introduction

Retrieval-based chatbots [1] are a typical kind of dialogue systems and have been widely used in many fields of the industry. As one of the most important part of retrieval-based chatbots, multi-turn response selection aims to identify the best response from a set of response candidates given a dialogue context, i.e., the conversation context. Previous works had focused on offering fine-grained text encoding and better interactions between response candidates and conversation contexts, by specially designed matching networks [2–4]. Some recent works [5–8] have also tried to build utterance-response matching models based on attention mechanisms and pre-trained languages, e.g., BERT, RoBERTa

© The Author(s), under exclusive license to Springer Nature Singapore Pte Ltd. 2024
D.-S. Huang et al. (Eds.): ICAI 2023, CCIS 2015, pp. 310–321, 2024.
https://doi.org/10.1007/978-981-97-0827-7_27

and ELECTRA, and achieved significant performance improvements in the multi-turn response selection. However, these works tend to give users the false positive answer and encounter inconsistencies between the context and the response, since they select the response by only considering the dialogue history without the guidance of knowledge grounding.

An example is given in Table 1, where the appropriate response selection should consider not only the conversation context, but also the background knowledge, i.e., personalized profiles. In order to better select proper response from candidates that is consistent with both context and background knowledge, some knowledge-grounded response selection models [9, 10] were also proposed. For example, Gu [9] extend the interactive matching network [3] to the dually matching version such that the context and knowledge match with the response candidate respectively. Hua [10] propose the pre-selection and post-selection strategies to detect the relevant parts of context and knowledge before they match with the response candidate. While Humeau [11] and Mazaré [12] pre-train BERT with millions of data extracted from Reddit and fine-tune it on downstream tasks to incorporate domain and task-specific knowledge.

Table 1. An example of knowledge-grounded dialogue from the Persona-Chat dataset. Red words demonstrate that the true response is related to the profile.

Background Knowledge	
A's profile:	**B's profile:**
I work part time at a pizza restaurant.	My mom taught me how to play the cello.
My major is business.	Everyday I go swimming.
I love the beach.	My favorite color is red.
I like to party.	I love going to concerts on the weekend.
Conversation	
A: Hi, how are you today?	
B: I am good. I just got done practicing the cello, my mom taught me.	
A: Do you enjoy playing it?	
B: Yes I do, I also go swimming everyday. How about you?	
A: I enjoy the beach so swimming is nice.	
B: The beach is nice, what's your favorite color?	
A: I am not sure really. You?	
B: Mine is **red**, do you like **concerts**? (True response)	
B: My favorite color is orange. (False response)	

These works have already demonstrated the effectiveness of using external knowledge, i.e., background knowledge and domain-specific knowledge, to better select responses. However, they still exist some following problems: (1) most of them neglect other knowledge or encode the context and knowledge separately without considering their internal relationships. (2) pre-training BERT-based matching models with extra external large corpus is computationally costly and time-consuming. (3) BERT-based matching models are unable to handle long sequences in knowledge-grounded dialogue due to the maximum length limitation (i.e., 512 for BERT), and thus might lose potential important information after roughly truncating the input sequences.

To address the above problems, the contributions of this paper can be summarized as follows:

(1) We design a consistency detection pre-training task for BERT to better capture the semantic coherence and consistency between the conversion and background knowledge, which yields a domain-adaptive BERT that contains domain-specific knowledge after pre-training on the task-specific dialogue corpus.
(2) We propose a BERT-based model BERT-KRS for knowledge-grounded response selection by taking the domain-adaptive BERT as the encoder and employing cross-attention mechanism to capture interaction between the context and background knowledge without truncating sequences.
(3) We conduct experiments on two widely used datasets, Persona-Chat [13] and CMU-DoG [14], and the results demonstrate that our model outperforms those strong baseline models and achieves new state-of-the-art performance.

2 Related Works

The most common practice are to build utterance-response matching models based on attention mechanisms and output a score indicating the adequacy of individual response candidates. More recently, Gu [8] have applied BERT into retrieval-based chatbots and obtained new state-of-the-art results. However, they mainly focus on designing fine-grained text encoding and building the relationship between dialogue context and response candidates without taking other knowledge into account. They also encode the context and knowledge separately without considering their internal semantic relationships.

To better incorporate knowledge grounding and improve the response selection performance, Zhang [13] present Persona-Chat that includes personal profiles as background knowledge, and propose a profile memory network to fuse profile information with the context. Zhou [14] release CMUDoG and employs movie-related documents as background knowledge. Zhao [15] propose a document-grounded matching network (DGMN) to select important matching information with the hierarchical interaction mechanism. Gu [3] propose a dually interactive matching network (DIM) to match the response candidate with the context and knowledge respectively. Hua [10] design selector modules to detect the relevant parts of the context and knowledge. Zhu [16] propose a document content selection network (CSN) to perform explicit selection of relevant document contents, and filter out the irrelevant parts.

To capture contextual information when the downstream task is biased towards a specific domain and different from the pre-trained corpora. Humeau [11] and Mazare [12] used Reddit as additional corpus to further pre-train BERT for learning more domain-specific knowledge. These works have demonstrate the importance of pre-training and domain adaptation could effectively improve the performance of knowledge-grounded response selection. However, they usually need to use extra large corpora, which are computationally costly and extremely time-consuming.

3 Model

3.1 Overview

As illustrated in Fig. 1, we first pre-train BERT on the training data of dialogue corpus with two domain adaptation tasks: masked language model (MLM) and consistency detection (CD). During the pre-training, a data augmentation (DA) method is also designed for increasing the amount of data. Then we propose a model BERT-KRS to take the domain-adaptive BERT as the encoder and matches response candidates with both context and knowledge.

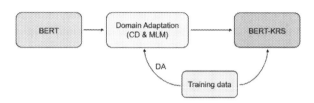

Fig. 1. An overview of our proposed model.

3.2 Task Definition

Given a dialogue dataset D, an example in D can be represented as (c, k, r, y), where $c = \{u_1, u_2,..., u_n\}$ denotes a conversation context with $\{u_i\}_{i=1}^{u_n}$ as utterances, $k = \{k_1, k_2,..., k_m\}$ denotes a collection of knowledge with $\{k_i\}_{i=1}^{u_m}$ as sentences. r is a response candidate, and $y \in \{0, 1\}$ is a binary label denoting whether r is a proper response for c and k. The goal of the task is to learn a matching model $g(c, k, r)$ from D, such that for a new context-knowledge-response triplet (c, k, r), $g(c, k, r)$ measures the degree of suitability of the response r to the given context c and the knowledge k.

3.3 Domain Adaptation

We propose a new domain adaptation task based on MLM, aiming to help the model better learn domain-specific knowledge.

Figure 2 shows the domain adaptation for knowledge-grounded response selection. The CD task can be regarded as a supplement of BERT without changing its original transformer structures.

For training our CD task, the positive examples are created by taking contexts and corresponding knowledge contents from the training corpus, while negative examples are created by pairing contexts with knowledge contents that are randomly sampled.

Formally, given a conversation context c and a collection of knowledge k, we set the binary label y_{cd} as 1 if they are matched, otherwise 0 for not matched. We formulate the input of BERT encoder as $x = \{[CLS], u_1, [EOU],..., u_i, [EOU],..., u_n, [SEP], k_1, [EOK],..., k_j, [EOK],..., k_m, [SEP]\}$, where [EOU] and [EOK] are two special tokens appended to the end of each utterance and knowledge sentence respectively.

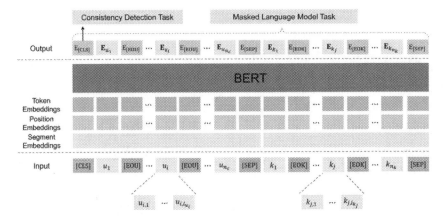

Fig. 2. Domain adaptation for knowledge-grounded response selection. u_i denotes the i-th utterance in the context and k_j is the j-th knowledge sentence.

The aggregated representation $E_{[CLS]} \in R^d$ is then fed to a single-layer perceptron for computing the final matching score via a sigmoid function as follows:

$$g_{cd}(c, k) = \text{sigmoid}(W_{cd} E_{[CLS]} + b_{cd}) \tag{1}$$

where $W_{cd} \in R^{1 \times d}$ and $b_{cd} \in R$ are trainable parameters. Finally, the objective of the consistency detection task is formulated as:

$$L_{cd} = -y_{cd}\log(g_{cd}(c, k)) - (1 - y_{cd})\log(1 - g_{cd}(c, k)) \tag{2}$$

Plus the MLM task, the learning objective of domain adaptation is the sum of the MLM loss L_{mlm} and CD loss L_{cd} as $L = L_{mlm} + L_{cd}$.

3.4 Model Architecture

We present the details of our proposed BERT-KRS model as shown in Fig. 3.

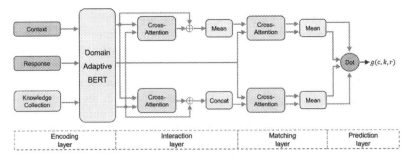

Fig. 3. The model architecture of BERT-KRS.

Encoding Layer

We use the above domain-adaptive BERT as the sequence encoder. Following previous works [11], we concatenate context utterances into a context sequence. After feeding the context, knowledge sentences and response to the BERT encoder, we can yield their hidden embedding representations $C \in R^{l_c \times d}$, $K = \{K_j\}_{j=1}^m$ and $R \in R^{l_r \times d}$ from the last layer of BERT, where l_c, l_r and d denotes the context length, the response length and the embedding size of BERT respectively. $K_j \in R^{l_{k_j} \times d}$ is the representation of the j-th sentence with the length of l_{kj}.

Interaction Layer

To further provide complementary information and enrich the representations of context and knowledge, we propose an interaction network for capturing their global features based on cross-attention mechanism.

Given C and $K = \{K_j\}_{j=1}^m$, we use two separate cross-attention networks to compute the word-word similarity between context and each knowledge sentence. For example, the similarity between C and the j-th sentence K_j is computed as $M_j = C \cdot K_j^T \in R^{l_c \times l_{k_j}}$. After normalization with *softmax* function, we combined them in a weighted way and yield the j-th knowledge-aware context representation C_j' and the j-th context-aware knowledge representation K_j' respectively, formulated as:

$$C'_j = softmax(M_j) \cdot K_j, K'_j = softmax(M_j^T) \cdot C, \tag{3}$$

A linear layer with RELU activation function and the residual connection is then adopted to yield the fused representations as:

$$C_j^* = max\left(0, C_j'W_1 + b_1\right) + C, K_j^* = max\left(0, K_j'W_2 + b_2\right) + K_j \tag{4}$$

where W_1, W_2, b_1 and b_2 are all trainable parameters.

In the same way, all fused context representations $\{C_j^*\}_{j=1}^m$ and knowledge representations $\{K_j^*\}_{j=1}^m$ can be obtained. We stack $\{C_j^*\}_{j=1}^m$ together as $C^* \in R^{m \times l_c \times d}$, and yield the final output $\overline{C} \in R^{l_c \times d}$ of the context in the mean-pooling way as $\overline{C} = mean(C^*)$. All knowledge sentence representations $\{K_j^*\}_{j=1}^m$ are also concatenated as the final output of knowledge, denoted as $\overline{K} = [K_1^*, K_2^*, \ldots, K_j^*] \in R^{l_k \times d}$ where $l_k = \sum_{j=1}^m l_{k_j}$ is the total length of the knowledge content.

Matching Layer

After using cross-attention to separately build the matching representations between context and the response, and the matching representations between knowledge content and the response, we can provide discriminative features and significant information to further decide their matching degrees. Formally, the aligned word-level matrix $M_{cr} = R^{l_c \times l_r}$ between the context and the response is defined as $M_{cr} = \overline{C} \cdot R^T$.

We further employ M_{cr} to build the response-aware context representation $\widetilde{C}$ and the context-aware response representation $\widetilde{R}_c$ respectively as follows:

$$\tilde{C} = softmax(M_{cr}) \cdot R, \tilde{R}_c = softmax(M_{cr}^T) \cdot \overline{C} \tag{5}$$

Similarly, we also use the above operations to obtain the response-aware knowledge representation $\widetilde{K}$ and the knowledge-aware response representation $\widetilde{R}_k$ respectively. To aggregate these matching representations on the sentence-level, we adopt mean-pooling strategy to yield the corresponding vectors $\hat{c} = mean(\widetilde{C}) \in R^d$, $\hat{r}_c = mean(\widetilde{R}_c) \in R^d$, $\hat{k} = mean(\widetilde{K}) \in R^d$ and $\hat{r}_k = mean(\widetilde{R}_k) \in R^d$.

Prediction Layer

We take the concatenation of $\hat{c}$ and $\hat{k}$ as the final query vector $x_f = [\hat{c}, \hat{k}] \in R^{2d}$, and the concatenation of $\hat{r}_c$ and $\hat{r}_k$ as the final response vector $y_f = [\hat{r}_c, \hat{r}_k] \in R^{2d}$. Then the final matching score is obtained with dot product operation $g(c, k, r) = x_f \cdot y_f$.

Learning

Let θ denote the parameters of the model and $f(\cdot)$ denote the softmax function. Then, θ would be optimized by minimizing the cross-entropy loss on D, which is defined as:

$$L(\theta) = -\sum_{(c,k,r,y) \sim D} y \log f(g(c, kr)) \qquad (6)$$

4 Experiment and Analysis

4.1 Experimental Settings

We evaluate our BERT-KRS model on two publicly used datasets, Persona-Chat [13] and CMUDoG [14]. For Persona-Chat, we use "revised" and "original" to indicate the different versions of the dataset. Since CMUDoG does not contain negative responses, we randomly sample 19 negative responses for each utterance as [10, 15]. Statistics of these two datasets are shown in Table 2.

Following [3, 10, 15], we utilize $R_n@k$ ($k = 1, 2, 5$) as evaluation metrics with $n = 20$ and $k \in \{1,2,5\}$, that is 1 in n recall at k, it gets 1 if a golden response is positioned in the k selected responses and 0 otherwise for these two datasets, n is equal to 20.

Table 2. Statistics of Persona-Chat and CMUDoG.

Statistics	Persona-Chat			CMUDoG		
	Train	Dev	Test	Train	Dev	Test
Total number of conversations	8939	1000	968	2881	196	537
Total number of turns	65719	7801	7512	36159	2425	6637
Average turns per conversation	7.5	7.8	7.8	13.5	13.3	13.1
Average length per utterance	11.6	11.9	11.8	17.3	18.6	16.9
Average number per knowledge collection	4.5	4.5	4.5	31.8	31.8	31.7
Average length per knowledge sentence	7.3	7.7	7.3	28.3	23.8	23.5

We use uncased BERT$_{base}$ as the basis for further pre-training. In the stage of domain adaptation training, we set the maximum lengths of the conversation, knowledge sequence and batch size as 330, 100 and 14 for Persona-Chat, and 400, 100 and 12 for CMUDoG respectively. In the fine-tuning stage, the default settings about the maximum length of the context, that of the response, that of the knowledge sentence, the maximum number of sentences in a knowledge collection and batch size are 310, 25, 25, 5 and 20 for Persona-Chat, as well as 360, 30, 30, 20 and 12 for CMUDoG. Optimizer uses Adam with the learning rate 3e−5. All experiment results except ours are cited from their original studies.

4.2 Baseline Models

We compare BERT-KRS with the following two kinds of baseline models:

Non-BERT-Based Models

(1) **Starspace** [2], Profile Memory [1] and KV Profile Memory [1] produce the fused representation between the context and knowledge, and calculate the cosine similarities between the fused representation and response candidate.

(2) **DGMN** [15] extracts matching information through a hierarchical interaction mechanism.

(3) **CSN-Sent** and **CSN-word** [16] use a document content selection network (CSN) to perform explicit selection of relevant document contents, and filter out the irrelevant parts from the perspectives of sentence-level and word-level respectively.

(4) **DIM** [3] and RSM-DCK [10] adopt a dual matching architecture, which lets response candidates interactive match with the context and knowledge respectively. RSM-DCK is the state-of-the-art model on Persona-Chat and CMUDoG.

BERT-based Models: Bi-encoder [11] and **Poly-encoder** [11] adopt BERT as the encoder to produce text representations. Both of them match the response candidate with the concatenation of the context and knowledge sentences.

4.3 Results and Analysis

The experiment results are shown in Table 3.

From Table 3, we can see that BERT-KRS significantly outperforms all baseline models over all metrics. Compared with RSM-DCK that is the current best baseline on two datasets, BERT-KRS achieves respectively 2.4% and 4.9% absolute improvements on the original and revised Persona-Chat, and 2.9% absolute improvement on CMUDoG with respect to $R_{20}@1$ metric. For other evaluation metrics, BERT-KRS also obviously outperform RSM-DCK. On all datasets, improvement on the revised Persona-Chat is greater than that on the original Persona-Chat and CMUDoG. It is because BERT-KRS can better capture the semantic similarity relationship between revised profiles and conversations rather than only considering trivial word overlap.

BERT-based models (i.e., Bi-encoder and Poly-encoder) outperform most non-BERT-based models, which indicate that BERT can provide much richer representations and context information for response selection. However, well-designed matching architectures, i.e., DIM and RSM-DCK, outperform BERT-based models, which demonstrate the importance of sufficient interaction. Since BERT-KRS takes full advantages of both BERT and attention-based interaction, it thus consistently outperforms all strong baseline models on both Persona-Chat and CMUDoG over all evaluation metrics, indicating the universality of our proposed model.

Table 3. Evaluation results on the two datasets. Numbers in bold indicate the best performance for the corresponding models on specific metrics.

Models	Persona-Chat						CMUDoG		
	Original Persona			Revised Persona					
	$R_{20}@1$	$R_{20}@5$	$R_{20}@10$	$R_{20}@1$	$R_{20}@5$	$R_{20}@10$	$R_{20}@1$	$R_{20}@5$	$R_{20}@10$
Starspace	49.1	60.2	76.5	32.2	48.3	66.7	50.7	64.5	80.3
Profile Memory	50.9	60.7	75.7	35.4	48.3	67.5	51.6	65.8	81.4
KV Profile Memory	51.1	61.8	77.4	35.1	45.7	66.3	56.1	69.9	82.4
DGMN	67.6	80.2	92.9	58.8	62.5	87.7	65.6	78.3	91.2
DIM	78.8	-	-	70.7	-	-	78.5	88.4	96.4
CSN-sent	77.5	88.8	96.8	70.1	83.4	95.1	70.1	82.5	94.3
CSN-word	78.1	89.0	97.1	71.3	84.2	95.5	69.8	82.7	94.0
RSM-DCK	79.6	90.2	97.4	71.8	84.9	95.5	79.2	88.8	96.6
Bi-encoder	77.0	87.8	96.0	69.1	82.4	93.8	78.0	87.7	96.3
Poly-encoder	77.9	88.8	96.6	69.0	83.0	94.6	66.4	78.0	87.3
BERT-KRS	**82.0**	**91.5**	**97.7**	**76.7**	**88.3**	**96.6**	**82.1**	**91.3**	**97.6**

4.4 Ablation Analysis

We conduct ablation experiments for investigating the effects of domain adaptation and different parts of the matching architecture, as well as the importance of the context and knowledge. The experiment results are shown in Table 4.

Table 4. Evaluation results on the two datasets. Numbers in bold indicate the best performance for the corresponding models on specific metrics.

Models	Persona-Chat						CMUDoG		
	Original Persona			Revised Persona					
	$R_{20}@1$	$R_{20}@5$	$R_{20}@10$	$R_{20}@1$	$R_{20}@5$	$R_{20}@10$	$R_{20}@1$	$R_{20}@5$	$R_{20}@10$
BERT-KRS	**82.0**	**91.5**	97.7	**76.7**	**88.3**	**96.6**	82.1	91.3	97.6
w/o domain adaptation	81.2	91.2	**97.8**	75.8	87.7	96.4	79.6	89.8	97.2
w/o interaction	80.4	90.9	97.6	75.9	87.8	96.4	81.1	90.5	97.5
w/o matching	76.0	87.8	96.3	70.1	83.1	94.7	80.4	90.4	97.3
w/o knowledge	66.4	80.7	92.5	67.8	81.2	93.3	79.3	89.2	96.8
w/o context	47.2	57.7	71.0	32.3	43.2	60.2	54.2	69.8	84.9

From Table 4, we can see that domain adaptation is necessary because removing it would bring obvious performance drops on both datasets over most metrics. Domain adaptation performs better on CMUDoG than on Persona-Chat, because it allows to deal with longer texts on CMUDoG and thus enables the model to capture more complete in-domain information.

When removing the interaction layer or the matching layer, the performance drop significantly, which confirms the necessity and effectiveness of these two layers. We can also observe that the conversation context plays a more important role than the knowledge collection for knowledge-grounded response selection, because the performance drops more when the context is removed. The reasons lie in the fact that conversation contexts provide the primary semantic information for response selection, while the background knowledge offers auxiliary information. It is worth noting that removing the knowledge collection causes more performance degradation on Persona-Chat than on CMUDoG, because the positive response is more closely related to background knowledge on Persona-Chat.

4.5 Effectiveness of Domain Adaptation

To further compare the effects of different domain adaption methods, we conduct experiments on CMUDoG. In the experiment situation, we concatenate context utterances and knowledge sentences to a long sequence and take them as the input of BERT for domain adaptation. Compared with using average sequence length on the Persona-Chat, the sequence length on CMUDoG exceeds the maximum length limitation of BERT, which as result requires truncation and thus might lose some potential information. To address this issue, we propose a data augmentation method.

Concretely, given a conversation context $c = \{u_1, u_2, ..., u_n\}$ and a knowledge collection $k = \{k_1, k_2, ..., k_m\}$, we transform them to $\{U_q\}_{q=1}^{S_c}$ and $\{K_p\}_{p=1}^{S_k}$ respectively, where U_q denotes the concatenation from u_i to u_{i+t}, and Kp denotes the concatenation from k_j to k_{j+d}. s_c and s_k denote the number of utterance segments and knowledge segments respectively. We perform all possible combinations and formulate the input as $x = [[CLS], U_q, [SEP], K_p, [SEP]]$. In this way, we can not only keep the complete conversations and background knowledge, but also can incorporate those training data for domain adaptation. Experiment results are listed in Table 5.

Table 5. Effectiveness of different domain adaption methods on CMUDoG.

Models	CMUDoG		
	$R_{20}@1$	$R_{20}@5$	$R_{20}@10$
BERT-KRS	**82.1**	**91.3**	**97.6**
w/o CD	81.0	91.1	97.3
w/o MLM	79.0	88.7	96.8
w/o DA	79.7	89.7	97.5

As shown in Table 5, we can see that CD and MLM, and the data augmentation method DA are all useful as removing any of them leads to a performance drop with the impact rank of MLM > DA > CD on the $R_{20}@1$ metric. The combination of CD and MLM can encourage the model to take the coherence and consistency relationship between conversation and background knowledge into consideration and thus learn more domain-specific knowledge. When removing DA, the performance drops, indicating that incorporating domain-specific corpus for domain adaptation training is effective, since DA can enable the model to better learn contextual representations and capture task-specific knowledge.

5 Conclusions

We propose a BERT-based model for knowledge-grounded response selection with respect to retrieval-based chatbot by designing a domain adaptation task to yield domain-adaptive BERT for capturing the coherence and consistency relationship between conversations and knowledge, taking the domain-adaptive BERT as the encoder and designing an attention-based matching network to strengthen their interactions before they match with response candidates.

Experimental results on two benchmarks demonstrate that our model outperforms those current strong baseline models and achieves new state-of-the-art results. In the future, we plan to develop a knowledge-grounded response selection model with self-supervised learning for capturing and utilizing more training knowledge and features.

References

1. Zhang, Z.S., Li, J.T., Zhu, P.F., et al.: Modeling multiturn conversation with deep utterance aggregation. In: Proceedings of the 27th International Conference on Computational Linguistics, pp. 3740–3752. Association for Computational Linguistics, Stroudsburg (2018)
2. Wu, Y., Wu, W., Xing, C., et al.: Sequential matching network: a new architecture for multi-turn response selection in retrieval-based chatbots. In: Proceedings of the 55th Annual Meeting of the Association for Computational Linguistics (ACL 2017), pp. 496–505. Association for Computational Linguistic, Stroudsburg (2017)
3. Gu, J.C., Ling, Z.H., Liu, Q.: Interactive matching network for multi-turn response selection in retrieval-based chatbots. In: Proceedings of the 28th ACM International Conference on Information & Knowledge Management, pp. 2321–2324. ACM Press, New York (2019)
4. Lu, J.Y., Zhang, C.B., Xie, Z.Y., Ling, G., Zhou, T.C., Xu, Z.L.: Constructing interpretive spatio-temporal features for multiturn responses selection. In: Proceedings of the 57th Annual Meeting of the Association for Computational Linguisticspp, pp. 267–275. Association for Computational Linguistics, Stroudsburg (2019)
5. Tao, C.Y., Wu, W., Xu, C., et al.: Multi representation fusion network for multi-turn response selection in retrieval-based chatbots. In: Proceedings of the Twelfth ACM International Conference on Web Search and Data Mining, pp. 44–50. ACM Press, New York (2019)
6. Yuan, C.Y., Zhou, W., Li, M.M., Lv, S.W., Hu, S.L.: Multi-hop selector network for multi-turn response selection in retrieval-based chatbots. In: EMNLP-IJCNLP 2019, pp. 111–120. Association for Computational Linguistics, Stroudsburg (2019)
7. Lu, J.Y., Ren, X.C., Ren, Y.Z., Liu, A., Xu, Z.L.: Improving contextual language models for response retrieval in multi-turn conversation. In: Proceedings of the 43rd International ACM SIGIR Conference on Research and Development in Information Retrieval, pp. 1805–1808. ACM Press, New York (2020)
8. Gu, J.C., et al.: Speaker-aware BERT for multi-turn response selection in retrieval-based chatbots. In: Proceedings of the 29th ACM International Conference on Information & Knowledge Management, pp. 2041–2044. ACM Press, New York (2020)

9. Gu, J.C., Ling, Z.H., Zhu, X.D., Liu, Q.: Dually interactive matching network for person-alized response selection in retrieval-based chatbots. In: Proceedings of the 2019 Confer-ence on Empirical Methods in Natural Language Processing, pp. 1845–1854. Association for Computational Linguistics, Stroudsburg (2019)

10. Hua, K., Feng, Z.Y., Tao, C.Y., Yan, R., Zhang, L.: Learning to detect relevant contexts and knowledge for response selection in retrieval-based dialogue systems. In: Proceedings of the 29th ACM International Conference on Information & Knowledge Management, pp. 525–534. ACM Press, New York (2020)

11. Humeau, S., Shuster, K., Lachaux, M., Weston, J.: Poly-encoders: transformer architec-tures and pre-training strategies for fast and accurate multi-sentence scoring. In: Eighth International Conference on Learning Representations. OpenReview.net, Ethiopia (2020)

12. Mazaré, P.E., Humeau, S., Raison, M., Bordes, A.: Training millions of personalized dialogue agents. In: Thirty-Second AAAI Conference on Artificial Intelligence, pp. 2775–2779. AAAI Press, Arizona (2018)

13. Zhang, S.Z., Emily, D., Jack, U., et al.: Personalizing dialogue agents: I have a dog, do you have pets too?. In: Proceedings of the 56th Annual Meeting of the Association for Computational Linguistics, pp. 2204–2213. Association for Computational Linguistics, Stroudsburg (2018)

14. Zhou, K.Y., Prabhumoye, S., Black, A.W.: A dataset for document grounded conversations. In: Proceedings of the 2018 Conference on Empirical Methods in Natural Language Processing, pp. 708–713. Association for Computational Linguistics, Stroudsburg (2018)

15. Zhao, X.L., Tao, C.Y., Wu, W., et al.: A document-grounded matching network for response selection in retrieval-based chatbots. In: the 28th International Joint Conference on Artificial Intelligence, pp. 5443–5449. AAAI Press, Arizona (2019)

16. Zhu, Y., Nie, J.-Y., Zhou, K., Du, P., Dou, Z.: Content selection network for document-grounded retrieval-based chatbots. In: Hiemstra, D., Moens, M.-F., Mothe, J., Perego, R., Potthast, M., Sebastiani, F. (eds.) Advances in Information Retrieval. LNCS, vol. 12656, pp. 755–769. Springer, Cham (2021). https://doi.org/10.1007/978-3-030-72113-8_50

An Assembly Instruction Question Answering System Based on Knowledge Graph

Yu Zhang, Liangzhi Gan[(⊠)], and Xueyu Pi

School of Electrical Engineering and Automation, Jiangsu Normal University, Xuzhou 221116, China
2392852551@qq.com

Abstract. Due to the widespread use of computers, assembly language is indispensable in operating computer software. Learning assembly language is particularly important. However, it requires strong logical and spatial imagination, making it difficult to learn and master. In order to enhance the efficiency of learning assembly language, this paper proposes the use of Chinese language to learn assembly language and designs a Knowledge Graph-based Assembly Language Question Answering (KG-ALQA) System. Taking the assembly language of the 8051 microcontroller as an example, firstly, we construct a knowledge graph of 8051 assembly language. Secondly, We construct a KG-ALQA system that utilizes templates to enable Chinese-to-8051 assembly language question answering (QA) functionality. Finally, to address the problem of template limitations in the KG-ALQA system, We propose a joint Transformer model to enhance the flexibility and accuracy of the system. Experimental results show that utilizing knowledge graph (KG) and deep learning techniques can effectively achieve Chinese-to-8051 assembly language QA functionality.

Keywords: Assembly Instruction · Question Answering System · Knowledge Graph

1 Introduction

The presence of computers is ubiquitous and computer operations rely on programming. Therefore, mastering programming languages is particularly important. As programming languages evolved from low-level to high-level, assembly language, although considered low-level, is closest to machine language. It can directly control the processor, with concise code, high efficiency, and easy manipulation [1]. In addition, skilled programmers proficient in assembly language can use disassembly to uncover bugs that high-level languages may not detect, so learning assembly language is essential. However, many beginners in assembly language lack spatial imagination, analytical skills, solid theoretical knowledge, and a thorough understanding of assembly language. Additionally, they lack practical experience, leading to poor learning outcomes [2]. We propose a method of generating assembly language code from natural language, converting Chinese into assembly language. By utilizing the expressive nature of Chinese, learners can focus on the logic and structure of assembly language.

© The Author(s), under exclusive license to Springer Nature Singapore Pte Ltd. 2024
D.-S. Huang et al. (Eds.): ICAI 2023, CCIS 2015, pp. 322–334, 2024.
https://doi.org/10.1007/978-981-97-0827-7_28

In the early stages, template-based code generation techniques were employed to generate repetitive code. However, due to the limited flexibility of template-generated code, researchers explored the use of traditional statistical machine translation (SMT) techniques to generate code. With the rapid development of artificial intelligence, attention shifted towards neural network-based automatic code generation. Yin [3] proposed a sequence-to-sequence model that generates abstract syntax trees, using tree structures as intermediate states in the code generation process to convert natural language into syntactically correct code. OpenAI introduced the Copilot [4] model, which uses natural language descriptions as comments to optimize code generation through docstrings and function name tuning. The model expands source text information using specialized blank tokens to generate the most relevant code. These methods are suitable for generating common and frequently used code, but they have limitations in specific domains.

To achieve code generation from Chinese to assembly language, this paper proposes treating Chinese and 51 assembly language as "QA pairs" and designs a QA system that takes Chinese input questions and generates assembly language output. To address the challenge of varied Chinese expressions with similar meanings, this paper suggests structurally storing 51 assembly language in a knowledge graph, serving as the knowledge source for the QA system. This knowledge graph represents addresses, registers, and other elements involved in assembly language as nodes, with data transfer relationships as edges, and instruction symbols as supplementary node attributes, enabling the QA system to accurately understand input statements based on capturing key information, irrespective of the expression format. Additionally, to overcome limitations in knowledge graph-based QA systems due to the size of the knowledge base, this paper trains a Transformer-based machine translation model, combining two models to achieve flexible and efficient QA between Chinese and 8051 assembly language.

2 Related Work

Converting Chinese into assembly language through a QA approach can be seen as a form of code generation. As programming languages developed, understanding and using them became a long-standing question for individuals seeking to work in related fields. Researchers started exploring how to convert natural language into programming languages or vice versa. Generating assembly language from natural language requires accurate understanding of the content and intent of the natural language description. This aligns with the idea of QA systems, which analyze the semantics of questions to understand user intent and provide correct answers. Therefore, this paper adopts a QA form to achieve the conversion from Chinese to assembly language and designs a QA system based on the assembly language knowledge graph.

Early QA systems, also known as expert systems, focused on specific domains and mainly relied on rule-based and template matching methods to answer domain-related questions. However, due to the limitations of rules and templates, these systems had low utilization of information and limited applicability. The emergence of the internet brought new breakthroughs to QA systems. Researchers utilized a vast amount of high-quality data from the internet for QA systems, which can be classified into document-based QA and knowledge-based QA, based on the source of information [5]. Document-based QA systems retrieve answers by searching network documents, while knowledge-based QA systems rely on high-quality knowledge graphs.

Since Google introduced the concept of knowledge graphs in 2012, there have been many large-scale, high-quality knowledge graphs, such as Freebase and DBpedia in English and domain-specific knowledge graphs in Chinese medicine. The accurate data and efficient retrieval capabilities of knowledge graphs have made knowledge-based QA systems a popular research direction. Fader [6] introduced a small set of templates to answer simple questions involving a single triple. For complex questions, Yin [7] proposed a pipeline approach to generate query graphs in multiple stages. Frankenstein [8] analyzed over 60 knowledge-based QA systems and proposed a pipeline framework based on knowledge graphs, which can be improved using neural network techniques for any module within the framework to enhance the overall performance of the QA system. Nowadays, knowledge-based QA systems have been widely applied in search engines like Baidu and Bing, as well as intelligent voice assistants like Siri and Xiaodu.

3   Method

3.1   Construction and Storage of Knowledge Graphs

The construction techniques of knowledge graphs can be divided into two main approaches: top-down and bottom-up [9]. There are three types of raw data: structured, semi-structured, and unstructured. Entities and relationships in a KG are typically expressed in triple form and stored in resource description frameworks (RDF) or graph databases. This paper adopts the top-down construction technique, extracting entity information related to assembly language from reference materials and manually annotating triples. It organizes the unstructured Chinese sentences describing the functionality of 51 assembly instructions into structured data, serving as the corpus for constructing the assembly language KG, which is stored in Neo4j.

Triples have two basic forms: (Entity1, Relationship, Entity2) and (Entity, Attribute, Attribute Value). Based on the characteristics of assembly language and the Chinese descriptive statements, we employ the triplet form of (Entity1, Relationship, Entity2). In this form, Entity1 represents the head entity, Entity2 represents the tail entity, and the Relationship represents the connection between the two entities. In constructing the corpus for the KG of 51 assembly language, various registers or memory locations, i.e., operands in the instructions, are treated as entities, and the data transfer relationship, i.e., the opcode, is considered as the relationship between entities. For example, given the sentence "Copy the content of register R0 to the accumulator" the entities in the sentence are "register R0" and "accumulator" and the relationship between the entities is "copy". Therefore, the triple form of this sentence is ("register R0", "copy", "accumulator"). An example of a partial knowledge graph dataset is shown in Table 1.

Due to the uniqueness of each entity and relationship, specific attributes can be extended to accurately identify relevant entities and relationships from the text. Utilizing their attribute information helps enhance the model's reasoning and comprehension capabilities. Some entity attributes and relationship attributes are presented in Table 2 and Table 3, respectively.

By organizing the knowledge graph corpus, this paper stores it in Neo4j. Neo4j is a type of graph database that enables the storage, management, and visualization of entities and relationships in a KG. Additionally, the Cypher language is utilized to query and

Table 1. Partial Example of a Knowledge Graph Dataset

Head Entity	Relationship	Tail Entity
register R0	input/send/copy	accumulator
direct address unit 01H	input/send/copy	accumulator
immediate operand #03H	input/send/copy	accumulator
accumulator	input/send/copy	direct address unit 04H
direct address unit 33H	input/send/copy	indirect address register R0
external data memory	input/send/copy/MOVX	accumulator
indirect address register @R0	half-byte swapping/XCHD	accumulator
direct address unit 00H	exchange/swap/XCH	accumulator
register R3	add/append	accumulator

Table 2. Partial Entity Attribute

ID	Operand	Symbol	Addressing	Type
1	accumulator	A	register	internal RAM
2	register R0	R0	register	internal RAM
3	direct address unit 00H	00H	direct	internal RAM
4	indirect address register @R0	@R0	register	internal RAM
5	immediate operand 00H	#00H	immediate	internal RAM
6	address register	DPTR	register	internal RAM
7	external data memory	@DPTR	register indirect	external RAM

Table 3. Partial Relationship Attribute

ID	Opcode	Symbol	Type
1	input/send/copy	MOV	internal RAM data transfer instruction
2	input/send/copy/MOVX	MOVX	external RAM data transfer instruction
3	input/send/copy/MOVC	MOVC	program storage data transfer instruction
4	swap the high four digits with the low four digits	SWAP	half-byte data exchange instruction
5	half-byte swapping/XCHD	XCHD	half-byte data exchange instruction
6	exchange/swap/XCH	XCH	byte data exchange instruction
7	plus one	INC	increment instruction

search the KG stored in Neo4j, serving as an important component in the design of the QA system in this paper. Figure 1 illustrates the stored results of the knowledge graph for partial 51 assembly instructions in Chinese. In the figure, nodes represent operands, and edges represent data transfer relationships between nodes.

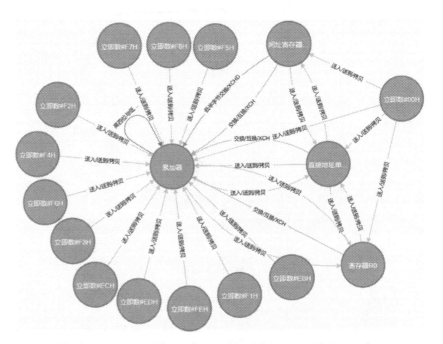

Fig. 1. Chinese Knowledge Graph of Partial 51 Assembly Instructions

3.2 Knowledge Graph-Based Assembly Language Question Answering (KG-ALQA) System

Different from QA systems that typically handle interrogative statements ending with question marks, such as "what", "why", "how", etc., the designed KG-ALQA system in this paper accepts declarative statements as input. For example, input statements like "Transfer the content in register R0 to the accumulator and convert it into a 51 assembly instruction" or simply "Transfer the content in register R0 to the accumulator" are expected to yield a response such as "MOV A, R0" which is the corresponding 51 assembly instruction. This approach is similar to machine translation. Therefore, in this paper, the question and answer modeling in the QA system is treated as a translation process. The symbolic representation of the assembly instruction's opcode and operand is considered as the attributes of nodes and edges in a KG. By parsing and retrieving the KG, the input statement is translated into the corresponding symbols, forming a correct expression, which serves as the output answer statement, i.e., the assembly instruction.

The QA system designed in this paper consists of four main components: question preprocessing, question classification, template matching, and knowledge base querying.

As shown in Fig. 2. The process of achieving question answering involves parsing the input statement, extracting entities and keywords from the question, matching the question keywords to corresponding question templates, retrieving relevant nodes and edges within the KG, and finally employing the appropriate answer template to form a correct expression that outputs the assembly instruction.

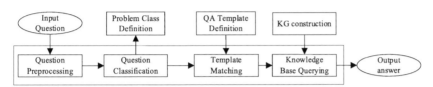

Fig. 2. Diagram of the KG-ALQA System

The question preprocessing component involves parsing the input statement to determine whether it contains question keywords and relevant entities. The format of assembly instructions is [label:] opcode [operand][,operand][,operand], where the opcode specifies the specific operation and nature of the instruction, serving as an essential component in an instruction. The operand can represent either a specific data value or the address where data is stored. There can be multiple operands or none at all in a single instruction. Therefore, based on the characteristics of assembly instructions, this paper treats the opcodes which represents the functionality as keyword phrases and the operands as entities. It constructs separate dictionaries for opcodes and operands that cover all the instructions. These dictionaries are used for keyword and entity detection to preprocess the input statements.

The question classification process involves categorizing the input statement based on the recognized question keywords and entities. The classification of questions primarily relies on the symbols of opcodes, where different opcodes represent different question categories. It is important to note that the symbols of opcodes are influenced not only by the type and addressing mode of the operand but also by the natural language expression of the opcode itself. For example, different natural language expressions such as "copy", "transfer" or "move" correspond to the same opcode symbol "MOV". Therefore, question classification needs to consider both the semantic meaning of the words and the type of operand. In this paper, a dictionary-based question classifier is constructed, defining over a hundred question types. When the question keywords and operand types meet the requirements, the question will be automatically assigned to the corresponding question category (Table 4).

Template matching involves setting question answering templates based on question categories and the characteristics of assembly language. These templates are used to match query results, thereby generating correct statements that comply with the requirements of assembly instruction syntax and format, which implement the functionality of system that utilizes natural language queries to access 51 assembly instructions. To enhance the general applicability of the templates, this paper labels all entities as "operand", assigns semantic entity identifiers and utilizes relevant entity attributes to establish question answering templates (Table 5).

Table 4. Question Categories (Partial)

Question Categories	Question Keywords	Operand Type
MOV_INRAM	input	internal RAM
	send	internal RAM
	copy	internal RAM
MOVX_OUTRAM	input	external RAM
	send	external RAM
	copy	external RAM
MOVC_OUTROM	input	external RAM
	send	external RAM
	copy	external RAM
SWAP_INRAM	swap the high four digits with the low four digits	internal RAM
	exchange the low four bits with the high four bits	internal RAM

Table 5. Question Templates (Partial)

Question Categories	Question Instance	Answer Templates
MOV_INRAM	copy the contents from register R0 (operand1) to the accumulator (operand2)	MOV opeand2.symbol,operand1.symbol
MOVX_OUTRAM	copy the contents from the external RAM's indirect address register R0 (operand1) to the accumulator (operand2)	MOVX operand2.symbol,operand1.symbol
MOVC_OUTROM	copy the contents from the addressing unit (operand1) with DPTR as the base address and A as the displacement to the accumulator (operand2)	MOVC operand2.symbol,operand1.symbol
SWAP_INRAM	swap the upper four bits of the accumulator (operand1) with the lower four bits of data	SWAP operand1.symbol

Knowledge base queries involve searching within the knowledge graph using Neo4j's query language, Cypher. For each question type, corresponding Cypher statements have been defined as query templates. These templates are used to search for nodes and edges that meet the criteria, and retrieve their properties. Finally, utilize the py2neo toolkit, which connects Cypher with Python, to return the query results. The retrieved results are then populated based on the answer templates to generate the final 51 assembly instructions.

3.3   Joint Knowledge Graph and Transformer-Based Assembly Language Question Answering (KGT-ALQA) System

In the previous section, when the question is contained within the KG, the KG-ALQA system performs exceptionally well. It has fast response times and can provide clear and accurate answers to assembly instructions. However, the KG is primarily constructed through manual annotation and has limited capacity. It may face difficulties in handling flexible and varied natural language inputs, meaning that it may not be able to retrieve answers when the input statement cannot be correctly recognized. To address the limitations of the KG-ALQA system, we consider combining the KG with a Transformer-based machine translation model [10]. By leveraging the deep learning and strong generalization capabilities of neural networks, the entire model can improve its accuracy and flexibility when dealing with complicated natural language inputs.

The KGT-ALQA system consists of two question answering methods. In most cases, the KG-ALQA system is used to retrieve answers. However, when the QA system encounters a recognition failure and outputs "query failed, please retype", the trained Transformer-based machine translation model is invoked to parse the input statement and obtain the question answering results. The question answering process of the dual-model is illustrated in Fig. 3.

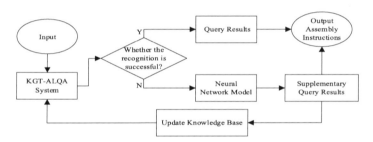

Fig. 3. Dual-model question answering process.

The Transformer-based machine translation model is constructed using the fundamental structure of the Transformer, employing an encoder-decoder framework. It consists of six encoder layers and six decoder layers, each layer comprising multi-head attention mechanisms and fully connected layers. The encoder is responsible for reading the word vector sequence of the input sentence, along with corresponding positional encodings, to obtain hidden vectors. The decoder takes the hidden vectors from the encoder and word vectors of the assembly instructions as inputs, decoding the dependency relationship between the input sentence and the assembly instructions. After computation, the decoder's output is passed through a linear layer to restore the word vectors and utilizes a softmax layer to calculate the positional weights of each symbol in the assembly instructions. Finally, the model employs a greedy search algorithm to select the symbol with the highest weight as the question and answering result.

The question answering results obtained through the Transformer model can compensate for the shortcomings of KG queries. On the other hand, the knowledge-based approach can accurately identify entities and relationships in complex sentences, addressing the high processing cost and potential loss of semantic information in the Transformer model when dealing with syntactically complex statements. Therefore, in this paper, the question answering statements obtained through the Transformer model are analyzed and processed using the techniques described in Sect. 3.1 to extract entities and attributes. These are then used to supplement and update the KG's knowledge base, enhancing the answering capability of the KG-ALQA system. Experimental results confirm that this combined question answering approach, which supplements knowledge graph query results with a Transformer model, expands the knowledge base and addresses the high cost and low efficiency of using a Transformer model through the knowledge graph. It achieves good performance in natural language and assembly language question answering tasks.

Table 6. Partial Bilingual Sentence Pairs Dataset

Source Sentence in Chinese	Target Sentence
copy the contents from register R0 to the accumulator	MOV A,R0
copy the data from direct address unit 00H to the accumulator	MOV A,00H
move the immediate value 03H into the accumulator	MOV A,#03H
move the data from indirect RAM to the direct address unit	MOV direct,@Ri
move the 8-bit immediate value to the indirect RAM unit	MOV @Ri,#data8
exchange the lower nibble between indirect RAM and the accumulator	XCHD A,@Ri
swap the contents between the register and the accumulator	XCH A,Rn
copy the contents from the addressing unit with DPTR as the base address and A as the displacement to the accumulator	MOVC A,@A + DPTR
add the content of indirect RAM with carry to the accumulator	ADDC A,@Ri

4 Experiments

4.1 Dataset

To better validate the effectiveness of the model, in addition to the Transformer model, a classic Seq2Seq model is also used for comparative experiments. For these neural network models, we construct a bilingual sentence dataset consisting of Chinese and 51 assembly language for training and validation purposes. The construction of bilingual sentence pairs involves using the functional description of 51 assembly instructions as the source sentence in Chinese and the corresponding 51 assembly instructions as the target sentence. To expand the dataset, we replace similar words in the Chinese source sentences based on the similarity of Chinese vocabulary, focusing on the 174 basic instructions. Additionally, we include all registers or address units related to certain instructions to form bilingual sentence pairs, further enhancing model training. The training set contains 1500 samples, and the test set contains 500 samples. The bilingual sentence dataset is shown in Table 6.

4.2 Experimental Environment and Parameter Settings

The experiments in this paper primarily rely on the Pytorch deep learning framework, with an RTX3050 GPU, Cuda11.0, and Windows 11 operating system. The parameter settings for the neural network model are shown in Table 7.

Table 7. Model Parameters

Parameter Name	Value
word vector dimension	256
learning rate	0.001
Dropout	0.1
Batch Size	1
optimizer	Adam
iterations	50

4.3 Experimental Results and Analysis

In this paper, comparative experiments were conducted on a KG-ALQA system, a Transformer-based assembly instruction translation system, and a Seq2Seq-based assembly instruction translation system. The experimental effectiveness of a KGT-ALQA system was also tested. The training processes for the Transformer model and the Seq2Seq model are illustrated in Fig. 4. After approximately 35 iterations, the loss rate of the Transformer model training stabilized, reaching close to 0. On the other hand, the loss rate of the Seq2Seq model training stabilized at around 57% after 30 iterations. Obviously, the Transformer model exhibited better convergence in training compared to the Seq2Seq model.

As this study involves language conversion between two languages, similar to machine translation, the performance of the three models was evaluated using the BLEU metric to assess the quality of generating 51 assembly instructions. The BLEU score ranges between 0 and 1. The results are presented in Table 8.

From Table 8, it can be observed that compared to the Seq2Seq model, the Transformer model exhibits lower loss rates and higher BLEU scores. Even the KG-ALQA system achieves a BLEU score of 100%, indicating that the generated assembly instructions are completely accurate. This can be attributed to the question answering template approach employed by the KG-ALQA system. When the input question correctly matches the question answering template, the assembly instruction response obtained through knowledge graph querying will yield the correct output format, which is in

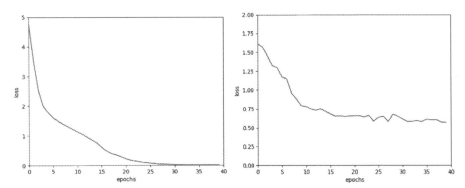

Fig. 4. Training Process of Transformer and Seq2Seq Models. The left image shows the training process of the Transformer model, while the right image depicts the training process based on the Seq2Seq model.

Table 8. Comparative Results of the Models

	Transformer	Seq2Seq	KG-ALQA
Loss(%)	0.706	57.03	
BLEU(%)	97.27	68.6	100

contrast to neural network models that are more susceptible to semantic and syntactic influences.

Based on the results of experimental comparisons, the Transformer model outperforms the Seq2Seq model, hence combining the knowledge graph and Transformer model to implement Chinese-to-assembly instruction question answering. This allows the system to generate answers with correct formats and content even when faced with different expressions conveying the same semantic meaning. The effectiveness of the question answering is illustrated in Fig. 5.

```
model init finished ......
```
输入：将直接地址单元5EH中的内容拷贝到累加器中
回答： MOV A,5EH
输入：外部数据存储器中的内容送入累加器
回答： MOVX A,@DPTR
输入：将以DPTR为基址、A为变址的寻址单元的内容送入累加器
回答： MOVC A,@A+DPTR
输入：将直接地址单元5EH和累加器中的内容交换
回答： XCH A,5EH
输入：将累加器的高四位和低四位数据进行交换
回答： SWAP A
输入：将间址寄存器R0与累加器进行低半字节交换
回答： XCHD A,@R0
输入：累加器的内容加一
回答： INC A

Fig. 5. Effectiveness of the Question Answering System

5 Conclusion

To address the difficulties in understanding the logical structure and the low learning efficiency of assembly language, this paper proposes a knowledge graph-based question and
answering system for assembly instructions that utilizes natural language descriptions of
the functional content. Additionally, it combines the use of a Transform-based assembly
instruction translation model to complement the limitations of the question-answering
system in terms of knowledge base size and question-answering templates. Experimental results demonstrate that the proposed model effectively achieves the conversion from
Chinese to assembly language, ensuring both the correct format of assembly instructions
and accurate symbol representation.

Furthermore, the approach of using natural language for program writing has gar-
nered extensive attention from researchers, with large-scale language models currently
being employed to facilitate the conversion between different natural languages and
programming languages. This paper aims to design a lightweight language model for
converting Chinese into assembly instructions, while also preparing it for future integration into assembly language compilation software. In the future, there are plans to
expand the variety of programming languages and the scale of datasets, and to enhance
the model's structure and parameters, in order to develop a natural language model that
is applicable to a wider range of programming languages.

References

1. Shen, X., Sun, K., Ju, J.: Computer programming languages and their applications. Electron.
   Technol. **51**(09), 12–14 (2022)
2. Jiang, S.: A brief discussion on the implementation plan of the course "51 single-chip
   microcontroller technology and applications." Inf. Syst. Eng. **5**, 161–164 (2022)
3. Yin, P., Neubig, G.: A syntactic neural model for general-purpose code generation. arXiv
   preprint arXiv:1704.01696 (2017)

4. Pearce, H., Ahmad, B., Tan, B., et al.: An empirical cybersecurity evaluation of GitHub copilot's code contributions. ArXiv abs/2108.09293 (2021)
5. Guo, Z.: Knowledge-Based Question Answering System in the Field of Artificial Intelligence. Dalian University of Technology (2022)
6. Fader, A., Zettlemoyer, L., Etzioni, O.: Paraphrase-driven learning for open question answering. In: Proceedings of the 51st Annual Meeting of the Association for Computational Linguistics (Volume 1: Long Papers), pp. 1608–1618 (2013)
7. Yih, S.W., Chang, M.W., He, X., et al.: Semantic parsing via staged query graph generation: question answering with knowledge base. In: Proceedings of the Joint Conference of the 53rd Annual Meeting of the ACL and the 7th International Joint Conference on Natural Language Processing of the AFNLP, pp. 1321–1331 (2015)
8. Wu, P., Zhang, X., Feng, Z.: A survey of question answering over knowledge base. In: Knowledge Graph and Semantic Computing: Knowledge Computing and Language Understanding: 4th China Conference, CCKS 2019, Hangzhou, China, 24–27 August 2019, Revised Selected Papers 4, vol. 1134, pp. 86–97. Springer, Singapore (2019). https://doi.org/10.1007/978-981-15-1956-7_8
9. Zhang, J., Zhang, X., Wu, C., et al.: Survey on knowledge graph construction techniques. Comput. Eng. 48(03), 23–37 (2022)
10. Tan, Z., Wang, S., Yang, Z., et al.: Neural machine translation: a review of methods, resources, and tools. AI Open 1, 5–21 (2020)

Improving Natural Language Inference with Residual Attention

Shanshan Yu[1], Jindian Su[2(✉)], Xiaobin Ye[3], and Dandan Ma[3]

[1] College of Medical Information Engineering, Guangdong Pharmaceutical University,
Guangzhou 510006, Guangdong, China
[2] College of Computer Science and Engineering, South China University of Technology,
Guangzhou 510641, Guangdong, China
sujd@scut.edu.com
[3] Guangdong Unicomm, Guangzhou 440100, China

Abstract. Natural language inference (NLI) aims to identify the logical relationship between a premise and a corresponding hypothesis, which requires the model should have the ability of effectively capturing their semantic relationship. Most of the existing transformer-based models tend to concatenate the premise and hypothesis together as the input of the model and capture their relationship through multi-head self-attention mechanism, which as a result might only consider their plain context-sensitive relationship and neglect the potentially mutual impacts of their contextual semantics. To better model the relationship between the premise and hypothesis, we propose a new transformer-based model RAN4NLI that consists of a sequence encoder based on pre-trained language model for encoding the input semantics and an interaction network based on residual attention for further capturing their relationship. We utilize residual attention for combining multi-head self-attention and cross-attention information so as to strengthen the potential semantic relationship between the premise and hypothesis. Experiments conducted on two canonical datasets, SNLI and SciTail, demonstrate that our RAN4NLI achieves comparable performance with other strong baseline models.

Keywords: Natural Language Inference · Natural Language Processing · Pre-trained Language Model

1 Introduction

Natural language inference (NLI) is an important and challenging task in the field of natural language processing (NLP), aiming to automatically identify the logical relationship (e.g., entailment, neutral or contradiction) between a premise and a corresponding hypothesis [1]. It can be also regarded as a sentence-pair classification task that consists of a premise and a hypothesis. In the past few years, NLI has received more and more attention from researchers due to the rapid developments of deep learning technologies and its wide applications in various fields, e.g., search engine, dialogue robots and knowledge reasoning. Previous works mainly focus on using traditional neural networks,

D.-S. Huang et al. (Eds.): ICAI 2023, CCIS 2015, pp. 335–346, 2024.
https://doi.org/10.1007/978-981-97-0827-7_29

i.e., long short-term network (LSTM), convolutional neural network (CNN) and graphical network, to offer independent fine-grained sentence encodings for the premise and hypothesis, and then design task-specific models for further capturing their semantic relationship, especially with attention mechanism to jointly interpret and align the premise and hypothesis. Some typical works include TBCNN [2], ESIM [3], DR-BiLSTM [1], HBMP [4] and BiLSTM-Max [4].

Recently, some scholars have begun to build NLI models based on pre-trained language models, e.g., BERT [5], RoBERTa [6] and achieved significant performance improvements. Compared with previous works that use common neural networks and statically pre-trained word embeddings, pre-trained language-based models can offer better contextual sentence representations for premise and hypothesis, as well as their inner- and intra-relation through pre-training and fine-tuning on the downstream task-specific data. However, these works tend to concatenate the premise and hypothesis together as the input of the model, and capture their relationship through multi-head self-attention (MHA) mechanism based on masked language model (MLM), as well as other auxiliary pre-training objectives, i.e., next sentence prediction (NSP) or sequence order prediction (SOP). As a result, the relationship between the premise and hypothesis might not be fully exploited, since they paid more attention to the word-to-word relationship and superficial next or sequential relationship. For example, the meanings of each word in the premise might vary with respect to different hypothesis context, and vice versa for that of the hypothesis with respect to premise. In order to further explicitly model the relationships between the premise and hypothesis during the encoding and inference processes and prevent the network from the loss of relevant, contextual information, we propose a new transformer-based inference model based on pre-trained language model and residual attention. We demonstrate that the output sequence embeddings of pre-trained language model still contain some potentially important relationship information about the premise and hypothesis, and explicitly modeling them can obtain performance improvement gains.

The contributions of this paper can be summarized as follows:

(1) We propose a transformer-based model RAN4NLI for NLI that consists of a sequence encoder for encoding sentence representations and an interaction network for explicitly modeling the relationship between the premise and hypothesis.
(2) We use residual attention to exploit relevant and contextual information between the premise and hypothesis by combining the self-attention and cross-attention information.
(3) We conduct experiments on two widely used datasets, SNLI [7] and SciTail [8], and the results demonstrate that our model achieves comparable performance with strong baseline models without additional pre-training or external knowledge.

2   Related Works

One of the common practice for earlier works is using sequence-based recurrent neural network (i.e., LSTM, GRU) [1, 4, 9, 10] or convolutional neural network as encoders to build independent sentence-level representations for the premises and hypothesis based on the pre-trained distributive expressions, and combine them through a classifier. For example, Samuel [9] and Conneau [10] tried to improve sentence-level representations of premises and hypothesis by using stack-augmented parser-interpreter neural network (SPINN) and bidirectional LSTM (BiLSTM) with max pooling. Ghaeini [1] proposed a novel dependent reading bidirectional LSTM network (DR-BiLSTM) to efficiently model the relationship between a premise and a hypothesis during encoding and inference. They also introduced a sophisticated ensemble strategy to combine the models for final predictions. Aarne [4] proposed a hierarchy-like structure HBMP by taking BiLSTM layers with max pooling as the sentence encoder for iterative refinement. Another common practice is treating the hypothesis and premise sentences together and focusing on capturing their relationship through some special mechanisms, e.g. cross-sentence attention [3, 11]. They put more emphasis on the inference relations rather than on the internal semantics of the sentences.

In the past few years, more and more scholars have begun to apply various pre-trained language models into NLI task and gained significant improvements over the previous works that based on staticly pre-trained word embeddings. Parts of these works tend to directly replace the static word embeddings with contextualized ones and traditional neural networks with transformer-based ones. Both BERT [5], RoBERTa [6] have shown excellent performances on SNLI only through pre-training and fine-tuning, without relying on any additional task-specific networks.

Some other works try to design more complicated strategies (i.e., multi-task learning, adversarial training) or incorporate external knowledge. For example, Liu [12] proposed a multi-task deep neural network (MT-DNN) model based on BERT, while Zhang [13] propose a semantics-aware BERT (SemBERT) by incorporating explicit contextual semantics over a BERT backbone. Liu [12] proposed a general adversarial algorithm ALUM for training large neural language models, which regularizes the training objective by applying perturbations in the embedding space that maximizes the adversarial loss. Gajbhiye [14] present a new model for NLI, External Knowledge Enhanced BERT (ExBERT), to enrich the contextual representation with real-world commonsense knowledge from external knowledge sources and enhance BERT's language understanding and reasoning capabilities. Pilault [15] proposed a transformer-based architecture CA-MTL that consists of a new conditional attention mechanism as well as a set of task conditioned modules for facilitating weight sharing.

On the whole, the existing works have already verified that pre-trained language models can capture the relationship between the premise and hypothesis well with the help of their pre-training capabilities (i.e., MLM, NSP or SOP) and MHA. But we argue that it is still far from enough, because NSP or SOP only consider the next or sequential relationship of sentences, while MHA mainly focuses on the word-to-word relationship, which as a result might lose some potential semantic relationship information between the premise and hypothesis. Different from the above described works, we want to further investigate and analyze how to utilize MHA for better modeling the relationship between the premise and hypothesis based on the sequence output of pre-trained language models, instead of designing task-specific networks or incorporating external knowledge.

3 Our Method

3.1 Overview

In this section we propose a residual attention network-based model for NLI, named RAN4NLI for short. The overall structure of RAN4NLI is demonstrated in Fig. 1.

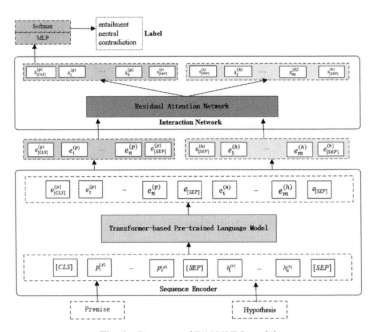

Fig. 1. Structure of RAN4NLI model.

Let $p = [p_1, ..., p_n]$ and $h = [h_1, ..., h_m]$ be a given premise consisting of n tokens and a hypothesis consisting of m tokens respectively, then the NLI task is to predict a label y that indicates the logical relationship between the premise p and hypothesis v, which is usually a triple label set $y = \{$entailment, contradiction, neutral$\}$ or $y = \{$entailment, neutral$\}$ in most of the relevant works.

3.2 Subsequence Encoder

We concatenate the premise $p = [p_1, ..., p_n]$ and hypothesis $h = [h_1, ..., h_m]$ as a pair of sentences and denote as $X = [CLS]p[SEP]h[SEP]$, where [CLS] and [SEP] are two special symbols padded to the beginning and the end of an input sequence respectively. Let *Transformer* denote the transformer encoder in the pre-trained language model that we use for sentence initializations and $E_X = \{e^{(p)}_{[CLS]}, e^{(p)}_1, ..., e^{(p)}_{[SEP]}, e^{(h)}_1, ..., e^{(h)}_{[SEP]}\} \in R^{L \times d}$ denote the corresponding sequence hidden embeddings of Transformer encoder from the last layer, as shown in Eq. 1. $L = n + m + 3$ is the total length of X, which is restricted to some maximum value (i.e. 512 or 1024). d is the dimension of the embeddings and usually set to 768 or 1024.

$$X = [CLS]p_1 \ldots p_n [SEP]h_1 \ldots h_m [SEP], E_X = Transformer(X) \qquad (1)$$

In many cases, the final hidden state corresponding to [CLS] is directly used as the aggregate sequence representation for final inference. To further explore the relationship between the premise and hypothesis, we split the hidden embedding sequence E_X into two parts, $E_X^{(p)} \in R^{(n+2) \times d}$ and $E_X^{(h)} \in R^{(m+2) \times d}$, that correspond to the premise p and hypothesis h, as shown in Eq. 2.

$$E_X^{(p)} = \{e_{[CLS]}^{(p)}, e_1^{(p)}, \ldots, e_n^{(p)}, e_{[SEP]}^{(p)}\}, E_X^{(h)} = \{e_{[CLS]}^{(h)}, e_1^{(h)}, \ldots, e_m^{(h)}, e_{[SEP]}^{(h)}\} \qquad (2)$$

Obviously, $E_X^{(p)}$ and $E_X^{(h)}$ encode the contextual semantic information for p and h respectively, as well as their superficial relationship via self-attention.

3.3   Interaction Network

After yielding the respective contextual embeddings of the premise and hypothesis, we further explore their relationship through an interaction network based on residual attention and the fully-connected feed forward layer.

Residual Attention Layer

The detailed structure of the residual attention layer is shown in Fig. 2. In each layer, we use two independent residual attention parts respectively to explore the mutual impacts of hypothesis as a sequence on the premise, and vice versa.

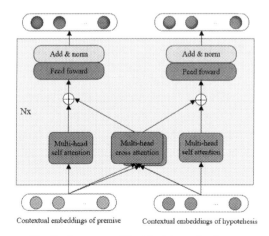

Fig. 2. Structure of Residual Attention Layer.

The residual attention layer consists of a multi-head self-attention module and a multi-head cross-attention module. We omit extensive formulations of the multi-head self-attention and recommend readers to get the details from [16]. We pay more attention to the cross-attention formulation [15].

Given a query $Q \in R^{d_k}$ and a key $K \in R^{d_k}$ of dimension d_k, and a value $V \in R^{d_v}$ of dimension d_v, we compute the dot products of the query with all keys, divide each by $\sqrt{d_k}$, and apply a softmax function to obtain the weights as Eq. 3 shows:

$$Attention(Q, K, V) = softmax\left(\frac{QK^T}{\sqrt{d_k}}\right) \tag{3}$$

Besides using traditional self-attention to calculate the similarities of the premise sentence itself, we also employ cross-attention mechanism to consider the similarities of each word in the premise with the contextual embeddings $E_X^{(h)}$ by taking $E_X^{(h)}$ as the key and the value, which is shown in Eq. 4.

$$S_s^{(p)} = Attention(E_X^{(p)}, E_X^{(p)}, E_X^{(p)}), S_c^{(p)} = Attention(E_X^{(p)}, E_X^{(h)}, E_X^{(h)}), S_X^{(p)} = S_s^{(p)} \oplus S_c^{(p)} \tag{4}$$

Similarly, we can yield the final representations of hypothesis as Eq. 5 shows.

$$S_s^{(h)} = Attention(E_X^{(h)}, E_X^{(h)}, E_X^{(h)}), S_c^{(h)} = Attention(E_X^{(h)}, E_X^{(p)}, E_X^{(p)}), S_X^{(h)} = S_s^{(h)} \oplus S_c^{(h)} \tag{5}$$

In this way, we can capture the relationship between the premise and hypothesis more comprehensively and deeply based on their plain context-sensitive embeddings output from transformer encoder in the pre-trained language model.

Feed Forward Layer

The contextual embeddings $S_X^{(p)}$ and $S_X^{(h)}$ from the residual attention layer are then fed to a feed forward layer that consists of a nonlinear transformation layer and a normalization layer *LayerNorm*, as shown in Eq. 6. W_1, W_2, W_3 and W_4 are trainable parameter matrices, b_1, b_2, b_3 and b_4 are the corresponding trainable bias values.

$$FFN(S_X^{(p)}) = LayerNorm(GeLU(S_X^{(p)}W_1 + b_1)W_2 + b_2)$$

$$FFN(S_X^{(h)}) = LayerNorm(GeLU(S_X^{(h)}W_3 + b_3)W_4 + b_4) \tag{6}$$

Finally, we select $S_{[CLS]}^{(p)}$ corresponding to [CLS] from the output $FFN(S_X^{(p)})$ as the aggregate sequence representation for final inference via a simple multi-layer perceptron (MLP) with a softmax function.

3.4 Loss

Following other works, we use the cross-entropy loss function as shown in Eq. 7 for RAN4NLI and train it to minimize the loss $Loss(\theta)$ given the input X.

$$Loss(\theta) = -\frac{1}{M} \sum_{i=1}^{M} \sum_{j=1}^{N} y_{ij} \, log(\hat{y}_{ij}) \tag{7}$$

where M denotes the sample number of each batch and N denotes the total number of labels in y. $\hat{y}_{ij}$ is the predicted label of the model and y_{ij} is the corresponding ground-truth label for the i_{th} input sentence-pair.

4 Experiment and Analysis

4.1 Datasets

We evaluate our models on two classical NLI datasets, Stanford Natural Language Inference (SNLI) [7] and SciTail [8].

SNLI is one of the most widely used entailment datasets that contains 570k human annotated sentence pairs, in which the premises are drawn from the captions of the Flickr30 corpus and hypotheses are manually composed for each relationship class (entailment, neutral, contradiction, and -). The "-" class indicates that there is no consensus decision among the annotators. We remove them during the training and evaluation, and use the same data split as other works [1, 3] to report comparable results.

SciTail is a textual entailment dataset that derives from treating multiple-choice question-answering as an entailment problem. The task involves assessing whether a given premise entails a given hypothesis or not. The hypotheses in SciTail are created from science questions, and the corresponding answer candidates and premises come from relevant web sentences retrieved from a large corpus.

The detailed information about SNLI and SciTail datasets are presented in Table 1.

Table 1. Statistical Information about SNLI and SciTail datasets.

Dataset	SNLI			SciTail		
	Train	Dev	Test	Train	Dev	Test
Total Number	550,152	10,000	10,000	23,596	1,304	2,126
Average Length of Premises	14.1			10.5		
Average Length of Hypothesis	8.3			6.9		
Numbers of Entailment, Neutral, and Contradiction Samples	190113, 189218, 189702			10,101, 16,925, -		

4.2 Experimental Settings

For fair comparisons, we use the pre-trained weights of $BERT_{large}$ and $RoBERTa_{large}$ to initialize our model. The dropout rate is set to be 0.1, the batch size is 24 and the number of residual attention layers is selected from $\{1, 2, 3, 4, 5, 6\}$. The maximum number of epochs is 3. Texts are tokenized using wordpieces with a maximum length of 128. We use the Adam method for optimization with the first momentum 0.9 and the second 0.999. The initial learning rate is $3e-5$. The warmup steps is 500 and the weight decays is 0.01. We report the accuracy as evaluation metric.

4.3 Results and Analysis

To verify the effectiveness of our RAN4NLI, we choose the following typical NLI models proposed in the past few years as the baseline models for comparison.

DR-BiLSTM (Ensem.) + Process [1] is a dependent reading bidirectional.

LSTM network (DR-BiLSTM) to model the relationship between a premise and a hypothesis during encoding and inference with sophisticated ensemble strategy to combine multiple models.

BERT [5] pre-trains on a bidirectional word-masking language modelling task, in addition to sentence pair prediction.

ExBERT$_{-BERT-large}$ + ConceptNet [14] is a BERT-based model that enriches the contextual representation with real-world commonsense knowledge from external knowledge sources.

MT-DNN [12] performs multi-task learning on the nine GLUE (General Language Understanding Evaluation) benchmark tasks, after BERT's pre-training.

SemBERT [13] is an improved language representation model that explicitly absorbs contextual semantics over a BERT backbone by incorporating contextual semantics from pre-trained semantic role labeling.

CA-MTL [15] is a transformer-based architecture consisting of a new conditional attention mechanism and a set of task conditioned modules that facilitate weight sharing.

ALUM [17] is a general adversarial algorithm for training large neural language models, which regularizes the training objective by applying perturbations in the embedding space that maximizes the adversarial loss.

The results on the SNLI and SciTail datasets are listed in Table 2, in which the results of the baseline models come from their published papers and - denotes that the results are not reported originally. The results indicated with * are based on our own experiments with the officially released pre-trained models. The bolded values represent the best results. Since there doesn't directly exist experimental results of RoBERTa$_{large}$ on SNLI and SciTail in other works, we conduct the experiments and take the average of the results from five epochs to decease uncertainty.

As shown in Table 2, we can conclude that:

(1) RAN4NLI outperforms most of the previous works on SNLI and SciTail, except for ALUM. Compared with DR-BiLSTM, BERT$_{large}$, ExBERT, RoBERTa$_{large}$, MT-DNN$_{BERT-large}$ and SemBERT$_{BERT-large}$, we achieve better performance obviously. Especially, RAN4NLI$_{BERT-large}$ outperforms both MT-DNN$_{BERT-large}$ and SemBERT$_{BERT-large}$ that also built on BERT-large model, without employing multi-tasking learning or additional structured information, which as a result confirm the effectiveness of further modeling the relationship between the premise and hypothesis.

(2) RAN4NLI$_{RoBERTa-large}$ performs better than CA-MTL$_{RoBERTa-large}$ on SNLI with marginal gains up to 0.6% points in accuracy, but worse than CA-MTL$_{RoBERTa-large}$ on SciTail by 0.5% points. Although they both use RoBERTa-large, CA-MTL use more complicated conditional attention mechanism and a set of task conditioned modules, as well as a new multi-task data sampling strategy, while our RAN4NLI only uses self-attention and cross-attention mechanisms, which is more intuitive and easy to understand. Additionally, we only consider single-task training instead of multi-task learning like CA-MTL, and achieve comparable performances, which as a result implies the importance of the interaction network in RAN4NLI.

Table 2. Statistical Information about SNLI and SciTail datasets.

Dataset	SNLI (Accuracy %)		SciTail (Accuracy %)	
	Dev	Test	Dev	Test
DR-BiLSTM (Ensem.) + Process	-	89.6	-	-
BERT$_{large}$	91.7	91.0	95.7	94.4
RoBERTa$_{large}$	92.1	91.5	96.2	94.8
MT-DNN$_{BERT-large}$	92.2	91.6	96.3	95.0
ExBERT$_{-BERT-large}$ + ConceptNet	-	91.5	-	95.2
SemBERT$_{BERT-large}$	92.3	91.6	-	-
CA-MTL$_{RoBERTa-large}$	92.4	92.1	97.8	96.8
ALUM$_{RoBERTa-large}$	93.1	93.0	97.4	96.3
ALUM$_{RoBERTa-large-SMART}$	**93.6**	**93.4**	**98.2**	**96.8**
RAN4NLI$_{BERT-large}$ (ours)	92.3	91.9	96.4	95.3
RAN4NLI$_{RoBERTa-large}$ (ours)	92.7	92.5	97.3	96.3

(3) ALUM$_{RoBERTa-large}$ and ALUM$_{RoBERTa-large-SMART}$ surpass RAN4NLI$_{RoBERTa-large}$, because they utilize a new adversarial training algorithm for large neural language models and continually pre-train RoBERTa based on RoBERTa-large model with a union set of Wikipedia, OPENWEBTEXT, and STORIES (total size 82 GB). Especially, ALUM$_{RoBERTa-large-SMART}$ also combines adversarial pre-training with adversarial fine-tuning and yields further gains. By contrary, we only directly fine-tune RAN4NLI on the downstream task without any further pre-training. Compared with ALUM$_{RoBERTa-large}$ that uses adversarial pre-training but standard fine-tuning, our RAN4NLI$_{RoBERTa-large}$ achieves almost comparable results, which implies the effectiveness of the interaction network. We believe that the performance would be further improved if we also incorporate the adversarial pre-training of ALUM.

4.4 Ablation Analysis

To analyze the major components that are of importance to RAN4NLI, we performed ablation study on SNLI and SciTail datasets.

Firstly, we study the effects of interaction network in RAN4NLI by comparing its performance with respect to several widely-used pre-trained language models, including: BERT$_{base}$, BERT$_{large}$, RoBERTa$_{base}$ and RoBERTa$_{large}$. We follow the default settings of these pre-trained models and only use their weights to initialize our RAN4NLI. When directly evaluating the performance of these pre-trained language models on SNLI and SciTail, we concatenate the premise and hypothesis into a sentence-pair and take the hidden embeddings of the [CLS] from the last layer of the transformer encoder for final classification via a MLP classifier. The comparison results are demonstrated in Table 3.

From Table 3, we can see that all RAN4NLI models outperform their corresponding pre-trained language models on both datasets, which demonstrate the necessity and

Table 3. Comparison of RAN4NLI with Different Pre-trained Language Models.

Dataset	SNLI (Accuracy %)		SciTail (Accuracy %)	
	Dev	Test	Dev	Test
BERT-base	91.0	90.8	94.3	92.0
RAN4NLI$_{BERT-base}$ (ours)$_e$	91.6	91.1	94.9	93.1
BERT-large	91.7	91.0	95.7	94.4
RAN4NLI$_{BERT-large}$ (ours)	92.3	91.8	96.4	95.1
RoBERTa-base	91.5	91.2	95.1	93.2
RAN4NLI$_{RoBERTa-base}$ (ours)	92.0	91.9	95.7	94.1
RoBERTa-large	92.1	91.9	96.2	94.8
RAN4NLI$_{RoBERTa-large}$ (ours)	**92.7**	**92.5**	**97.0**	**96.3**

effectiveness of the interaction network, since RAN4NLI only use it to further model the relationship between the premise and hypothesis based on the sequence outputs of pre-trained language models, without employing any other complicated strategies or extra models, i.e., multi-task learning or external knowledge. The results also demonstrate the generalization capability of interaction network since it is consistently effective for different pre-trained language models.

Secondly, we continue to investigate the influence of the number of residual attention layers. We consider the number ranges from 0 to 6 and only report the accuracy results on the test sets of SNLI and SciTail respectively. See Fig. 3 for the details.

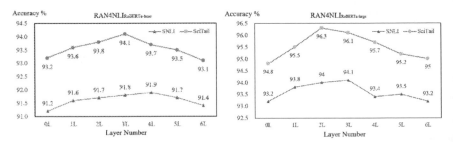

Fig. 3. Performances of RAN4NLI$_{RoBERTa-base}$ and RAN4NLI$_{RoBERTa-large}$ on SNLI and SciTail

From Fig. 3, we can see that:

(1) As the number increase from 0 to 1, the performances of both two RAN4NLI models improve obviously, which confirms the effectiveness of residual attention layers and it can enables the model to capture deeper potential relationship between the premise and hypothesis.

(2) The performance of RAN4NLI improves steadily as the number of residual attention layers increases, but decreases after it reach some threshold value. In our experiment

situations, we find different threshold values for different RAN4NLI models built on different pre-trained language models. For example, RAN4NLI$_{\text{RoBERTa-base}}$ achieves best performance when the number is 4 for SNLI and 3 for SciTail. According to our analysis, we think the reasons are mainly due to the different sizes of training data in these two datasets. As the layer number increases, the model has stronger abilities to explore more comprehensive relationship between the premise and hypothesis, but it also brings the side-effect of rapidly increasing numbers of the trainable parameters. Since SNLI has a much larger scale than SciTail, it can fine-tune RAN4NLI more fully and thus adding more layers can bring further improvement gains to some extent.

(3) After the number surpasses the threshold value, the performances of both RAN4NLI models decrease. We think the main reason might be due to the fact that more layers require more task-specific data for fine-tuning and the model tends to be over-fitting when training data are insufficient. Additionally, as the layer number increases, higher layer will contain more abstract relationship information about the premise and hypothesis, which might not be necessary for simple classification.

5 Conclusions

This paper proposes a transformer-based inference structure RAN4NLI by using a sequence encoder based on pre-trained language model to capture the input semantics and an interaction network based on residual attention to further model their relationship. Different from the common practice of pre-trained language models that use the hidden embeddings of $[CLS]$ for inference by concatenating the premise and hypothesis into a pair of sentences as input, we also consider the independent premise and hypothesis information and propose an interaction network based on residual attention to further explore their relationship. The experiments conducted on two widely used NLI datasets, SNLI and SciTail, demonstrate that our RAN4NLI achieves comparable performance with most of existing strong baseline models. The ablation analysis also confirms the effectiveness of the interaction network and the corresponding residual attention.

In our future work, we will continue to further explore how to strengthen the relationship modeling between the premise and hypothesis by pre-training with more effective objectives or external knowledge. We also want to investigate the roles of interaction network and residual attention more extensive and comprehensive.

References

1. Ghaeini, R., Hasan, S.A., Datla, V., et al.: DR-BiLSTM: dependent reading bidirectional LSTM for natural language inference. In: Proceedings of NAACL-HLT, pp. 1460–1469. Association for Computational Linguistics, New Orleans (2018)
2. Mou, L., Men, R., Li, G., et al.: Natural language inference by tree-based convolution and heuristic matching. In: The 54th Annual Meeting of the Association for Computational Linguistics, pp. 130–136. Association for Computational Linguistics, Berlin (2016)
3. Chen, Q., Zhu, X., Ling, Z.-H., et al.: Enhanced LSTM for natural language inference. In: The 55th Annual Meeting of the Association for Computational Linguistics (ACL 2017), pp. 1657–1668. Association for Computational Linguistics, Berlin (2017)

4. Aarne, Y., Anssi, J., Jorg, T.: Sentence embeddings in NLI with iterative refinement encoders. Nat. Lang. Eng. **25**(4), 467–482 (2019)
5. Devlin, J., Chang, M.W., Lee, K., Toutanova, K.: BERT: pre-training of deep bidirectional transformers for language understanding. In: NAACL-HLT, pp. 4171–4186. Association for Computational Linguistics, Minneapolis (2019)
6. Liu, Y.H., Ott, M., Goyal, N., Du, J.F.: RoBERTa: a robustly optimized BERT pretraining approach. arXiv preprint arXiv:1907.11692 (2019)
7. Bowman, S.R., Gauthier, J., Rastogi, A., et al.: A large annotated corpus for learning natural language inference. In: Proceedings of the 2015 Conference on Empirical Methods in Natural Language Processings, pp. 632–642. Association for Computational Linguistics, Portugal (2015)
8. Khot, T., Sabharwal, A., Clark, P.: SciTaiL: a textual entailment dataset from science question answering. In: 32nd AAAI Conference on Artificial Intelligence (AAAI 2018), pp. 5189–5197. AAAI Press, CA (2018)
9. Bowman, S.R., Gauthier, J., Rastogi, A., et al.: A fast unified model for parsing and sentence understanding. In: The 54th Annual Meeting of the Association for Computational Linguistics, pp. 1466–1477. Association for Computational Linguistics, Berlin (2016)
10. Conneau, A., Kiela, D., Schwenk, H., et al.: Supervised learning of universal sentence representations from natural language inference data. In: Proceedings of the 2017 Conference on Empirical Methods in Natural Language Processing, pp. 670–680. Association for Computational Linguistics, Copenhagen (2017)
11. Tay, Y., Tuan, L.A., Hui, S.C.: Compare, compress and propagate: enhancing neural architectures with alignment factorization for natural language inference. In: Proceedings of the 2018 Conference on Empirical Methods in Natural Language Processing, pp. 1565–1575. Association for Computational Linguistics, Stroudsburg (2018)
12. Liu, X., He, P., Chen, W., Gao, J.: Multi-task deep neural networks for natural language understanding. In: The 57th Annual Meeting of the Association for Computational Linguistics (ACL 2019), pp. 4487–4496. Association for Computational Linguistics, Berlin (2019)
13. Zhang, Z., Wu, Y., Zhao, H., et al.: Semantics-aware BERT for language understanding. In: Proceedings of the Thirty-Fourth AAAI Conference on Artificial Intelligence (AAAI 2020), vol. 34, no. (5), pp. 9628–9635. AAAI Press, CA (2020)
14. Gajbhiye, A., Moubayed, N.A., Bradley, S.: ExBERT: an external knowledge enhanced BERT for natural language inference. In: 30th International Conference on Artificial Neural Networks, pp. 460–472. European Neural Network Society, Switzerland (2021)
15. Pilault, J., Elhattami, A., Pal, C.: Conditionally adaptive multi-task learning: improving transfer learning in NLP using fewer parameters & less data. In: International Conference on Learning Representations (ICLR 2021). OpenReview.net, Vienna (2021)
16. Vaswani, A., Noam, N., Niki, P., et al.: Attention is all you need. In: 31st Conference on Neural Information Processing Systems (NIPS 2017), vol. 30, pp. 5998–6008. Currant Associates, CA (2017)
17. Liu, X.D., Cheng, H., He, P.C., et al.: Adversarial training for large neural language models. arXiv preprint arXiv.2004.08994 (2020)

Enhancing Reasoning Pathways for Power Defect Analysis Using CoT and Role-Play Prompt

Jimin Xu[1]([✉]), Xuemin Cao[1], Tao Li[1], Fei He[1], Shun Yao[1], Baolu Jia[1], Hongxu Zhang[2], and Shitao Cai[2]

[1] State Grid Shaanxi Electric Power Co., Ltd., Baoji Power Supply Company, Hengyang, China
30920231154367@stu.xmu.edu.cn
[2] School of Informatics, Xiamen University, Xiamen, China

Abstract. Modern Large Language Models (LLMs), such as ChatGPT, demonstrate exceptional capabilities in text classification and reasoning. The categorization of severity levels for descriptions of Power Defects and the inference of Defect Causes present an innovative and challenging task aimed at providing comprehensive and accurate reasoning pathways to power grid workers. In this study, a comparison is made among three Chain-of-Thought (CoT) prompting methods and the Role-Play prompting method using a Power Grid dataset. It is observed that the manually designed Manual-CoT method achieves the best results, with other methods showing significant improvements in classification accuracy and the coherence of reasoning pathways. This further highlights the potential for substantial enhancement of Large Language Models' reasoning abilities in specialized domains through expert-guided template pathways.

Keywords: Large Language Models · Chain-of-Thought · Role-Play Prompting · Power Defects

1 Introduction

Society's increasing emphasis on large-scale production and scientific technological advancement has led to a growing demand for electrical power supply. Consequently, higher requirements have been placed on the reliability and stability of key power grid components such as transformers. With the accelerated development of intelligence in the power industry, the functionality and automation level of power equipment continue to advance. In order to ensure the secure operation and cost-effectiveness of the power system, and to guarantee the smooth progress of daily life and production, the safe and efficient operation and maintenance of power equipment have gradually become a crucial issue. However, knowledge within the field of electricity is often highly specialized and intricate. Relying solely on human memory and analysis during the maintenance and operation of power equipment not only incurs significant time costs but also risks errors and omissions, which are insufficient to meet the real-time and efficient demands of

D.-S. Huang et al. (Eds.): ICAI 2023, CCIS 2015, pp. 347–356, 2024.
https://doi.org/10.1007/978-981-97-0827-7_30

power equipment diagnostics. Therefore, expediting the integration of power companies' grid production operations with artificial intelligence technology has become profoundly important.

Recent research has shown that Large Language Models (LLMs) have achieved significant success in complex comprehension and question-answering tasks when appropriately guided [1]. When tackling intricate reasoning tasks, LLMs often break down multi-step problems into intermediate steps and then generate answers gradually, demonstrating impressive performance [2]. Particularly noteworthy is the recent introduction of the "Chain of Thought" (CoT) approach, through which LLMs are capable of effectively addressing reasoning tasks, including arithmetic, commonsense, and symbolic reasoning.

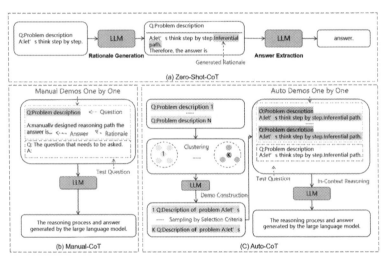

Fig. 1. Zero-Shot-CoT [3] (using the "Let's think step by step" prompt), Manual-CoT [1] (using manually designed demonstrations one by one), AutoCoT [4] (using clustering, select the central problem and build it automatically through "Let's think step by step") with structure of an LLM.

The CoT paradigm can be categorized into three main approaches as shown in Fig. 1. The first involves the Zero-Shot CoT paradigm, where a single prompt like "Let's think step by step" is added after the test question to encourage the reasoning chain within LLMs [3]. This zero-shot CoT paradigm is agnostic to tasks and does not require input-output demonstrations. Through zero-shot CoT, LLMs have demonstrated impressive zero-shot reasoning capabilities. The second paradigm is Manual CoT, achieved by designing reasoning demonstrations manually one by one. Each demonstration consists of a question and a reasoning chain, where the reasoning chain comprises a rationale (a sequence of intermediate reasoning steps) and an expected answer.In contrast, the third paradigm is Auto-CoT, used to automatically construct demonstrations with questions and reasoning chains. Auto-CoT consists of two main steps: firstly, clustering questions from a given dataset, and secondly, selecting a representative question from each cluster and generating its reasoning chain using simple heuristics and the zero-shot CoT prompt.

Specifically, Auto-CoT employs the prompt "Let's think step by step" to iteratively generate reasoning chains for demonstrations, step by step.

In addition to using prompts to generate chains of thought, employing roleplaying prompts with predefined scenarios is also highly valuable in specialized domains. Unlike the CoT prompt approach that enhances model quality, role-playing prompts focus on generating more targeted outputs by setting the perspective of specific roles within a scenario. While the CoT method primarily enhances reasoning capabilities by generating intermediate reasoning steps, role-playing prompts guide the model to adopt different angles when generating output by setting specific roles or perspectives. The objective of this approach is to enable the model to generate content more contextually, producing relevant answers based on the backgrounds and viewpoints of the designated roles. Through role-playing, the model can better grasp the context, leading to more accurate responses.

2 Related Work

2.1 Chain-of-Thought

CoT enhances the model's inference accuracy by guiding it to generate a series of natural language intermediate steps that lead to the ultimate answer. This concept draws inspiration from how humans utilize thoughtful thinking processes to undertake complex tasks. Experiment results from various LLMs, such as GPT-3 [5] and PaLM [6], indicate that CoT prompts can improve performance across a spectrum of arithmetic, commonsense, and symbolic reasoning tasks. Originally proposed by Wei et al., the "Manual-CoT" method requires the manual design of demonstrations to facilitate the generation of reasoning paths. In contrast, Kojima et al. introduced "Zero-Shot CoT", which employs a single zero-shot prompt to extract CoT from LLMs. By simply adding "Let's think step by step" before each answer, the "Zero-Shot CoT" demonstrates that LLMs are capable of zero-shot reasoning without the need for any manually constructed few-shot examples. Furthermore, Wang et al. proposed an "Auto-CoT" strategy for automatically constructing demonstrations for large language models. This involves sampling multiple outputs from various LLMs and aggregating them through majority voting. This strategy encourages the model to consider multiple CoTs when generating answers. However, generating a substantial number of reasoning paths is essential for optimal performance, which subsequently leads to increased computational costs.

2.2 In-Context Learning and Prompt

CoT prompts are closely related to In-Context Learning (ICL) [7]. ICL involves incorporating prompt examples into the input to enable LLMs to perform target tasks. ICL allows a single model to be versatile across multiple tasks without requiring gradient updates. Research directions aimed at improving ICL performance include: (i) dynamically retrieving training examples related to a given test input, which has become a popular practice [8]; (ii) incorporating finer-grained information, such as integrating task instructions [9]; (iii) adjusting LLMs' output probabilities to replace direct computation of target label likelihoods [10]. Although ICL has achieved success, studies [11] point out that the efficacy of ICL might significantly vary due to the choice of context demonstrations. Specifically, factors such as the format of prompts, like wording or the order of demonstrations, could lead to performance fluctuations [12] has even questioned the necessity of a direct input-output mapping for benchmarks, suggesting that using incorrect labels in examples only slightly diminishes performance. Aobo Kong [13] employed models like ChatGPT and Llama 2 to validate that role-playing prompts consistently outperform standard zero-shot methods on most datasets. The authors believe that role-playing prompts not only enhance contextual understanding but also serve as implicit triggers for Chain of Thought (CoT), thus elevating the quality of reasoning. By comparing our approach with zero-shot CoT techniques, which encourage models to think step by step, we further validate that role-playing prompts are capable of generating more effective CoTs, highlighting their potential to enhance LLMs' reasoning capabilities.

2.3 LLM in Specialized Domains

Currently, Large Language Models (LLMs) have shown immense potential across various domains like Natural Language Processing, Text Generation, and Machine Translation [14]. Examples such as GPT-3 and BERT have demonstrated wide-ranging applications in multiple fields [15]. In the domain of Natural Language Processing, they are employed for text generation, including the creation of articles, poems, and conversations, providing creativity and support for content creation [16]. In the realm of Machine Translation, they have enhanced the efficiency and accuracy of cross-lingual communication. Regarding sentiment analysis, they possess the capability to comprehend the underlying emotional tendencies in text, thereby playing a role in social media monitoring and brand reputation management [17]. Within intelligent assistants and virtual characters, they offer a more natural and smooth conversational experience, becoming the core of intelligent customer service [18]. In specialized sectors like healthcare and law, their applications are gradually being explored to aid in tasks like medical diagnosis [19] and legal document analysis [20]. Despite the rapid development of Large Language Models, challenges in their development exist due to concerns such as data privacy and training costs.

3 Methodology

Zero-Shot. When presented with a question describing a power grid defect phenomenon, the Large Language Model (LLM) will automatically provide a corresponding answer indicating the level of the power grid defect. However, this process tends to be relatively straightforward, resulting in lower accuracy.

Zero-Shot-CoT. Zero-Shot CoT doesn't require manually annotated task demonstrations; instead, it directly generates reasoning steps and then employs the generated CoT to derive the answer. In this approach, the LLM first generates reasoning steps prompted by "Let's think step by step" and subsequently derives the final answer prompted by "Therefore, the answer is." Specifically, for this question, when inputting a query describing a power grid defect phenomenon, the model automatically adds "Let's think step by step" to trigger CoT, generating intermediate reasoning steps to guide the Large Language Model (LLM) through complex reasoning. This enables the model to provide justifications for its decisions, thereby enhancing the quality of answers.

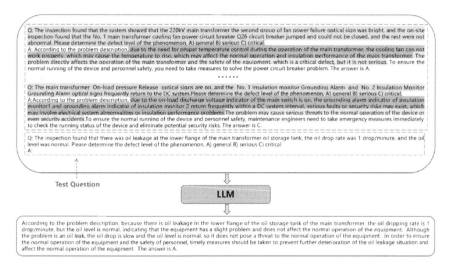

Fig. 2. Manual-CoT [1] (using manually designed demonstrations one by one) with example inputs and outputs of an LLM

Manual-CoT. The Manual Chain-of-Thought (Manual-CoT) is an approach within the context of CoT prompting. It utilizes manually designed demonstrations to prompt Large Language Models (LLMs). In this method, the process of human reasoning, known as the Chain of Thought, is explicitly embedded in the prompt message using a few-shot prompt format. This transformation modifies the prompt message from <input, output> to <input, chain of thought, output>. When the Chain of Thought is included in the prompt, it compels the LLM to produce the Chain of Thought output before presenting the final answer. From the perspective of conditional probability distribution, placing the Chain of Thought before the answer enhances the accuracy likelihood. Therefore, in experiments employing the Manual-CoT, the initial step involves the manual design of CoT prompts. These prompts consist of multiple sets of example demonstrations. Each demonstration set encompasses a question and a chain of reasoning. The chain of reasoning comprises a series of intermediate reasoning steps leading to the expected answer.

In the manual rationale design of this experiment, "due to" is first used to elicit the cause and impact of the problem. Subsequently, employing the phrase "The problem may cause" it segues into an evaluation of the severity of the issue based on its impact analysis. Finally, initiated by "To ensure the normal running of the device and personnel safety" it introduces solutions, recommendations, and concludes with a final statement. The Manual-COT also adheres to the crafted response format, explicating and elucidating the question to ultimately derive a conclusion, as shown in Fig. 2. During actual question-answering, when provided with a question describing a power grid defect phenomenon, the model processes both the prompt and the question as inputs. Subsequently, the LLM supplies both the reasoning steps and the answer. This approach highlights the importance of explicitly designed prompts, including the Chain of Thought, to guide the LLM's reasoning process effectively in order to generate accurate answers.

Auto-CoT. Auto-CoT constructs demonstrations by sampling a diverse set of questions and generating corresponding chains of reasoning. It primarily involves two steps: Question Clustering and Demonstration Sampling. In the Question Clustering step, the k-means algorithm is used to cluster questions into k clusters. For each cluster, questions are sorted based on their distance from the cluster center. In the Demonstration Sampling step, a corresponding demonstration is built for each cluster. In this experiment, the model performs Question Clustering using k-means, forming clusters of questions. For each cluster, a central question is selected, and the Zero-shot Chain-of-Thought (Zero-shot-CoT) method is employed to automatically construct examples with questions and chains of reasoning as prompt inputs. During actual question-answering, when presented with a question describing a power grid defect phenomenon, the model simultaneously takes the prompt and the question as inputs. Subsequently, the Large Language Model (LLM) provides both the steps of reasoning and the answer. This approach enables the model to engage in reasoning and answer questions without the need for manual design.

Role-Play Prompting. In traditional approaches, Role-Playing Prompts combine reasoning questions from different roles with a single prompt and involve a single interaction with a Large Language Model (LLM). Given our focus on specialized datasets in the power grid domain, the model has a significant advantage in decision support. To further guide the LLM and maximize its effectiveness, we upgraded the single-round interaction to a two-round dialog process.

This dual-round structure serves different purposes: the first round immerses the model into a specific role, while the second round prompts the model to provide answers based on predefined roles. In both rounds of dialogue, the introduction of the initial role aids the subsequent reasoning. Considering the challenge of fully controlling the quality of the initial response, we sample multiple responses in the first round to determine the optimal answer for handling various questions. By ensuring the optimal initial reaction, we concatenate the input and output of the first interaction with the reasoning question, generating a single prompt that guides targeted responses, as shown in Fig. 3.

In summary, our Role-Playing Prompt method follows a two-stage process as illustrated in Fig. 2: firstly, establishing the optimal role-immersive interaction for each task, and then guiding the model to respond to each reasoning question based on a specific role. Specifically, in this method, the first round of dialogue is used to elaborate on the role the model is assuming, namely, an expert role in the power grid domain. Such descriptions aid in shaping a deeper understanding of the model's framework and role characteristics. Subsequently, the second round of dialogue guides the model to respond to reasoning questions, as illustrated in the example in Fig. 3.

Through this approach, we enhance the model's understanding and integration into the professional background of the power grid domain. This dual-round dialog design enables the LLM to exhibit more expertise in reasoning and response, thereby providing stronger support for decision-making in the power grid domain.

Existing research indicates that both Role-Playing Prompt Learning and Chain of Thought approaches can enhance the accuracy and fluency of text generation. These techniques aid models in improving their grasp of language grammar and structure, while also bolstering model interpretability. As a result, models become more adept at comprehending context and tasks, leading to more precise and dependable outcomes in text generation and language model training. To compare which approach is better suited for datasets in the power grid domain, we conducted experimental validations.

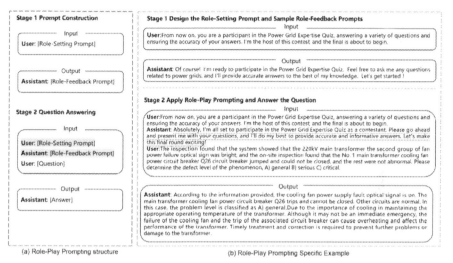

Fig. 3. Two stages of Role-Playing prompt program description. In stage 1, multiple role feedback prompts were sampled. In stage 2, select the optimal role feedback prompt (underlined in blue) to generate an answer. (Color figure online)

4 Experiment

4.1 Datasets and Tasks

To evaluate the effectiveness of Role-Play Prompting and CoT methods on an electric grid dataset, we conducted experimental research using a dataset of transformer defect descriptions provided by an electric grid company. The dataset comprises a total of 38,412 defect descriptions, 38,408 corresponding defect severity levels, and 273,294 corresponding classification criteria for defect severity. On average, each defect description is approximately 112.3 characters long, and the defect severity levels include general, serious, and critical. The average length of the classification criteria for each defect severity level is around 31.2 characters. We extracted a subset of 1,200 data samples from the dataset, following a 1:1:1 ratio for defect severity levels, and referred to this subset as "DataSet."

We designed two sets of experiments. In the first set of experiments, we employed four different CoT methods by invoking GPT-3 (gpt3-xl) to run on the DataSet and obtained results for comparative analysis. In the second set of experiments, we tested the Role-Play Prompting method and compared its results with those of the four CoT methods. Finally, we combined the results from both sets of experiments to draw conclusions. These experiments were conducted to investigate the effectiveness of the Role-Play Prompting method and various CoT methods on the electric grid dataset, aiming to gain a better understanding of their advantages and applicability.

4.2 Baseline and Evaluation

Baseline. We directly ran the dataset on GPT-3 (gpt3-xl) and used the generated results as a baseline for comparison with our model.

Evaluation. Automatic Evaluation: To assess the quality of generated answers, the accuracy (ACC) of the model on a test dataset is used as a criterion for model performance.

Manual Evaluation: In order to evaluate the quality of the model generation inference path, a manual evaluation was performed. Four graduate experts specializing in the field of power grids or computer science were recruited for annotation. These expert annotators were instructed to rate the answers based on four criteria: fluency, relevance [21].

Fluency: Which robot's response is smoother and more easily understandable?

Relevance: Which robot's reasoning for defect level classification is more closely aligned with the criteria?

Rating was conducted within a 5-star range, where 5 stars represent the highest rating. To evaluate the quality of generated responses, we followed the steps outlined below:

(i) 100 sets of data were randomly selected as evaluation samples.
(ii) The 4 graduate experts, specialized in power grids or computer science, were divided into 2 groups, each comprising two annotators.
(iii) Each group of two annotators independently annotated 50 questions.
(iv) For the same questions, the score differences between the two annotators in each group were calculated.
(v) Answers with score differences exceeding two points were re-annotated by the other group of annotators.
(vi) Steps (iv) and (v) were repeated until a sufficiently high level of agreement was reached (assessed using Cohen's Kappa coefficient [22]).

4.3 Results and Analysis

We chose the results of running the dataset directly on GPT-3 (gpt3-xl) as the baseline. The evaluation metrics include Precision (P), Recall (R), F1 score, and Accuracy (ACC). The comprehensive assessment results are shown in Table 1.

Among the evaluated methods, Manual-COT demonstrated the best performance on the dataset, achieving the highest accuracy, fluency, and relevance scores. Auto-COT also outperformed the baseline significantly across the three evaluation metrics. Role-Play Prompting and Zero-shot-COT exhibited similar accuracy and relevance, but the use of Role-Play Prompting led to smoother professional domain expressions. All four

Table 1. Experimental Results and Baseline Performance

Model	ACC	Fluency	Relevance
GPT-3(gpt3-xl)	0.533	3.56	3.12
Zero-shot	0.467	/	/
Zero-shot-cot	0.608	3.97	3.57
Manual-cot	**0.867**	**4.48**	**4.33**
Auto-cot	0.692	4.23	3.96
Role-Play Prompting	0.599	4.15	3.46

methods showed improvements over directly invoking GPT-3 (gpt3-xl). However, using Zero-shot directly resulted in lower accuracy than the baseline, and the lack of reasoning pathways in its responses made it challenging to assess fluency and relevance.

The success of Manual-COT aligns with expectations, as general large-scale models might exhibit biases in professional domain reasoning. For example, when facing defects like "more than 2/3 of the total amount of the silica gel hygroscopic discoloration part," a generic language model might, without contextual prompts, classify it as "serious" based on common data, even though it should be classified as "general." A manually designed demo could address such issues of misjudgment by pre-designing rationales to guide the model towards the intended outcome, thereby significantly improving accuracy. Moreover, guiding reasoning using phrases like "due to" "the problem may cause" and "ensure the normal running of the device and personnel safety" not only enhances the fluency of reasoning but also helps to lead to accurate conclusions.

5 Conclusion

This study, through the meticulous design of a manual demo tailored for the domain of power grids, along with a comparative analysis against four other methods on the dataset, unveils that the manually designed CoT significantly outperforms other models in terms of experimental results within the professional domain dataset. Furthermore, it validates the substantial enhancement in reasoning capabilities of large language models in the realm of power grids, as achieved by both CoT and Role-Play. These results notably emphasize the potential of applying CoT and Role-Play to specialized datasets. In summary, this research provides valuable insights for further deepening the application of large language models in domain-specific question answering and reasoning. It also underscores the crucial direction for the practical application of large language models, offering a significant reference for realizing their practical value.

Acknowledgement. This work is supported by the Research Funds from State Grid Shaanxi (SGSNBJ 00BYJS2311111).

References

1. Wei, J., et al.: Chain-of-thought prompting elicits reasoning in large language models. In: Advances in Neural Information Processing Systems, vol. 35, pp. 24824–24837 (2022)
2. Brown, T., et al.: Language models are few-shot learners. In: Advances in Neural Information Processing Systems, vol. 33, pp. 1877–1901 (2020)
3. Kojima, T., Gu, S.S., Reid, M., Matsuo, Y., Iwasawa, Y.: Large language models are zero-shot reasoners. In: Advances in Neural Information Processing Systems, vol. 35, pp. 22199–22213 (2022)
4. Zhang, Z., Zhang, A., Li, M., Smola, A.: Automatic chain of thought prompting in large language models. arXiv preprint arXiv:2210.03493 (2022)
5. Blair-Stanek, A., Holzenberger, N., Van Durme, B.: Can GPT-3 perform statutory reasoning? arXiv preprint arXiv:2302.06100 (2023)
6. Chowdhery, A., et al.: PaLM: scaling language modeling with pathways. arXiv preprint arXiv:2204.02311 (2022)
7. Radford, A., Wu, J., Child, R., Luan, D., Amodei, D., Sutskever, I., et al.: Language models are unsupervised multitask learners. Open AI Blog 1(8), 9 (2019)
8. Rubin, O., Herzig, J., Berant, J.: Learning to retrieve prompts for in-context learning. arXiv preprint arXiv:2112.08633 (2021)
9. Mishra, S., Khashabi, D., Baral, C., Hajishirzi, H.: Cross-task generalization via natural language crowdsourcing instructions. arXiv preprint arXiv:2104.08773 (2021)
10. Holtzman, A., West, P., Shwartz, V., Choi, Y., Zettlemoyer, L.: Surface form competition: why the highest probability answer isn't always right. arXiv preprint arXiv:2104.08315 (2021)
11. Liu, J., Shen, D., Zhang, Y., Dolan, B., Carin, L., Chen, W.: What makes good in-context examples for GPT-3? arXiv preprint arXiv:2101.06804 (2021)
12. Webson, A., Pavlick, E.: Do prompt-based models really understand the meaning of their prompts? arXiv preprint arXiv:2109.01247 (2021)
13. Kong, A.: Better zero-shot reasoning with role-play prompting. arXiv preprint arXiv:2308.07702 (2023)
14. Kaddour, J., Harris, J., Mozes, M., Bradley, H., Raileanu, R., McHardy, R.: Challenges and applications of large language models. arXiv preprint arXiv:2307.10169 (2023)
15. Ling, C., et al.: Beyond one-model-fits-all: a survey of domain specialization for large language models. arXiv preprint arXiv:2305.18703 (2023)
16. Mirowski, P., Mathewson, K.W., Pittman, J., Evans, R.: Co-writing screen plays and theatre scripts with language models: evaluation by industry professionals. In: Proceedings of the 2023 CHI Conference on Human Factors in Computing Systems, pp. 1–34 (2023)
17. Park, P.S., Schoenegger, P., Zhu, C.: Artificial intelligence in psychology research. arXiv preprint arXiv:2302.07267 (2023)
18. Shuster, K., et al.: BlenderBot 3: a deployed conversational agent that continually learns to responsibly engage. arXiv preprint arXiv:2208.03188 (2022)
19. Singhal, K., et al.: Large language models encode clinical knowledge. arXiv preprint arXiv:2212.13138 (2022)
20. Yu, F., Quartey, L., Schilder, F.: Legal prompting: teaching a language model to think like a lawyer. arXiv preprint arXiv:2212.01326 (2022)
21. Liu, S.: Towards emotional support dialog systems. arXiv preprint arXiv:2106.01144 (2021)
22. McHugh, M.L.: Interrater reliability: the kappa statistic. Biochemia medica 22(3), 276–282 (2012)

Author Index

Printed in the United States
by Baker & Taylor Publisher Services